Energiepolitik und Klimaschutz
Energy Policy and Climate Protection

Reihe herausgegeben von
L. Mez, Berlin, Deutschland
A. Brunnengräber, Berlin, Deutschland

Weltweite Verteilungskämpfe um knappe Energieressourcen und der Klimawandel mit seinen Auswirkungen führen zu globalen, nationalen, regionalen und auch lokalen Herausforderungen, die Gegenstand dieser Publikationsreihe sind. Die Beiträge der Reihe sollen Chancen und Hemmnisse einer präventiv orientierten Energie- und Klimapolitik vor dem Hintergrund komplexer energiepolitischer und wirtschaftlicher Interessenlagen und Machtverhältnisse ausloten. Themenschwerpunkte sind die Analyse der europäischen und internationalen Liberalisierung der Energiesektoren und -branchen, die internationale Politik zum Schutz des Klimas, Anpassungsmaßnahmen an den Klimawandel in den Entwicklungs-, Schwellen- und Industrieländern, die Produktion von biogenen Treibstoffen zur Substitution fossiler Energieträger oder die Probleme der Atomenergie und deren nuklearen Hinterlassenschaften.

Die Reihe bietet empirisch angeleiteten, quantitativen und international vergleichenden Arbeiten, Untersuchungen von grenzüberschreitenden Transformations- und Mehrebenenprozessen oder von nationalen „best practice"-Beispielen ebenso ein Forum wie theoriegeleiteten, qualitativen Untersuchungen, die sich mit den grundlegenden Fragen des gesellschaftlichen Wandels in der Energiepolitik und beim Klimaschutz beschäftigen.

Reihe herausgegeben von
PD Dr. Lutz Mez
Freie Universität Berlin

PD Dr. Achim Brunnengräber
Freie Universität Berlin

Weitere Bände in der Reihe http://www.springer.com/series/12516

Angela Pohlmann

Situating Social Practices in Community Energy Projects

Three Case Studies about the Contextuality of Renewable Energy Production

Springer VS

Angela Pohlmann
Faculty of Business, Economics
and Social Sciences
University of Hamburg
Hamburg, Germany

Dissertation at the University of Hamburg, Faculty of Business, Economics and Social Sciences, 2016

Funded with a full PhD scholarship by Hans-Böckler-Stiftung

Hans **Böckler**
Stiftung ▰▰▰

Energiepolitik und Klimaschutz. Energy Policy and Climate Protection
ISBN 978-3-658-20634-5   ISBN 978-3-658-20635-2   (eBook)
https://doi.org/10.1007/978-3-658-20635-2

Library of Congress Control Number: 2017963538

Printed on acid-free paper

This Springer VS imprint is published by Springer Nature
The registered company is Springer Fachmedien Wiesbaden GmbH
The registered company address is: Abraham-Lincoln-Str. 46, 65189 Wiesbaden, Germany

Acknowledgements

This work could not have been written without intense help, inspiration and advice from many people. First of all I want to thank the project members of the CDT, KEBAP, and the staff members of the IBA Hamburg, who all let me take part in their daily routines and who spent their time to contribute their knowledge, information, and insights to my thesis. This work is based on your friendliness, your knowledge, and your engagement. I also want to thank all the other people who were willing to give me interviews and take part in my research. I am extremely grateful to my first supervisor, Anita Engels! She has not only been the first person to believe that I could indeed write a PhD-thesis, but throughout the process she has supported me with adequate critique, insightful advice, practical support, and trust in my competencies. Many thanks also go to my second supervisor, Jörg Knieling who has always been willing to meet me and debate my ideas. Besides his competent expert advice, I am grateful for the many pragmatic recommendations he has made. Janette Webb has not only welcomed me in her team for some month, but has given me the opportunity to take part in all discussions, meetings, and event which took place at the time of my stay. Throughout my stay at the University of Edinburgh she furthermore has taken the time to intensely debate my research, my ideas, and the relevance and value of theoretical concepts with me. I also want to thank Janette Webb, Mags Tingey, David McCrone, and David Hawkey for actually making me feel like being part of the team, for talking things over with me, sharing their tea, coffee, and milk with me, and generally making me feel at home in Chisholm House. Thank you: Kerstin Walz, Sarah Debor, and, Lena Borlinghaus who kindly have each read, commented upon, and corrected parts of my thesis. Who have contributed ideas, were willing to discuss things over and over again, and whose work and patience definitely has improved this work. Clearly, this thesis could not have been written without the material and immaterial support from the Heinrich-Böll Stiftung. Getting a scholarship from the Heinrich-Böll Stiftung has secured my livelihood, enabled my fieldwork in Scotland, and my stay at the University of Edinburgh. Furthermore, and maybe even more importantly: being part of the "Cluster Transformationsforschung" has put me in contact with many brilliant and likeminded people.

Table of Contents

List of Abbreviations

ANT	Actor-Network-Theory
BImA	Bundes Immobilien Agentur
BTTP	Block-Type Thermal Power Station
BSU	Behörde für Stadtentwicklung und Umwelt Friends
BUND	of the Earth Germany
KEBAP	Kultur-und-Energie-Bunker-Altona-Projekt
KW	Kilowatt
CCF	Climate Challenge Fund
CCS	Carbon Capture and Storage
CDT	Comrie Development Trust
CES	Community Energy Scotland
CHP	Combined Heat and Power
CO2	Carbon Dioxide
DGRV	Deutscher Genossenschafts- und Raiffeisenverband
DTA	Development Trust Association
EEG	Erneuerbare-Energien Gesetz
FIT	Feed-in Tariff
IBA	International Building Exhibition
INES	International Nuclear Events Scale
IPCC	Intergovernmental Panel on Climate Change
kWh	kilo Watt per hour
KSB	Keep Scotland Beautiful
LTS	Large Technical Systems
MCA	Maximum Credible Accident
MoD	Ministry of Defence

PoW	Prisoner of War
RE	Renewable Energy
RHI	Renewable Heat Incentive
SCOT	Social Construction of Technology
SDHL	Scottish District Heating Loan Scheme
STS	Science and Technology Studies
UHUN	Unser-Hamburg-Unser-Netz Westray
WDT	Development Trust

Figures

Introduction

Bunkers and Prisoner-of-War (PoW) camps from the Second World War (WWII) are not typically associated with renewable energy production. During the war, bunkers and PoW camps were used to separate humans from their environment. Renewable energy production sites are usually approachable; they dissolve the geographical and social separation of people and energy production that characterizes conventional energy production (Walker et al. 2007: 68). WWII is associated with dictatorship and strict top-down power relations, while renewable energies are in many ways closely related to citizen initiatives and bottom-up processes (Rosenbaum/Mautz 2011: 411; likewise among others, Bergman et al. 2010; Mautz et al. 2008). Last but not least, WWII stands for a destructive past while renewable energy is equated with the future of low-carbon energy production. There is, however, an important similarity between bunkers and PoW camps and renewable energy production sites—both are associated with what were or are major threats to humankind. Bunkers and PoW camps sheltered people from the hazards of the World War—air raids as well as enemy combatants. Renewable energy sources derive from the aim to lessen the extent of anthropogenic climate change. Thus, both types of buildings are responses to major hazards for mankind, created by human conduct.

In the following, three buildings from the WWII which are transformed into renewable energy production sites serve as case studies for this thesis. The case studies are analysed with the research interest to gain insights into the contextuality of renewable energy and renewable energy production. This analysis of the case studies aims to contribute to an understanding of renewable energy and renewable energy production as complex, socially embedded phenomena. In Hamburg, a northern German city, two bunkers have been or are in the process of being refurbished to host renewable energy production technologies. In Comrie, a village in central Scotland, renewable energy production takes place in a former PoW camp. One of the two bunkers has been refurbished within the context of an international building exhibition (IBA). After restoration, the building now houses an energy production centre, a café, and an exhibition about the WWII and the history of the bunker. The energy production centre consists of a biomass boiler for burning woodchips, a gas biomethane-fired combined heat and power unit, a wood combustion system, and a solar thermal unit, as well as waste heat from an industrial plant. The heat produced by these technologies is fed into a 2000m³ buffer storage facility. The heat is distributed via a newly installed micro-grid that connects the bunker to about 1600 households.

The second bunker is going to be transformed into a culture-and-energy bunker (KEBAP) by the combined efforts of a cooperative and a charity organisation. Both organisations have their origins in the same citizen initiative. One part of the bunker is meant to provide room for the installation of a modular system comprised of a biomass combustion system for burning woodchips, a gas block-typed thermal power station (BTTP), and a buffer storage facility. Installation of solar thermal panels on the rooftop is being considered. Within the second part, cultural and social activities are projected to take place. The produced heat will be fed into Hamburg's long-distance heat grid and (virtually) distributed to customers all over the city. The income generated by selling the heat will be used to subsidize cultural and social activities in the second part of the building. These include but are not limited to urban gardening, music exercise rooms, yoga, a social kitchen, and a multipurpose room for exhibitions, neighbourhood meetings, and parties. At the moment the project is negotiating with Hamburg's government and the district government about conditions for purchasing the building.

In Scotland, a former PoW camp called the Cultybraggan Camp has been purchased by a community organisation—the Comrie Development Trust (CDT). Besides a multitude of different activities taking place at the camp, the CDT has installed photovoltaic panels and a biomass woodchip-based boiler to produce heat and electricity for the camp users. For this purpose, a micro heat grid has to be installed as well in the Camp. The facilities offered at the camp include a community orchard, allotments, a heritage centre, history re-enactment groups, small-scale entrepreneurs, a gym, and a music exercise room. The financial income from selling electricity and heat is fed back into the local community.

Using bunkers and PoW camps for renewable energy production alters the way these sites are made sense of. Initially, all the buildings were sites of military activities, more specifically WWII. While they continue to be memorials of war, the buildings have now become associated with ideas like climate change and sustainability. As mentioned above, anthropogenic climate change is one of the biggest challenges for today's human societies. Being created through human activities, the '*warming effect of greenhouse gases has the potential to trigger abrupt, large-scale, and irreversible changes in the climate systems*' (Stern 2006: 19). These changes in the climate system have effects on most ecological systems in the world. They will cause changes in sea levels and temperatures, draughts and desertification, melting of ice caps and glaciers, and the desiccation of swamps and marshes. The anticipated effects of these changes include the loss of biodiversity and an increase in natural hazards like storm-floods, tornados, and strong rains (Dunlap/Brulle 2015). Besides—and because of—the negative impacts on the flora and fauna, climate change and its effects will also affect human societies. Loss of arable and/or settled land and an increase in natural hazards will affect human lives and are likely to cause mass migrations and wars over resources (Smith 2007; Reuveny 2007). In particular, economically disadvantaged groups and individuals are going to suffer due to the effects of climate change (Adger 2006). Because of their lack of resources, they are more vulnerable and less resilient to the effects of climate change. Climate change thus *presents perhaps the most profound challenge ever to have confronted human social, political, and economic systems'* (Dryzek et al. 2011: 3; likewise Weyer 2010: 385).

While the effects of climate change are related to human societies, climate change is also caused by human activities. As Working Group 1 states in its contribution to the Fifth Assessment Report of the Intergovernmental Panel on Climate Change (IPCC), there are *'multiple lines of evidence that the climate is changing across our planet, largely as a result of human activities'* (Cubasch et al. 2013: 121).[1] More specifically, climate change is caused by the emission of greenhouse gases, most significantly carbon dioxide (CO_2), into the atmosphere. Most greenhouse gases result from the burning of fossil fuels like coal, gas, and oil. Fossil fuels are burned to produce energy. Modern societies require energy in the forms of electricity and heat and also for mobility and transport. The utilization of energy is a 'by-product' of human activities. People want to use, not energy per se, but their computers, cars, and other technical devices. *'Energy demand is the outcome of what people are* doing, *of the interlinking of practices and energy-intensive material arrangements'* (Urry 2014: 4). Furthermore, people do not use these technical devices in social isolation. Humans drive their cars on a daily basis because they need to get to work or in order to fulfil diverse social demands. They use a hoover and shower in hot water because of social conventions of cleanliness (Shove 2003). They switch on their computer as part of their social interactions, for work, and for information gathering. Taken together, *'energy in a variety of guises is bound up technically, economically, and politically with our societies, communities, and livelihoods in very diverse ways'* (Rutherford/Coutard 2014: 1354).

The social sciences have for a long time been marginalized in the scientific debates about climate change (Urry 2011; Brulle/Dunlap 2015; Rosa et al.

---

1   This thesis does not elaborate upon the scientific discussions about the factual existence or social reality of climate change. But see Steffen 2011; Boykoff et al. 2010; Norgaard 2006; Olausson 2009; McCright/Dunlap 2003.

2015). Arguing against purely natural science and technocratic understandings of climate change, social scientists have started to argue against '*the neglect of "society" in analysing current and future climate and resource processes*' and instead to '*bring society centrally into the analysis of climate change*' (Urry 2011: 2). Within this realm, sociology has become increasingly engaged in understanding the human dimensions of climate change, especially its social, institutional and cultural dynamics (Brulle/Dunlap 2015: 2). Social scientists who take part in political and scientific debates about climate change and carbon reduction argue that

> '[s]trategies of mitigation and adaptation cannot be accomplished by technical means alone but require a fundamental change in society and culture because climate change puts the traditional concept of industrial societies in itself in question, its technologies based on carbon, its economics, its principles of growth, and its ways of life' (Leggewie/Welzer 2010: 031009-2).

Instead of working on purely technical and natural science-based solutions, societies need to engage in radical, encompassing, and deliberate transformations. In order to transform today's societies it does not suffice to develop energy-efficient devices. In fact, '*there is clear evidence that improvements in efficiency often have failed to have substantial effects on the overall scale of environmental problems*' (Rosa et al. 2015: 33). Increases in energy efficiency are often thwarted by rebound effects (ibid.: 33; Frondel 2012). In addition to energy-efficiency measures, people's energy consumption patterns need to change in order to achieve significant carbon reductions. As it will be possible to neither eliminate carbon consumption absolutely, nor increase carbon efficiency sufficiently, these strategies need to be accompanied by an energy production approach that does not offset CO_2 or utilize non-renewable resources. Transforming today's societies means to fundamentally change the system of energy production. A main question is if and how renewable energy production can be incorporated into the existing system, or if the low carbon society requires a radical break with the existing structures and institutional infrastructures (Weyer 2010: 385)?

Energy efficiency, reduction of energy consumption, and renewable energy production are the key mitigation strategies to decrease carbon emissions and transform present-day societies into low- or no-carbon societies (Diesendorf 2011: 561). Within social sciences, researchers so far have studied different aspects of energy efficiency (Kousky/Schneider 2003; Rutland/Aylett 2008; Slocum 2004), energy consumption (Shove/Pantzar 2005; Ehrhardt-Martinez et al. 2015; Urry 2010), and energy production and distribution (Webb 2015; Bolton 2011). Additionally, different scholars have developed approaches to theoretically explain transformation processes (among others Geels 2004, 2010;

Markard/Truffer 2008; Geels/Schot 2007; Berkhout et al. 2003). Last but not least, social scientists have analysed the existing barriers to transformations and developed ideas to overcome them (Foxon/Pearson 2008; Burch 2010). These scholars have identified path dependencies (Wise et al. 2014), lock-in effects (Unruh 2000), social power (Avelino/Rotmans 2009), and individuals' cognitive barriers to prevent or act upon climate change (Lorenzoni et al. 2007; Slocum 2004) as obstacles.

Social scientists have acknowledged and even emphasised the complexities of the causes and consequences of climate change (Dryzek et al. 2011: 4pp.; Rotmans/Loorbach 2009; Norberg/Cumming 2008). While recognising this complexity, many studies in the field of '*transformation research*' (WBGU 2011) are characterized by an underdeveloped estimation of complexity with regard to the transformation processes themselves. Many models and theoretical approaches, for example, draw on a linear concept of transformation processes (for example Westley et al. 2011; Moore et al. 2014) or transition pathways (Foxon/Pearson 2011; Geels/Schot 2007). This simplistic understanding can also be seen in terms like 'the great transformation'. Different scholars have repeatedly criticized the failure to integrate complexity within transformation research (Engels 2015; Fuchs/Hinderer 2014; Shove/Walker 2007). Instead of talking about 'the' transformation, it would be more advisable to use the plural form—thus acknowledging the complexity of actors, events, processes, and outcomes (Engels 2015).

This thesis will make visible some of the complexities that are part of transformative processes. More precisely, it studies the complexities and heterogeneities that occur among three rather similar local renewable energy projects. In fact, the projects have a lot of similarities. Besides being situated in buildings from the WWII, each project uses woodchip biomass boilers and solar and/or photovoltaic panels to produce renewable heat. Furthermore, no site is used exclusively for energy production. Instead, all three projects combine energy production with facilities for social and/or cultural activities. Last but not least, all three projects explicitly aim to contribute to the development of low-carbon, renewable, and decentralized energy systems. Using examples of only one type of actors—local renewable energy production projects—within transformations, the analysis of the case studies focuses on heterogeneities and complexities between the three projects. It is analysed how renewable energy and the production of renewable energy are made sense of in different ways, not only by the different projects, but also by different members of each project.

A number of studies have already shown the existence of a range of dissimilarities among local renewable energy projects. According to these studies, projects can be differentiated according to whether the owning, running, and/or profiting organisations are private businesses, public institutions, or '*grassroots initiatives*' (Seyfang/Smith 2007). Grassroots initiatives, again, might vary, for example with regard to the organisational form chosen. Organisations as diverse as charities, cooperatives, and community development trusts have been identified (Seyfang et al. 2014). Community engagement is another differentiating factor. Projects vary with regard to the involvement of citizens in planning as well as profiting from local renewable energy projects (Walker/Devine-Wright 2008). Bauwens (2016) found very heterogeneous motivations among individuals participating in community renewable energy projects. Also, projects have attained very different levels of 'success' (Kunze 2011). While some projects are able to install even more energy production unities than had originally been anticipated, other projects die somewhere along the way. Studies have also analysed variances between renewable energy projects from different national backgrounds (Schreurs 2008; Breukers/Wolsink 2007). These studies have, for example, shown how national institutions and policies influence the realization of renewable energy projects. Most of the differentiating factors analysed for renewable energy projects are rather descriptive. Also, many studies are inspired by the aim to produce applicable results (for example, Seyfang et al. 2014; Schneidewind/Singer-Brodowski 2014).

In studying how energy and energy production are made sense of in these projects, this thesis has transferred one of the most basic sociological questions onto the issue of renewable energy. It aims to tackle two fallacies inherent to all existing studies on local renewable energy projects. First, this thesis addresses the inherent idea within all existing studies—the assumption that energy is 'somehow just there'. To counter this common idea, this thesis analyses how energy is produced through specific human activities. These activities take place in specific social situations. As such, energy is an outcome of particular social contexts. More precisely, as a specific type of human activity, renewable energy production is always embedded into the social contexts in which these activities are enacted. It is a meaningful part of the context in which it is enacted, and is both shaped by the context and shapes it.

Second, this thesis aims to challenge the lack of awareness that energy and the process of energy production are socially situated and thus have dissimilar meanings in different projects. By not acknowledging this difference, scholars tend to homogenize local renewable energy projects. Different ideas about energy and energy production within one project are overlooked. Challenging this inherent homogenization of renewable energy (production), this thesis is interested in the complex and dynamic understanding of energy and energy production within certain contexts.

By focusing on the contextuality of sense-making within local renewable energy projects, this thesis realises one of the main principles of qualitative social research (Hollstein/Ullrich 2003: 36). According to this principle, sense-making is always embedded in certain contexts and can only be understood in relation to these contexts. Studying energy and energy production as outcomes of contextualized, complex, and heterogeneous sense-making requires a theoretical approach that is able to not only uncover these complexities, but also explain how heterogeneous positions are negotiated within the projects. Taking serious insights from postmodern thinking, negotiation is understood as a dynamic, constantly contested process (Clarke 2003).

In order to analyse how energy and energy production are understood in the projects, this thesis uses a practice theoretical approach. Scholars have already applied practice theories to environmental, specifically low-carbon, issues. Studies have, for example, looked at social practices and how they change with regard to the environment (Hargreaves 2011) and energy consumption (Shove/Walker 2014; Strengers 2012; Røpke 2009; Shove/Pantzar 2005). The interplay of social practices and smart technologies such as smart homes (Wilson et al. 2015; Strengers 2012); various domestic technologies (Hargreaves et al. 2011), or smart grids (Naus et al. 2014; Nyborg/Røpke 2013) have also been analysed with practice theoretical approaches. Specifically, with regard to warmth, scholars have analysed the relationships between heat comfort and social practices (Gram-Hanssen 2010) as well as the social practices in the context of cooling (Strengers/Maller 2011).

Among the growing number of practice theories, I have adopted the approach developed by Theodore Schatzki. Analysing energy production with his concept of social sites '*directs attention to how practices and arrangements causally relate, how arrangements prefigure practices, how practices and arrangements constitute one another, and how the world is made intelligible through practices*' (Schatzki 2010: 146). His concept offers a way to analyse how energy and energy production are positioned in, related to and have meaning for other elements in a specific context. Though it takes complexity and dynamic developments into account, his approach, however, does not specifically enable researchers to focus on these. Furthermore, he does not offer any theoretical tool to study how sense-making practices of different actors are negotiated in a certain context. To create a theoretical approach which is able to grasp this research interest, his concept is combined with Adele Clarke's situational analysis.

Adele Clarke's situational analysis is a method/theory package based on grounded theory. It provides a methodology that focuses on negotiations of different actor groups. Members of different '*social worlds*' negotiate, bargain with, or even coerce other social worlds in '*social arenas*' (Clarke 1991). Analysing these instances of negotiation with the tools of situational analysis draws attention to the complexities and instabilities that are part of every such situation. Combining practice theory with situational analysis offers a way to relocate the focus from the reproductive and routine aspects of embodied practices onto their negotiated and situated character in a specific situation or context.

Specifying the aforementioned research interests in accordance with the methodological premises and the theoretical framework, the first research question is:

How are energy production activities shaped by the specific context within which they are enacted? This question can be differentiated into two sub-questions: What specific motivations, ideas, knowledge, and intentions are underlying the production of energy? How do other elements of these contexts influence the production of renewable energy?

Energy production, however, is assumed to not only be shaped by but also shape the context within which it is produced. Consequently, the second research question is:

How does energy production shape the context within which it is enacted? This question can also be split into two sub-questions. How is energy production related to other activities within a certain context? How does it influence the human actors, material artefacts and local environments of the context?

These research questions are answered through a qualitative analysis of the three projects briefly described in the beginning of this introduction. Qualitative fieldwork in these projects was conducted between the beginning of 2012 and the end of 2013 and consisted of participant observation, interviews, and the collection of grey literature.

This thesis is structured as follows. Chapter 1 describes the state of the art of research on local renewable energy projects. It situates this thesis within the scholarly discussions of a certain scientific community and defines its thematic limits and boundaries. This chapter is followed by an explanation of the methodological premises, the research design, and the methods employed for the research. The description of the methodological premises is motivated by principles of qualitative social research and essentially aims to make transparent the why and how of this thesis. Deriving from the methodological premises, Chapter 3 develops the theoretical framework by first providing an introduction to practice theoretical ideas and concepts. Subsequent to the general overview, Section 3.5 details Theodore Schatzki's practice theory approach and provides an explanation as to why his approach best suits the research interests. As is elaborated, Schatzki's approach does not suffice to realise the specific research interest and methodological premises of this thesis. It is combined with Adele Clarke's situational analysis, described in Section 3.6. The last section of this chapter (3.7) details how the two approaches are combined and operationalized in order to analyse the three case studies. In the second part of this thesis, these cases are individually described and analysed. The conclusion summarizes the results and refers back to the research interest and questions.

1   State of the art

Different theoretical and empirical strands of literature have contributed to the development of this thesis. This chapter introduces those studies and theoretical works that have influenced my thinking and contributed ideas to the analysis and interpretation of the three case studies. Subsequent to some introductory remarks about the scope and limits of the literature review, this chapter provides an overview to the socio-technical science publications that have contributed to the analysis of renewable energy production from a theoretical perspective. The largest part of this review presents those empirical works that have studied new actors and activities in the field of decentralized and renewable energy production. The chapter concludes with an identification of existing research gaps and an explanation of how this thesis fits into the existing body of research.

1.1   Scope and limits

On the most general level, this thesis contributes to the body of research on transformation to low-carbon societies. So far, social sciences have had a marginal role in research on the causes and consequences of climate change (Brulle/Dunlap 2015: 2; Welzer et al. 2010). One reason is that most social sciences only started to engage with climate change in the beginning of the 1990s, at a time when the natural sciences had already developed sophisticated climate models (Brulle/Dunlap 2015: 4). Scholars have pondered the reasons for sociology's missing interest in climate change. One explanation is seen in the division of nature and 'the social' which underlies sociology. In order to legitimate the new discipline, early theorists had to separate sociology from other disciplines, especially the dominant natural and physical sciences. The differentiation between nature and society was enabled by Enlightenment's understanding of nature as '*primordial, autonomous, and mechanistic*' (Goldman/Schurman 2000: 564). Since then sociologists have focused on what was perceived as 'social facts', while 'nature' was left to the natural sciences (Dunlap/Catton 1979: 244). Since the 1970s, however, environmental degradation increasingly forced sociologists to recognize that nature and society are inseparable (Dunlap/Catton

1979). Especially climate change '*makes a mockery of the premise that society and nature are separate and mutually exclusive*' (Beck 2010: 256).

As a result of these historical developments, sociologists since the 1990s have started to take part in debates about climate change (Shove 2010). By the time social scientists started to argue the social factor in climate change, both the causes of climate change and the strategies to tackle it, had already been framed as natural science or technological issues (Brulle/Dunlap 2015). Increasingly, however, recognition has set in that

> '[t]he phenomenon of global warming is also driven by processes that cannot be adequately analyzed by physicists, geologists, or meteorologists alone. The dynamics of anthropogenic climate change is not merely a question of natural processes but first and foremost a question of economy, society, and culture' (Leggewie/Welzer.: 31009-1).

Social scientists argue that '*the drivers of anthropogenic climate change are deeply rooted in the routines of everyday life and the social structure of modern societies*' (Brulle/Dunlap 2015: 4; likewise Rosa et al. 2015: 32). What needs to change is not only the energy efficiency of technological artefacts, but also the ways they are used by and within societies. Likewise, scholars argue that it does not suffice to provide information about the most sustainable options to individuals and expect them to change their behaviour. Instead,

> 'the "problem" of human behaviour which leads to emissions needs to be placed within the wider contexts where social practices are undertaken. Norms and values shape practices, and so do infrastructures, institutional arrangements and systems of governance' (Moloney et al. 2010: 7615).

Acknowledging the social background of climate change also means to understand transformations to low-carbon societies as socio-technical processes (among others Leggewie/Welzer 2010; WBGU 2011). This nearly paradigmatic change in the perception of climate change has had effects on the existing literature. While technical and business studies still make up the main part of publications—representing the often diagnosed '*technological bias*' in discourses on transition (among others Sovacool et al. 2014, 2015)—the number of studies focusing on or at least including insights from social sciences has increased.

An important aspect in the socio-technical transformations towards low-carbon societies are changes in the energy system. Social scientists have criticized the '*blind spot*' in '*conventional techno-economic thinking*' (Lutzenhiser/Shove 1999: 217; likewise Sovacool 2014; Sovacool et al. 2015) concerning necessary changes in the systems of energy production, distribution, and consumption. These scholars argue that energy research has so far in most parts downplayed the role of human dimensions in energy production, distribution, and consumption. Sovacool and his colleagues state that '*energy advocates, the climate change community, and related policymakers need to recognize that energy production, consumption, and policy are both social and technical domains*' (Sovacool et al. 2015: 95). A greater degree of scholarly interest has to be generated for understanding the interplay of the technical and social components in the energy system (Sovacool 2014: 26).

Scholars from different disciplines within the social sciences have offered ideas about and insights into transformations to low-carbon societies. Geography, among others, has contributed with extensive debates about the change of land usage in the context of new infrastructure for renewable energy production in rural settings (among others Gailing/Röhring 2015) and the conflicts arising due to these transitions (Murphy 2010). Urban planners have provided insights into infrastructural changes that support mitigation and adaptation strategies in urban settings (Fröhlich 2011; Knieling et al. 2011; Condon et al. 2009). Apart from other important issues, an interest in the question of governance and formal participation instruments has originated from the political sciences (among others Stephan et al. 2015; Goldthau 2014; Kern/Bulkeley 2009; Betsill/Bulkeley 2006). International relations and treaties are another important topic that has been highlighted by political scientists with regard to low carbon transformations (Luterbacher/Sprinz 2001). Historians have offered insights into the progress of various transition processes (Geels 2011, 2010, Grin et al. 2010b; Markard/Truffer 2008).

Sociologists have worked on a broad range of topics in the realms of climate change and low-carbon transformations. Among many other issues, sociologists have studied the involvement of different social actors (Sommer/Schad 2014; Schaefer Caniglia et al. 2015; Dunlap/McCright 2015; Voss/Schildhauer 2016), the role of social institutions and how they (need to) change (Dunlap 2010; Buttel 2010), and the practices of energy consumption (Shove/Pantzar 2005; Nyborg/Røpke 2013; Shove/Walker 2014). Furthermore, sociologists have analysed discourses on climate change (Weingart et al. 2000; Reusswig 2010), aspects of environmental and/or climate justice (Harlan et al. 2015), and the influence of existing and emergence of new markets (Engels et al. 2008; Engels 2010; Perrow/Pulver 2015).

As this short overview illustrates, the existing wealth and diversity of literature cannot be captured in one chapter. Instead, this review focuses on those bodies of research that are directly relevant to the research question and interest. Only studies that centre on energy production activities at the local level are included. While acknowledging that transformations to low-carbon societies are multi-level governance processes (among others Betsill/Bulkeley 2006;

Kern/Bulkeley 2009), in which local, regional, national, and international levels are interconnected, neither the multitude of relevant social levels nor their interactions are included here. For the same reason, research publications dealing with global or international issues do not feature in this chapter. This overview concentrates on the topic of the thesis—the local production of renewable energy. Thus, despite being inextricably related to renewable energy production, the issues of conventional energy production, energy grids, energy efficiency, or energy consumption are not part of this chapter. To further limit the scope of this review, only publications that provide insights into the German and the British context are presented. Publications from other national contexts are included only if they are significant for the scientific discussion of energy transitions in Germany or the UK.

Important contributions on the other hand, not only originate from scientific institutions. Publications by political organisations like the report from the German Advisory Council on Global Change ('*Wissenschaftlicher Beirat der Bundesregierung Globale Umweltveränderungen*' or WBGU) or the Energy White Paper in the UK and publications from Federal Ministries or different federal state institutions are relevant to the political and scientific debates in both countries. Publications from civil society institutions have also contributed to the British and German transformation discourses. Among these are environmental organisations like Friends of the Earth or Greenpeace, funding organisations like the Heinrich-Böll Stiftung or Community Energy Scotland, and organisations as varied as the umbrella organisation for cooperatives in Germany ('*Genossenschaftsverband Deutschland*') and the German association for cooperatives ('*Deutscher Genossenschafts- und Raiffeisenverband*' or DGRV), as well as community organisations in Great Britain (Development Trust Association Scotland). Publications from these organisations feature in this literature review if they contribute to the process of finding, developing, and answering the research questions.

1.2 Theoretical approaches to renewable energy production

One subfield in the social sciences that has frequently been dealing with issues of energy production and distribution is science and technology studies (STS). Within STS, the socio-technical branch in particular is '*concerned with explaining how social processes, actions, and structures relate to technology*' (Mackay/Gillespie 1992: 658). While the socio-technical approach is not a unified set of theories but rather a research agenda (Bolton 2011: 37), a common concern of all socio-technical approaches is the rejection of technical determinisms

(Mackay/Gillespie 1992: 658pp.; Bolton 2011: 37). They are further united '*by an insistence that the "black box" of technology must be opened, to allow the socio-economic patterns embedded in both the content of technologies and the processes of innovation to be exposed and analysed*' (Williams/Edge 1996: 866). In essence, socio-technical approaches are interested not in the social impacts of certain technologies or innovations but in the range of social factors that form the basis of the design and implementation of technologies (ibid: 865).

When the '*new sociology of technology*' (Bijker/Pinch 1987: xiv) came into existence in the 1980s, three approaches constituted this new field of research: the social construction of technology (SCOT) approach, the systems concept, later renamed as large technical systems (LTS), and Bruno Latour's actor-network theory (ANT) (ibid. xiv pp.). Here, I focus on the concept of SCOT. While the concept of LTS implies

> 'that large-scale technologies—such as electricity networks, railroads, telecommunications grids, roads and automobiles, and sewage systems—weave together technical artefacts, organisations, institutional rule systems and structures, and cultural values' (Sovacool 2014: 24pp.).

and thus fits my research interests, the approach is—as the name suggests—dedicated to explaining large technical systems, while my focus is on small and medium-sized energy production systems. Ideas from ANT that are important for this thesis are included in both Adele Clarke's situational analysis and Theodore Schatzki's concept of social practice.

SCOT approaches derive from the motivation to combine social constructivist approaches from the study of science with the study of technology (Pinch/Bijker 1984). Adapting the idea '*that there is nothing epistemologically special about the nature of scientific knowledge: it is merely one in a whole series of knowledge cultures*' (ibid: 401) to the study of technology, the implication is to treat '*technological knowledge in the same symmetrical, impartial manner that scientific facts are treated within the sociology of scientific knowledge*' (ibid: 405). Technological artefacts are not understood to be outcomes of quasi-natural evolutionary processes, but as social constructs. When Michel Callon (1986) analysed the controversies and negotiation processes evolving around the development of electric vehicles in France, he showed that the technology was not the outcome of a selection process of the technologically best options, but that a controversy involving different actors' heterogeneous aims, ambitions, resources, ideas, and power access crucially shaped the technology. Analysing the changing patterns of renewable energy implementation in the UK, Walker and Cass argue that in order to understand the mutual construction of technologies and the society, studies should '*focus on the relationships between an object and surrounding actors*' (Walker/Cass 2007: 459).

This last quote describes one of my main research interests. Technologies do not exist in social isolation; they are related to the social contexts within which they are invented or used. Throughout the analysis of the three cases, readers constantly encounter the idea of SCOT approaches that energy production technologies, their meaning, and their usage are related to the social context within which they are created or used. However, while STS approaches have inspired and influenced this thesis, they focus on social aspects that are directly related to certain technologies. They start with a certain technology by identifying the social elements or aspects that have shaped or are shaped by it. STS concepts—like the ANT (Schatzki 2010: 135), however, do not provide any conceptual tools to analyse *how* technologies are related to the surrounding actors or contexts. This is the aim of this thesis. While STS has contributed the general idea of studying the social contextuality of renewable energy technologies and thus has significantly shaped the research interest, it is not used for the analysis of the three case studies.

1.3 Research on civil engagement in local energy production

The second field of transformation research relevant for this thesis are empirical studies, studying the activities of different actors in transformation processes. My research interest and case studies relate this thesis to those studies that deal with the engagement of civil society actors in energy production. The subfield of studies on public engagement in energy production constitutes the research community into which this thesis is most intensely embedded.

Research on civil engagement in local energy production can roughly be differentiated into two streams. The first (and older) of these is interested in studying civil protests against energy projects. A large part of this stream consists of acceptance and risk-perception studies. The second stream, which has recently emerged in the social sciences, is distinctly interested in new actors and activities related to the production of renewable energy. This second stream of research has been inspired by the increasing participation of actors from civil society in energy production (among others Seyfang/Smith 2007; Kunze 2011; Becker et al. 2012; Becker et al. 2013; Seyfang et al. 2013; Fuchs/Hinderer 2014; Schmid et al. 2016). These studies are interested in the different aspects of those projects, in which citizens become the producers of renewable energy. Before explaining the main issues in these two fields, I describe some fundamental aspects of the history and actual political processes related to (renewable) energy production in Germany and the UK. Providing this background information is crucial to understanding the historical and social contexts within which conflicts against and engagement of citizens in energy production occur.

Different social processes have contributed to the growing number and diversity of actors engaged in energy production. Technological developments are an important aspect of this development. The technical and economic structures of the existing energy systems in Germany and the UK date back to the beginning of the 20th century and are still largely in existence today (Mautz et al. 2008: 11). Technical innovations from the end of the 19th century enabled the production and utilization of energy in large scales, which could also be transported over long distances (Urry 2014: 4). Conventional energy production utilizes resources like coal, gas, or uranium, which can only be extracted and processed in an economically viable manner when organised in centralized production units (Fuchs/Hinderer 2014: 354). The produced energy is transported via energy grids to distant consumers. Energy production thus takes place at those places that are economically most feasible for centralized energy production, at sites that are often geographically distant from the places of energy consumption (Mautz et al. 2008: 11pp.; Kocka 1990: 18). Furthermore, conventional energy systems are characterized by a high degree of market concentration. In both Germany and the UK, an oligopoly of few large energy providers dominates these nations' energy systems (Mautz et al. 2010: 11pp. for Germany; Winskel 2007 for the UK). While people consume energy on an everyday basis, they are not involved in the processes of energy production. These factors have contributed to the creation of a '*significant spatial and psychological distance between energy generation and use*' (Walker et al. 2007: 68).

By the middle of the 1970s, more and more people became frustrated with the ecological and social effects of the by then conventional energy system (Mautz et al. 2008: 33pp.). Inspired by ideas of '*soft energy*' (ibid: 33), members of the new ecological social movement started to puzzle over alternative ways to produce decentralized, inclusive, and environment-friendly energy (ibid: 34). Prototypes of wind turbines, biogas plants, and solar panels were developed in mostly non- or semi-professional settings (ibid: 44pp.). Within a short period, these technologies spread throughout society. While these new technologies were an important technical background for the decentralized production of energy, institutional factors have been found to be even more important than the technical features of energy production (Schmid et al. 2016: 272). Social scientists from both the UK and Germany have found that the processes of liberalization and privatization in the late 1980s and 1990s constituted a disruptive process (Winskel 2007: 184) that opened up opportunities for new market entrants (for the UK see for example Winskel 2007; Walker et al. 2007; for Germany see

Schmid et al. 2016). In Great Britain the Labour government throughout the 1990s despite liberalization still supported large scale private-sector led models of energy production. This changed in 2010 when the new coalition government was formed. One of the key policy aims of the new government was to reduce the influence of central government and devolve power to local communities (Eagle et al. 2017: 55). In 2011 the 'Localism Act' was passed, which—among other things—gave communities the right to protect and better control assets in their locality (ibid: 56).

At about that time, the increasing necessity to tackle climate change and other environmental issues induced a change in energy policy in both countries. Both countries have set more or less ambitious targets for carbon reduction and renewable energy production. The German targets envisage a 40% reduction of carbon emissions by 2020 and 80–95% by 2050, compared to emission from a 1990 baseline (www.bmub-bund.de, 27.08.2016). In order to achieve these targets, the German national government aims to increase the percentage of renewable energy to 40–45% by 2020 and 55–60% by 2035 (ibid.; Klemisch 2014: 154). Great Britain likewise has decided to reduce carbon emissions by 80% by 2050 (www.gov.uk, 27.08.2016.). According to the national renewable energy action plan, Great Britain aims to produce 20% of overall energy consumption from renewable sources by 2020.

Also contributing to changes in policy and practice of energy production, especially in Germany, is the active engagement of civil society actors against conventional energy production, particularly against nuclear- and coal-based energy production (Mautz et al. 2008). Activities of the large energy providers have increasingly become subject to local, regional, national, and (in the case of nuclear waste transports) even transnational criticism and conflicts. Most conflicts in the energy sector develop around infrastructural projects. The biggest conflicts, in terms of number of actors involved, concern activities like pithead mining, mine dumping, nuclear waste, and related issues like the evacuation of villages (Schumann et al. 2010; Moss et al. 2014; Becker et al. 2016). Additionally, new technologies, like fracking and carbon capture and storage (CCS) are heavily contested both generally and with regard to specific testing sites (Schumann et al. 2010: 54; Rost 2015).

While many studies have found high levels of public acceptance for renewable energy, the construction of renewable energy facilities has also caused numerous conflicts (among many others van der Horst 2007; Becker et al. 2012; Becker et al. 2016). Among renewable energy projects, wind farms are the most controversial technology (a conclusive review of the literature on protests against wind farms is given in Devine-Wright 2005). The level of civil society protest against wind parks varies among different countries, with anti-wind organisations being more common in the UK than in Germany or Denmark (Toke 2007: 167). Solar parks and biomass projects have also given rise to local conflicts (Becker et al. 2012: 45; Otto/Leibenath 2013), albeit on a much smaller scale than wind farms. Technologies differ in terms of their environmental, social, and economic impact, depending on the natural source being used and the way it is being used. Protests against a wind-farm will be based on other arguments than protests against a biomass plant (Devine-Wright 2007: 8). Infrastructural projects related to energy supply have also been subject to protest. A number of conflicts grow around the construction of new overhead networks for the transport and distribution of renewable energy (Zimmer et al. 2012; Neukirch 2014). Lately, initiatives aiming for re-communalization of energy utilities, for example in Hamburg and Berlin, have aroused a lot of scientific interest in Germany (Blanchet 2015).

Protests against energy projects are a key obstacle facing renewable energy projects, whereby *'[l]ocal people have been identified as the chief influence on planning outcomes in places like the UK where population density is high and where landscape issues loom large'* (Toke 2007: 168; likewise Wüstenhagen et al. 2007). Protest against development plans were found to be particularly strong if communities had no or only limited opportunities to take part in the planning and realization of energy projects, and especially *'if major utility companies were seen as making money at the community's expense'* (Walker et al. 2007: 71; likewise Warren/McFadyen 2010). The phenomenon of citizens being supportive of renewable energy in principle while at the same time heavily opposing the construction of the necessary facilities in their own vicinity has come to be known as 'NIMBYism' (not in my backyard). The term, however, has often been found to be used as a strategy to de-legitimize local protest against energy projects (Devine-Wright 2005; Ek 2005; van der Horst 2007). It homogenizes the activities of very different groups that may be opposing particular energy projects or aspects thereof for very different reasons. Despite the differences between individual protest activities, citizen initiatives have been found to play a major role in all these conflicts (Schweizer-Ries et al. 2013: 24; Neukirch 2014: 20). They are able to mobilize the necessary resources as well as to organise and focus the different aims and interests of heterogeneous actor groups.

The recent developments in political thinking described above have led to the creation of certain policy instruments, the most prominent of which are feed-in tariffs (FITs), which have significantly influenced the production of renewable energy by individuals and small producer organisations in both countries. When the FIT in Germany was redesigned to conform to the Renewable Energies Act (EEG) in 2000, this led to *'an increase in the share of renewables in electricity generation from 3% in 1990 to 26% in 2014'* (Schmid et al. 2016: 264). By setting definite prices for each kilowatt–hour of energy produced from renewable sources which are fed into a grid, FITs have established a financially sound and trustworthy base for small-scale energy producers, who can be assured of getting a certain price for the coming years. As such, FITs contribute to the emergence of new actors, among them individuals, households, communities, or co-operatives, in energy production, *'creating a highly diverse actor landscape'* (Moss et al. 2014: 4; likewise Becker et al. 2012: 43) in the Energiewende. In fact, citizen energy organisations *'Bürgerenergie'* (among many others Kress et al. 2014) play a major role in the German Energiewende. Citizen energy groups own up to 47% of the installed renewable energy capacity in Germany, while the conventional private energy companies only have ownership of about 12 percent (trend:reserach 2013: 69).

Like in Germany, civil society renewable energy projects in Great Britain are historically based on alternative energy initiatives from the 1970s (Walker/Devine-Wright 2008: 497; Seyfang et al. 2013: 978). Traditionally, the British energy system has been characterized by a high degree of centralization and dominance of large private energy companies since its reorganisation around nuclear power in the 1950s (Winskel 2007: 184). In the late 1990s, however, a new governmental discourse emerged, which emphasised the potential benefits of a more localized and distributed energy generation and participation of local actors (Walker et al. 2007: 68). This new discourse was further established by the influential 2003 Energy White Paper. In this document, supporting community energy was mentioned as a way to increase the percentage of renewable energy production and thus to meet the government's carbon reduction targets. Also, supporting community energy provided a strategy for the national government to circumvent European directives about energy incentives (ibid.). In their interviews with community energy project leaders in rural areas, Walker and colleagues found that a further important impetus for policy programmes was

> 'related not to energy policy objectives per se, but to the social and economic outcomes that could be derived from community RE [renewable energy, AP] projects. In the context of a narrative of "countryside in crisis" and the urgent need for rural regeneration, renewable energy projects were seen as a way to provide new sources of income and employment for communities suffering from agricultural decline, depopulation and economic collapse' (Walker et al. 2007: 73).

Walker and colleagues conclude that the policy discourse in Great Britain is not driven by a climate change or local sustainability agenda; rather it coalesces around the notion of community (ibid: 74). Recent policy measures seem to have benefited community energy projects in the UK. *'However, commitment towards large scale generation facilities is more widespread within the UK's governance regime of energy policy'* (Nolden 2013: 545). Especially key sociotechnical features, such as market support and planning arrangements, still favour large enterprises and keep community energy projects at the margins (Strachan et al. 2015: 96; Markantoni 2016: 157). In 2013, only about 10% of the electricity generated from renewable sources originated from community energy projects (Nolden 2013: 543). In 2014 the Department of Energy and Climate Change (DECC) released its Community Energy Strategy, which sets out the foundations for local engagement in energy production (Markantoni 2016: 157). As a consequence, community energy is gaining momentum in the UK, albeit on a still rather low level. In Germany on the other hand, the energy transition is mainly a project involving citizens, communities, and small private companies. Only 5% of renewable energies are produced by the Big Four energy companies (trend:reserach 2013).

Political programmes and schemes have generally been found to be an important prerequisite for local renewable energy projects. The increasing number of decentralized, local, and renewable energy projects has motivated a growing number of scholars to study the prerequisites, organisation, and outcomes of local renewable energy projects from a social science perspective (among others Kunze 2011; Li et al. 2013; Moss et al. 2014). In particular, projects based on or involving civil society actors have recently received growing scholarly interest. While Kunze in 2011 only found a very limited number of scientific publications on these types of renewable energy projects in Germany, since then the scientific 'database' on these actors has been growing steadily. In the British discussion, first publications about community energy and grassroots initiatives date back to the beginning of the millennium (among others Hinshelwood 2001; Leaney et al. 2001) and have been proliferating since around 2007. Different terms have been used to describe decentralized energy production projects dominantly based on civil society actors. *'Bottom up'* (Leibenath 2013) and *'grassroots'* (Seyfang/Haxeltine 2012; Seyfang et al. 2014; Hargreaves et al. 2013; Middlemiss/Parrish 2010) are commonly used terms. While these terms emphasise on civil society actors, they imply a homogenizing perspective of the different actors involved. Both terms associate these projects with local actors from the civil society. In all three case studies analysed in this thesis, however, different types of actors were found to play important roles in the planning and realization processes. Civil society projects cannot be realised without consent, permission, and/or support from public authorities. Also, 'professional' actors like technical advisors or financial counsellors have contributed significantly to all three projects. Likewise, actors from the civil society have taken part in the municipal case study. Maybe even more importantly, official actors have worked extra hours and were inspired by private motivations and convictions when working on the project. Consequently, the terms grassroots or bottom-up (or conversely top-down) are not used in this thesis. Another common term that is used by not only scholars but also practitioners is '*community energy*' (Walker et al. 2007; Walker/Devine-Wright 2008; Hinshelwood 2001). Here, I follow the definition of community energy given by Walker and Devine-Wright, who identified two key dimensions underlying community energy projects.

> 'First, a process dimension, concerned with who a project is developed and run by, who is involved and has influence. Second, an outcome dimension concerned with how the outcomes of a project are spatially and socially distributed—in other words, who the project is for; who it is that benefits particularly in economic or social terms' (Walker/Devine-Wright 2008: 498).

According to these two dimensions, community energy projects are projects in which '*communities exhibit a high degree of ownership and control, as well as benefiting collectively from the outcomes*' (Seyfang et al. 2013: 978).

Community ownership is organised differently in Germany and the UK (for Germany see Moss et al. 2014; for the UK see Seyfang et al. 2014). Energy cooperatives have a long tradition in Germany; the first electricity cooperative was founded at the end of the 19th century (Yildiz 2013: 177). In the last few years, energy cooperatives have become the fastest growing form of citizen energy production organisation in Germany (DGRV 2016). Since 2009, the number of energy cooperatives has increased significantly, with more than 100 new cooperatives being registered in the DGRV between 2009 and 2013. The insecurities arising from the amendment of the EEG in 2014 has caused a drop in the number of foundations, with 'only' 40 newly registered cooperatives in 2015 (ibid: 5). While most energy cooperatives produce electricity, cooperatives also produce heat—they own and run both electricity and heat grids (ibid: 7). About 800 cooperatives are active in the energy sector nowadays. As cooperatives have to act economically, their engagement in the energy sector is partly related to the creation of new institutions, regulations—not only the EEG, but also regulations concerning cooperatives—and technologies through which small-scale energy production has become a financially viable activity (Klemisch 2014: 152pp.). Other reasons are seen in the social learning and communication of best practice examples (ibid: 152). As cooperatives have to ensure a safe investment for their members, they are found to be one of the most insolvency-safe organisational forms (AAE 2013: 5). From a participatory perspective, one of the main advantages of cooperatives is that the amount of money needed to buy shares and thus become a member is normally quite affordable. They consequently enable individuals or groups with limited financial means to become actively involved in and profit from renewable energy production (Ott/Wieg 2014: 830). Furthermore, cooperatives often plan and realise renewable energy projects in cooperation with residents, local policymakers, public authorities and organisations, and regional banks (AAE 2013: 4). Scholars have found energy cooperatives to be important actors for the German energy transition, which combine environmental, economic, and social aspirations in their activities (Yildiz 2013; Ott/Wieg 2014; Klemisch 2014; Elsen 2014).

In Great Britain, community energy projects have been organised in a range of different types '*including voluntary organisations, cooperatives, informal associations, etc., and partnerships with social enterprises, schools, businesses, faith groups, local government, or utility companies*' (Seyfang et al. 2014: 24. likewise Clark/Chadwick 2011). Unlike in Germany, the cooperative model is not very common in Scotland. In 2012, only about 12% of the community requirement was produced by cooperatives. Most community energy projects in Scotland instead are organised in the legal form of development trusts (Harnmeijer et al. 2012).

Both in Germany and the UK, the scale of most community energy projects is small or medium, ranging from single photovoltaic arrays to medium biomass or wind farm projects (Nolden 2013: 544). In a large-scale quantitative survey of community energy projects in Britain, Seyfang and colleagues found that, with regard to technologies used,

> 'the field was clearly dominated by installations of solar photovoltaic renewable technologies (71%). The next most common types were solar thermal (23%), ground source heat pumps (22%), onshore wind (20%), air source heat pumps (16%), biomass (14%), and hydroelectric power (14%)' (Seyfang et al. 2013: 983).

Especially micro-hydro power has expanded drastically in the UK (Bracken et al. 2014). Such specific numbers could not be found in the German context. In energy cooperatives, the dominating technology for community energy projects is solar photovoltaic (Yildiz 2013: 173). Wind and biomass, as sources of energy production are respectively the second and third most planned key investment activities for 2015 (DGRV 2016: 8).

Besides civil society organisations, local governments are another important player in local renewable energy production. Many studies have underlined the relevance of local public authorities in climate change-related activities. Local governments have many tools at their disposal for implementing climate policy. They have legislative and executive control of many of the factors related to the production of renewable energy production, including land use planning and regulation, building permission and emission rights, and (sometimes) the supply networks (Kousky/Schneider 2003: 370; likewise

Betsill/Bulkeley 2006: 143). While the 'passive' role of local governments in the production of renewable energy has been acknowledged in many scientific publications, active engagement—for example in the form of municipal utilities—has not received much scholarly interest so far. In the British context, this can be attributed to the fact that for the last century, local authorities played only a very limited role in the provision of energy services (Hawkey et al. 2013: 22). After the number of municipal utilities declined for most part of the last century in Germany (Kunze 2011: 158), different communities have recently started to either re-communalize formerly privatized utilities or to establish new ones (Becker et al. 2013). Studies have found that reasons for local authorities to become actively engaged in renewable energy production might include:

> 'local social and economic benefits including mitigating pollution, retaining greater proportions of energy payments in local economies, reducing energy costs and cost fluctuations for residents, businesses and public sector organisations, and contributing to regeneration' (Hawkey et al. 2013: 22).

For realising and running renewable energy projects, local governments might collaborate with private or civil society organisations (Mautz et al. 2008: 105; Kunze 2011; Webb 2012; Becker et al. 2013: 24).

In both Germany and the UK, most of the new actors and renewable energy production sites are situated in rural and semi-rural areas and regions (Kunze 2011; Moss et al. 2014: 2). In Britain, 65% of community energy projects are rurally located while 23% are in urban areas and 12% in suburbs (Seyfang et al. 2013: 981). This situatedness of renewable energy projects is mirrored in literature. By far, most studies are based on rural or semi-rural cases (among many others Nolden 2013; Kunze 2011; Seyfang et al. 2013; Hinshelwood 2001; an urban focus on the other hand is taken by Webb 2012; Hawkey et al. 2013).

Regardless of the chosen organisational form, civil society groups are faced with a range of hurdles and challenges when planning and realising community energy projects (Walker 2008; Rogers et al. 2008). Generally, community energy projects *rely on people with limited power, limited resources, and limited ability to influence others*' (Middlemiss/Parrish 2010: 7559). They often lack necessary resources and support (Hinshelwood 2001: 96). Time, finance, and expertise are especially scarce resources for community energy projects (ibid: 104pp.; Gubbins 2007).

In order to realise a community energy project, local communities need to mobilize different resources (Bomberg/McEwen 2012; Middlemiss/Parrish 2010). The mobilization of financial resources is a crucial aspect of the realization of renewable energy projects (Seyfang et al. 2013: 984). The high upfront investment necessary to purchase renewable energy production technologies makes it more or less impossible for communities to realise renewable energy projects without financial support. The mobilization of financial resources is difficult because of the limited amount of financial support opportunities and the high degree of competition between groups to access the existing funds and grants. In particular, communities lacking the specialist knowledge, information, or time resources needed to produce a successful application will face difficulties in raising the necessary financial resources. Within the process of applying for funds, the choices available and the conditions of prospective funding organisations influence the structure and planning of projects (Hinshelwood 2001: 96). While this shaping effect is often beneficial in terms of the project becoming more feasible or realizable, it might also result in communities compromising some of their ideals or principles in order to access funding (ibid: 101). '*The ideal situation of raising enough money from donors to carry out community ideas without compromise is rare*' (ibid: 104). Furthermore, the financial power of external organisations might lead to a transfer of the community leadership and control over the project from community actors to funding organisations (ibid: 104). The ambivalent role of external bodies is apparently recognized by many project members. In a survey among community energy projects in Cumbria, Gormally and colleagues found that while most respondents preferred locally owned energy projects (2014: 921), the necessity for an external driver or body in order to overcome difficulties in terms of knowledge, skills and motivation was also recognized by most people (ibid: 922).

In addition to the provision of financial schemes like FITs, community energy projects require further institutional support (Nolden 2013: 544). One major requirement is the availability of land. Without access to appropriate sites, the installation of renewable energy production installations is just not possible. Energy projects might own or lease land. In Scotland, communities owning the land and the assets on it have been found to be more confident, ambitious and strategic in with their development plans (Callaghan/Williams 2014: 660). If institutional support through public authorities is low or absent, getting planning consent or building permission from local authorities can become a major obstacle (Gubbins 2007: 82). More generally, Eagle and colleagues argue that the success of the UK Government's localist policy is diminished by a persistent conservative push for economic individualism and neoliberalism, which prevents or hinders co-operation (Eagle et al. 2017: 59pp.). Without adequate central governmental structure, funding and the creation or support of platforms for community energy projects to share information and advice are limited (ibid: 61). Top-down support for community energy, especially through funding and advice from experts, were found to be crucial for creating and catalysing motivation (Markantoni 2016: 167).

Opposition within communities and hostility from local residents can severely hinder or even render impossible the realization of renewable energy projects (Middlemiss/Parrish 2010: 7559). Networks and social capital hence are an important resource for community energy projects (Seyfang et al. 2013: 985pp.; likewise Devine-Wright et al. 2001; Kunze 2011; Hinshelwood 2001; Allen et al. 2012). The existence of social capital, a collective identity (Bomberg/McEwen 2012), or trust (Walker et al. 2010) among the members of the community is both an important prerequisite for and an outcome of community energy projects. Community participation not only defines community energy projects but is also crucial for keeping them alive (Hielscher et al. 2013: 143). Public participation in community energy projects increases social capital within a community, as members from different backgrounds and different ways of life are brought together (ibid.). Nevertheless, pre-existing social cohesion underpins community energy projects rather than resulting from them (Devine-Wright et al. 2001; Walker et al. 2010; Haggett et al. 2013). Partnerships with established local organisations have also been found to have positive effects. Besides providing access to information, support, and knowledge, partnerships with already established organisations improve a project's credibility and legitimacy both locally and externally (Hinshelwood 2001: 106). Proving credibility and legitimacy to external actors is particularly relevant to the search for financial support or investment. Not only partnerships with established institutions, but also social networks, for example with policy makers, are relevant for the mobilization of financial resources (Brickmann et al. 2012: 76). Planning and realising a community energy project often requires partnerships and collaborations with professional technical organisations, support institutions, and public authorities at different levels and across different sectors (Hinshelwood 2001: 105). These are for example necessary for the mobilisation of expert knowledge. Expertise, however, is not only used but also contested within community energy projects (Bracken et al. 2014).

Many studies focus on the role of individual participants in community energy projects. Community energy projects are largely dependent on the mobilization of volunteers (among others Callaghan/Williams 2014; Gormally et al. 2014; Bauwens 2016). Generally, actors from within a community are differentiated from external actors like support agencies (Hinshelwood 2001). Scholars have found that members of community energy projects from within the community might have very different motivations for their participation (Bauwens 2016). Gormally and colleagues, however, in a survey conducted in three different community energy projects found that *'respondents ranked their motivations in the following order: financial (1), fuel security (2), helps reduce climate change (3), influence over siting of the development (4) and community partici-pation (5)'* (Gormally et al. 2014: 923). Generally, existing studies support the importance of individuals with the necessary competencies, resources, and com-mitment to mobilize resources and convince other community members or out-side actors; such individuals are essential for the realization of the project (Som-mer/Schad 2014). These *'change agents'*—actors who actively support and pro-mote ideas or concepts—have received a lot of scholarly interest (Mautz et al. 2008; Heins/Alscher 2013; WBGU 2011; Sommer/Schad 2014). With regard to initiating and realising energy projects, additional types of actors like innova-tors, transformers, the mainstream, and laggards have been analytically differ-entiated (Hinshelwood 2001). The concept of change agents and other actors can relate to individuals, groups, or organisations. If community energy projects are successfully realised, they might themselves become change agents for the en-ergy transition (Grin et al. 2010a; Heins/Alscher 2013).

External actors like support agencies often function as *'intermediaries'* (Seyfang et al. 2014; Hargreaves et al. 2013). Intermediaries are organisations that provide communities with advice, information, networks, and funding or support their resource mobilisation strategies (Eagle et al. 2017: 64). While on the one hand *'passing down'* resources to community energy projects, these in-termediaries also help to distribute best practice examples to other local groups or *'scale-up'* concepts or ideas (Hargreaves et al. 2013: 11). While congrega-tions of community energy projects as a collective actor and/or other intermedi-ary organisations are thought to have the potential to put pressure on local or national governments (Smith et al. 2016), intermediaries so far have been found to not provide the relevant tools for growth (Hargreaves et al. 2013; Seyfang et al. 2014).

If community energy projects are successfully realised, they are found to be beneficial to the local community (Cass et al. 2010). In fact, environmental, economic, and social sustainability and development of the community are key motivating factors for community energy projects (Hinshelwood 2001; Rogers et al. 2008). Decentralized energy production provides communities and regions with an opportunity to create a financial income for themselves (among others Middlemiss/Parish 2010; Kunze 2011; Bomberg/McEwen 2012; for a critical view see Munday et al. 2011). If communities produce their own energy, they are often able to distribute energy to local residents at a lower price than con-ventional energy companies (Moss et al. 2014). Local ownership of small scale energy generation thereby secures accessibility and affordability of energy and hence contributes to energy equity within a community by alleviating energy poverty (Forman 2017). Community energy projects hence might play an active role in the enactment and realization of energy justice (ibid.; Fuller/Bulkeley

2013). Furthermore, community energy projects can potentially produce financial overhang when the community sells excess capacity to the grid (Fuchs/Hinderer 2014: 362). They produce capital for the community; this revenue might be magnified further, depending on the degree to which money is spent locally and thus contributes to the local economy (Callagahn/Williams 2014: 659). According to recent research from Community Energy England (2015), many community energy groups are able to turn relatively modest public funding into far larger sums of private investment, which in turn is used for public benefit services and institutions. This potential of community energy schemes is especially valuable in areas experiencing withdrawal of public spending and services (Adams/Bell 2015: 1474). Also, community energy schemes have been found to generate additional income, for example by contributing to an area's tourist value (Kunze 2011). Economic benefits from community energy schemes can induce social improvements if the income is spent for public benefit, for example to improve local infrastructure or to support other community projects (Bomberg/McEwen 2012: 442). Other social benefits to the community include employment, training, and education opportunities for local residents (Brickmann et al. 2012; Berlo/Wagner 2011: 237). When appraising the social benefits of community energy projects, however, it should be noted that most projects in Scotland are realised in less deprived areas (Haggett et al. 2013). Likewise, in Germany, the number of energy cooperatives is much smaller in economically deprived regions, especially in the new eastern states of Germany (Klemisch 2014: 160). Furthermore, within a community, inhabitants might not only gain from energy schemes. Adams and Bell found equity questions and conflicts to emerge about questions of who will be able to buy into community energy schemes and who will profit from them on the one hand and who will suffer from decreasing property prices, views or loss of other amenities by for example a wind turbine (2015: 1480pp.).

With regard to sustainability, community energy projects have been found to successfully raise awareness for and knowledge about environmental issues, especially climate change, to introduce or improve sustainable consumption patterns, to increase acceptance of renewable energy projects, and to contribute to local, regional, and national carbon or renewable energy targets (Hinshelwood 2001: 98; Rogers et al. 2012; Seyfang/Haxeltine 2012; Seyfang et al. 2014). More specifically, according to Allen and colleagues, community energy schemes are able to motivate a range of other sustainable efforts in a locality (2012: 272). While the evidence is ambiguous with regard to changing consumption patterns, Rogers and colleagues (2012), in their case study about a wood-fuel heating scheme in the Lake District, found that four more biomass systems were installed in the area after the community had successfully realised the scheme. One explanation for this effect is that community energy projects offer new ways for individuals to perceive and interact with systems of energy generation and supply (Cass et al. 2010; Rogers et al. 2012: 239). The fact that community energy projects have developed a more holistic approach to energy than most conventional projects is especially relevant (Hielscher et al. 2013). In community energy projects, energy production is often interlinked with energy consumption campaigns or other environmental, social, and economic issues. Community energy thus offers a way to overcome the geographical and psychological isolation of energy production, which is an outcome of the conventional energy system and which is seen to have contributed to heedless consumption of energy (Rogers et al. 2012: 239). Community energy projects not only make energy production 'visible' to local residents by situating it close to their day-to-day life, but actively involve citizens as participants in the production process or as beneficiaries of the outcomes (Peters et al. 2010). Local actors are provided with the opportunity to transform from passive consumers to active producers of their own energy supply. Dissolving the boundaries between producers and consumers, citizens active in community energy projects become '*prosumers*' (Toffler 1980; Schleicher-Tappeser 2012: 69; Huener/Bez 2015; Ellsworth-Krebs/Reid 2016). Community energy projects not only create visible pilot projects and engage local residents, but also—through information and consultation meetings—introduce experts to the public and publish information about the background, the technological, environmental, and social aspects, and the success potential of the installation. This

> 'sets community energy initiatives apart from individual domestic, business, or public sector installations and from large-scale commercial developments. Although the latter may include consultation events, these are usually run on developers' rather than communities' terms, designed to close, not open up, discussion' (Rogers et al. 2012: 243; likewise Cass et al. 2010).

Furthermore, by engaging individuals in a common endeavour, community energy projects provide a meaningful context for individual action on sustainability (Rogers et al. 2012: 240). This is particularly important because issues like climate change and energy transitions are large scale and distant and might make individual engagement seem futile. Integrating individuals in a shared context for action could assure individuals that their activities are meaningful. Engaging citizens in community energy projects could thus contribute to energy behaviour change in individuals (Heiskanen et al. 2010; Middlemiss 2008). On the other hand, community energy groups might emerge as a collective actor, influencing policy and business (Smith et al. 2016: 409). While many community energy groups are wary of being defined as political actors, they have been found to contribute to raising political awareness. More or less all community energy projects sooner or later '*encounter impediments arising from social structures inherent to regimes. Influence is seen arising through the shared discussion, awareness, reflection, and points of action towards these social structures*' (ibid: 412).

Internal evaluations of a project's success might be very different from the perspective of an outsider—political, financial or scientific—actor. While scholars often evaluate a project's success with regard to CO_2 reduction, produced energy, or the financial income generated by these projects, the project members might be more interested in the successful realization of aims like participation, development of capacities, and realization of the project in general (Hinshelwood 2001: 101). Most scholars who have studied local projects, especially community energy projects, have found that energy production is often not the only aim of these projects (Seyfang et al. 2013; Schmid et al. 2016; Strunz 2014; Fuchs/Hinderer 2014).

The literature on community energy deals with many different aspects of these types of local projects. However, certain research gaps do exist. Most of the existing studies about community energy focus on electricity producing projects. In Germany, 32.6% of the electricity in 2015 was generated by renewable sources, while only 13.2% of heat was produced through renewable technologies (www.erneuerbare-energien.de, 27.08.2016). The situation in the UK is not much different. The small number of publications on renewable heat projects reflects the asymmetrical reality of renewable heat production. Also, there are hardly any studies dealing with renewable energy production projects in urban settings. As with heat, this trend reflects the current empirical situation in which most renewable energy is being produced in rural settings (Gailing/Röhring 2015: 32). By considering two urban projects and one rural project, this thesis aims to reduce this research gap. Additionally, this thesis studies local renewable energy projects in different national settings. While the projects are not directly compared with one another, this thesis nevertheless integrates and reproduces knowledge about the two national settings and thus contributes to the increasing awareness about aspects of German and British energy transitions.

The most important research gap in the community energy field of study concerns the heterogeneity of community energy projects. While most studies mention differences between community energy projects, for example with regard to the organisational structure, these have not been the focus of research so far. Heterogeneity among community energy projects has in fact often been downplayed or ignored in order to differentiate these projects from more conventional types, or to develop typologies of different actor types within these projects. Studies that acknowledge the differences between community energy projects (Seyfang et al. 2013; Walker/Devine-Wright 2008; Kunze 2011; Bauwens 2016) do not pay specific attention to the differences about the role and meaning of energy and energy production in these projects. Heterogeneity has so far only been seen as a specific topic with regard to the reasons for, type, quantity, and quality of community involvement (Walker/Devine-Wright 2008; Bauwens 2016). This thesis addresses this research gap by analysing the highly specific ways in which energy and energy production are embedded into particular contexts. Thereby, I not only focus on differences between the projects but also pay attention to the ambivalences, heterogeneities, and complexities within each project. The aim is to show not only the distinctiveness of local renewable energy projects, but—more specifically—the situatedness and contextuality of energy and energy production.

2 Methodology, research design, and methods

An important principle of 'good' scientific work is that researchers should keep their methods of data collection and analysis transparent in order to explain and account for how they have arrived at their results (Silverman 1994: 145). A second scientific principle regarding methods and methodology is that these have to be defined by the research interest and question (Silverman 2013: 11; likewise Flick 2009: 15pp.). For this PhD project, the research design, methodology, and methods had originally been chosen to fit the research interest of studying the (re)production of social inequality in local renewable energy projects. In the course of the analysis and interpretation of the material gathered during the fieldwork process, however, the original research interest and questions changed, due to the insights and ideas developed during the research process. Of course, these insights were shaped by the chosen research design, methodology, and methods. These created the type of information gathered and insights accumulated. Research interest, methodology, and methods have thus mutually influenced each other throughout this PhD project.

This chapter provides a short overview of the methodology, research design, and methods used to gather, analyse, and interpret empirical material on local renewable energy projects. Each section begins with an explanation of the theoretical background of methodology, research design, and methods. Subsequently, each section also describes their practical application in this PhD project. The chapter starts with an introduction into the methodology of qualitative social research. While grounded theory is the methodological approach that has played the biggest role in shaping the processes of data gathering and analysis, I do not venture too deeply into grounded theory in this chapter. Instead, grounded theory, in its postmodern version developed by Adele Clarke, is described as a method for analysis in Section 2.4 and as a theoretical approach in the subsequent chapter on theory.

2.1 Methodology

A methodology is defined as '*the basic belief system or worldview that guides the investigator*' (Guba/Lincoln 1994: 105). These belief systems or worldviews naturally impact the researcher and his or her assumptions and findings.

After spending most of its history at the margins of the social sciences (for an historical overview see Flick 2009; Mey/Mruck 2014, for an overview on actual developments see Knoblauch 2014), qualitative social research has now become part of the sociological mainstream (Mey/Mruck 2014: 13pp.; likewise Hitzler 2014: 56pp.).[2] 'Qualitative research' is an umbrella term for a variety of theoretical approaches, methods, and methodologies (among others Guba/Lincoln 1994: 105; Hollstein/Ullrich 2003; Mey/Mruck 2014; Kardorff 1995: 5; Kleining 1995). After introducing the core principles of qualitative social research, grounded theory, which has had the biggest methodological impact on the processes of data gathering and analysis, is introduced.

It is the intention of this thesis to analyse energy production as a contextualized social practice. The aim is to illustrate that energy production can be subject to very different meanings, intentions, and values depending on the context in which it is realised. I thus aim to understand and describe energy production projects like '*Sinngebilde*' (entities of sense) (Soeffner 2006) through the subjective and interactive processes of sense-making or meaning-giving. This research interest corresponds with qualitative social research methodology, which is based on '*the fundamental epistemological axiom of the interpretative paradigm*' (Hitzler 2014: 62, translation AP). The interpretative paradigm understands reality not as an objectively given pre-social entity, but as an outcome of the actors' sense-making processes. The subjects in a qualitative paradigm are not expected to construct reality in isolation. Instead, people's perceptions, behaviour, and practices are always accompanied by interpretations (Soeffner 2014: 36), which have in turn been shaped by socially objectified constructions of reality (Berger/Luckmann 1991 [1966]). While qualitative researchers reject the idea of an objective world, they recognize objectification of worlds.

---

2   The development of qualitative social research has been accompanied by extensive debates between adherents of the qualitative and the quantitative paradigm. One important point in these debates concerns the question of the scientific nature of qualitative methods. Quantitative scholars were critical about the ability of quantitative social research to meet the requirements of validity, reliability, and objectivity. Qualitative scholars have argued that instead of trying to fit qualitative methods to quantitative logics, the scientific principles need to be interpreted according to the core concepts of the interpretative paradigm. The scientific principles for qualitative social research are replicability and transparency of the research process. Moreover, methods need to be adequate with regard to the research interest and questions. For conclusive argumentation, see Przyborski/Wohlrab-Sahr 2014: 21pp.; Meyer/Meier zu Verl 2014.

'Indeed, they contend that some objectification is essential if human conduct is to be accomplished. [...] Objectivity exists, thus, not as an absolute or inherently meaningful condition to which humans react but as an accomplished aspect of human lived experience' (Prus 1996: 247).

According to the interpretative paradigm, reality is not objectively given but is interpreted and reconstructed by individuals who engage in sense-making processes. Qualitative research aims to reconstruct the constitution of social realities by reference to social actors' sense-making processes (Hitzler 2014: 64; Hollstein/Ullrich 2003: 35pp.). Seemingly objective circumstances within this paradigm are understood as objectified constructs (Hitzler 2014: 65). This approach not only underlines the subjectivity of human experience and existence (Silverman 2013: 6) but also aims to show that '*apparently obvious features of the social world depend upon complex social phenomena and processes*' (ibid: 7).

Deriving from this fundamental epistemological axiom, qualitative social research is interested in the reconstruction and interpretation of subjective sense-making (Hitzler 2014: 56; Kardorff 1995: 4). Qualitative research aims for '*methodisch kontrolliertes Fremdverstehen*' (Przyborski/Wohlrab-Sahr 2014: 14pp.), meaning a methodically controlled understanding of the other. It aspires to capture and understand social realities by reconstructing sense-making processes that are implicit in the social actors' day-to-day life (Przyborski/Wohlrab-Sahr 2014: 12; Lamnek 1995: 61). The reconstruction of sense-making can be differentiated into the reconstruction of subjective, social, or objective sense (Hollstein/Ullrich 2003). The aim of this thesis is neither to solely understand the subjective sense-making of individuals nor to reconstruct the underlying invariant objective structures (ibid: 37). Instead, I focus on reconstructing social sense-making processes, or the negotiation of orders, by actors in shared social contexts (ibid.). While being interested in these shared forms of sense-making, taking a qualitative approach means to seriously consider the variety of perspectives that are always present in each situation or context (Flick 2009: 16). Taking a qualitative methodological stance means to be aware of the continuing mutual relation and tension between subjective sense-making and socially shared senses (Przyborski/Wohlrab-Sahr 2014: 11).

To reconstruct social sense-making, qualitative research needs to seriously consider two principles. First, qualitative research needs to take into account the contextuality of sense-making (Hollstein/Ullrich 2003: 36). Sense-making is always embedded in certain contexts and can be understood only in relation to these contexts. Furthermore, to enable methodically controlled understanding of the other, researchers need to understand sense-making processes in their relation to the subjective context of their research participants (Przyborski/Wohlrab-

Sahr 2014: 17). The principle of contextuality accordingly requires researchers to enable research participants to present issues and describe phenomena using their own systems of relevance (ibid.).

Second, scholars need to be aware that by reconstructing the social actors' first-order constructions of reality, they engage in second-order constructions of reality (ibid: 13). While this reconstructive principle does not entail complete relativity of qualitative social research, it enjoins researchers to be sensitive to the fact that understanding is always preliminary, especially if it is based on pre-existing ideas about the research object. Qualitative social research realises this sensitivity through the principle of openness. Taking the principle of openness seriously entails not '*an empty head, but an open mind*' (Pulla 2014: 19; likewise, Richardson/Kramer 2006: 501) towards new issues and understandings coming up during the research process and their relation to theoretical concepts. How these principles and their different aspects are realised methodically is explained later (2.3 Methods).

Qualitative social research furthermore understands the process of data gathering, analysis, and recording of research results as an interactive and interpretative process. Researchers do not uncover an objectively existing reality, but engage in interaction with their research objects and subjects in interactive processes of interpretation. Being bound by their own subjective sense-making, researchers are able to interpret what they find only in terms of their own sense constructions.

2.2 Research strategy

Research strategies are not methods, but transform methodologically derived theoretical standards into instructions for practical research (Lamnek 1995: 5). Within social sciences, research strategies entail approaches like surveys, experiments, histories, and case studies (Yin 1981b: 59). Case studies have been called the '*implicit companion*' (ibid: 58) of qualitative research. The close relationship between qualitative research and case studies stems from the fact that case studies are best equipped to realise the core concepts of qualitative research, particularly context-sensitivity, day-to-day life, and openness to new and unforeseen information and insights.

Case studies are defined as '*scholarly inquiry that investigates a contemporary phenomenon within its real-life context, when the boundaries between phenomenon and context are not clearly evident; and in which multiple sources of evidence are used*' (Dooley 2002: 335pp.; see likewise Yin 1981b: 59). According to the quote, case studies are context-sensitive. Context-sensitivity derives from the fact that, unlike surveys, case studies do not research a phenomenon with regard to a number of preconceived variables for which they try to gather as many instances as possible. Instead, case study research '*emphasize[s] the study and contextual analysis of a limited number of events or conditions and their relationships*' (Dooley 2002: 337). Unlike experiments, which '*deliberately divorce a phenomenon from its context*' (Yin 1981b: 59) by studying the phenomenon in the artificially constructed context of the laboratory, case studies observe phenomena in their real-life context. Unlike the other two research strategies, case studies also aim to observe the everyday reality of a certain phenomenon. While case studies often study individuals, groups of people—like associations, families, or organisations—are also common entities in case study research (Lamnek 1995: 5pp.).

As two of the most renowned case study researchers summarize, critiques of case studies mainly state that '*this way of doing research (1) should be used at the exploratory stages, (2) leads only to unconfirmable conclusions, and (3) is really a method of last resort*' (Yin 1981a: 97), or on a more general level that '*[o]ne cannot generalize on the basis of an individual case; therefore, the case study cannot contribute to scientific development*' (Flyvbjerg 2006: 220). While this discussion is not detailed here (for more details see Yin 1981a; Flyvbjerg 2006), it is argued that these critiques generally derive from a quantitative logic in which generalization is understood to be crucial. Quantitative scholars look for representative samples. Representativeness refers to both the quantity and quality of the sample, which should be sufficiently large (in relation to the real population of phenomenon) and characterized by the same dispersion of characteristics as the original population. In most cases, quantitative research tries to ensure the representativeness of samples through random statistical sampling (Lamnek 1995: 21pp.). Qualitative research employs another strategy and logic of sampling. Theoretical sampling (Glaser/Strauss 1973 [1967]) is based on the idea that in order to develop theoretical concepts, cases should be selected in such a way that they provide complex, differentiated, profound, and even divergent information about the phenomenon of interest (Lamnek 1995: 22). In other words, cases in qualitative social research should be selected in such a way that they enable the development of more complex, differentiated, and profound theoretical concepts.

Aiming to gather as much information and dimensions about a case as possible means that most case studies use different methods. Case studies are generally open for all methods and techniques of social research (Yin 1981b: 58pp.; Lamnek 1995: 7). Using a set of methods ensures that case studies do not remain in the sphere of everyday life, but instead enable methodically controlled understanding. The case study approach in itself does not identify methods to be used for data gathering or analysis. Instead, methods have to be adapted to the respective research interests and questions. The next chapter introduces the methods chosen for this research project. First, however, this chapter explains the process of finding and choosing the three case studies on which the results within this PhD project are based.

The selection of case studies happened through a mixture of research interest and coincidence. I was interested in studying how renewable energy production is influenced by social processes and conditions. At that time my focus was specifically on conditions and processes related to questions of social power and social inequality—a focus that was abandoned during the research process. Through internet research, I learned about a decentralized renewable energy project being realised in Hamburg. The project was meant to be carried through by a daughter organisation of the Hamburg city government, and intended to contribute innovative solutions to urban planning in the context of climate change. At more or less the same time, through private contacts I got to know about the existence of another project which also decidedly aimed to create a decentralized renewable energy project. Especially interesting was the fact that both projects were intended to be realised in old bunkers from WWII and so far had developed a similar technological concept. With regard to material and technological aspects, the two projects consequently were comparable to one another. At the same time, however, the projects differed with regard to the participating actors as well as their ideas, histories, and organisational background and structure.

After learning about these two projects, I refined my research question to ask how social conditions—especially aspects of power—would influence and be (re)produced in these two projects. I chose the case study approach for studying the two projects due to the necessity of gathering complex and context-sensitive information about the project's daily activities, as I was interested in understanding how social conditions influenced and shaped different aspects of the daily organisation of renewable energy production in the two projects. Also I wanted to study how actors in the two projects handled and either reproduced or resisted these conditions in their daily activities. It was assumed that the different backgrounds of the projects would imply that different social conditions affect the projects, and that the different background would affect the handling of these conditions in the projects.

After starting fieldwork in both projects and also the preliminary analysis, I became conscious of polarizing the two projects. Specifically, I had begun to ascribe all differences between the two projects to their different organisational background and structure. In order to curb this tendency, I decided to include a third case that would force me to stop dichotomizing my findings, and enable a more complex and differentiated view. The idea was to find a project that would be comparable to the two cases while providing a different context of local renewable energy production. Based on these requirements, I decided to look for a suitable project in another national—and thus legislative, social, cultural, and political—context. The search was limited by the fact that in order to ensure basic comparability, the project needed to be a European project in which local heat would be produced from renewable sources. Furthermore, in order for me to be able to grasp interactions, members needed to speak German or English. These limitations confined the search to northern European countries. After some research, I came across a Scottish project in which a PoW camp from WWII was reused to produce renewable heat for the local community. Like the two German projects, the CDT uses a mixture of solar panels and biomass boilers. On the other hand, not only was the project realised in another national context, but it also had a different social background compared to the two German cases, and thus fitted into the established case study design. It was possible to apply the same research interest and questions to the 'new' project while realising the principle of theoretical sampling in so far as the CDT provided different and 'new' characteristics, information, and insights to the study.

2.3 Methods

Methods should not only derive from the research interest but '*must be fitted to a predetermined methodology*' (Guba/Lincoln 1994: 108). Methods used to gather and analyse empirical material have to adhere to the principles of qualitative social research and enable the reconstruction of sense-making processes and practices in the three case studies. Methods for gathering empirical material have to provide contextualized information about sense-making processes, allow for openness—especially for unexpected issues or aspects—and enable the research to unfold as an interactive process. Methods of analysis likewise have to enable contextualized interpretation of sense-making processes and help to maintain the researcher's openness to new interpretations or theoretical assumptions as far as possible. In addition, the methods need to allow an emic understanding of the projects. In other words, methods need to enable an understanding of the projects '*from the point of view of the people who participate*' (Flick et al. 2004: 3). In accordance with the adopted theoretical framework, methods have to provide insights into the everyday practices within which social sense is created in the three projects.

This section explains the qualitative research methods of semi-structured interviews, participant observation, and analysis of grey literature. These were chosen in light of the requirements sketched above. The theoretical underpinnings of these methods are only briefly described.[3] Most of this section is dedicated to explaining how the methods were practically employed throughout the fieldwork process.

Interviews have been called the '*royal road*' of qualitative social research (Lamnek 1995: 35, translation AP). The main reason is that information from interviews can be gathered in a *status nascendi* (ibid.), i.e. in their '*natural context*'. Interviews are able to capture events or aspects that are distanced in time and space and thus not directly observable (Spittler 2001: 4). Furthermore, information from interviews—if they have been tape-recorded—are relatively authentic and inter-subjectively approachable (if interview quotes are used in the text). Using interview quotes in the written text allows readers to directly compare interview quotes and their interpretations. This provides a high degree of intersubjective control (ibid.).

Interviews enable the capturing of people's linguistic accounts of their interpretative frames (Lamnek 1995: 61). Unlike any other method, qualitative interviews provide participants with an opportunity to explain their own account of an issue and thus their interpretations of reality (ibid.). By allowing people to unfold their own frames of reference in an interview, researcher might get '*access to the meanings people attribute to their experiences and social worlds*' (Miller/Glassner 2010: 133) while being aware of the socially constructed status of reality. Subjective perspectives and their relatedness to objectified social senses can be revealed in qualitative interviews (Strübing 2013: 86). By confronting the researcher with interviewee's sense-making of the research topics, qualitative interviews not only realise the principles of openness and contextual information, but also provide emic accounts of the research topic.

In order to enable qualitative interviews to unfold these potentials, researchers have to abstain from forcing interviewees or their answers into predefined categories (Lamnek 1995: 61). To prevent predetermination by the researcher, open questions should be formulated and researchers should refrain from closing answer options. By being asked questions that do not suggest a certain way of answering or force people to choose from existing answering categories, interviewees are enabled to present their own understandings of an issue and its relations to other issues. Interviewees are given the opportunity to introduce issues and ideas that have not been anticipated by the researcher.

---

3   Only those aspects deemed relevant for this research are described. For a more nuanced insight into the different methods used, please refer to the cited literature.

All in all, I conducted 34 qualitative interviews. Most interviews (15) were with participants of the Scottish CDT and members of related institutions. In the context of the German citizen initiative KEBAP, 13 interviews with 14 people were conducted, while 12 interviews were held with 12 people who were involved in different ways with the German municipal project IBA. Four interviews were about both KEBAP and the IBA. Being situated in the same city, some actors who were engaged in renewable energy would know and be willing to talk about both projects. The difference in numbers is largely due to project-specific aspects. In the IBA, only three people were directly working on the project. The additional interviews were conducted with two local residents, one representative from a collaborating institution, two employees of Hamburg Energie, the municipal utility which realised and now operates the energy project, two former politicians, and one leading representative of the municipal public administration. In KEBAP, about 15 people were regular members at the time of the fieldwork. Of these, nine were formally interviewed. With most of the other members, I was able to have intense informal talks about the project at least once. Because of a manifested conflict in the project at the time of my entrance and throughout the fieldwork, two participants were not willing to take part in an interview; one person left Germany and another left the project for private reasons before interviews could take place. Three other people did not want to take part in an 'official' interview. In addition to the members, interviews about KEBAP were conducted with one representative from the district administration, a technical scientist who supports the project as a technical advisor in his private life, one former politician, and one representative of a potentially collaborative energy provider. For the Scottish case study, all four staff members of the CDT were interviewed. Also, eight of the then 10 board members found time to give an interview. Additionally, one group leader who was not a board member at that time[4] was formally interviewed. Additional interviews were conducted with representatives of the Land Fund and the Climate Challenge Fund. Another employee of the Climate Challenge Fund and an employee from Resource Efficient Scotland agreed to have tape-recorded meetings with the CDT.

Before the interviews started, I obtained permission for the interview from the interviewees. Permission was given either in written form or verbally. While written consent was mainly requested and provided in the IBA (see Appendix I for an example), interview permission was always verbal in both the Scottish and the German citizen project. Interviewees thereby '*agreed to be interviewed for a predetermined length of time, at a particular venue, on a particular topic, and under clear conditions of confidentiality*' (Legard et al. 2003: 147). The principle of confidentiality requires that the interviewer anonymize all personal

---

4   Most group leaders are either also staff or board members.

data, delete the audio tape after transcription, keep personal data separate from information used, use all given information only for scientific purposes, and not pass on any private information to third parties. These terms of confidentiality are in agreement with ethical standards of scientific research.

Most interviews took about an hour and were held with one person at a time, either in project offices or in private settings at the interviewees' homes. In one case, an interview took place with two people at the same time because of schedule restrictions. The interview provided highly valuable information and insights. The experience with this two-person interview showed that focus groups may have been a valuable additional method to 'uncover' similar or heterogeneous ways of sense making among members. It might, however, also have been the case that individuals—especially the ones who were not so outspoken—would have felt more pressure to agree to the explanations, descriptions, and estimations of dominant participants. Within the individual interview setting, most interviewees did mention or extensively describe their individual critical views about certain aspects of the respective project. It is doubtful whether they would have done so (at least so extensively) in a focus group setting. In the two-person interview, the two people were not only members of the same project but also friends. Having a good and friendly relationship allowed these participants to voice dissenting opinions and ideas without fearing impediment in their future cooperation in and for the project. The 'lack' of focus groups is somewhat offset by the circumstance that all three projects allowed me to take part in (and often even tape-record) internal discussions and meetings with external actors (see participant observation further below). Through these meetings and discussions, I could experience and register how the participants negotiated their different ideas and interests in situ.

Depending on the interviewee and situation, interviews would drift between more or less structured models. While interviews with project participants were relatively unstructured, interviews with representatives from official institutions tended to be more structured. This difference was mainly due to the fact that representatives of institutions were mostly questioned about certain aspects of the project under study. Interviews with politicians were mainly focused on the political aspects of a project, while interviews with employees of the municipal utility or technical advisors of the projects were mainly about aspects of the technologies.

For all interviews, however, an interview guideline or '*topic guide*' (Legard et al. 2003: 141), had been prepared, in which key topics and issues to be covered during the interview were listed. Both the structure of topic guides and their handling throughout the interview were flexible, '*to permit topics to be covered in the order most suited to the interviewee, to allow responses to be fully probed and explored and to allow the researcher to be responsive to relevant issues raised spontaneously by the interviewee*' (ibid: 141). The intention behind this approach is to allow interviewees to frame issues, topics, and their ideas about them in their own way, i.e. to remain in their own frame of reference. Enabling interviewees to do so is particularly crucial with regard to the research interest and question. In accordance with these, issues to be covered in the interviews were a) the specifics and individualities underlying the seemingly similar practices of renewable energy production, and b) the ambiguities and heterogeneities that had to be negotiated in each project. The intention was to uncover project-specific and individual ideas, intentions, aims, and motivations. Finding out about ambiguities and heterogeneities meant that topic guidelines needed to vary between different participants within one project. Guidelines were to be adapted for each interviewee, in order to best capture the specific roles, commitments, and activities of an individual. Also, guidelines were adapted to account for the areas of expertise of an individual. While, for example, group leaders were asked questions about specific actual group activities and their relation to the project's overall intentions, long-term members of the projects were specifically motivated to describe the beginnings and early developments of the projects.

All interviews were transcribed by professionals. German and English interviews were transcribed by different people who were both native speakers in the respective language of the interview. The transcripts were proofread and formatted and controlled for thorough anonymization of all interviews (see Appendix II: Guideline for transcription). All transcripts were analysed and have contributed to this thesis.

The second main research method employed in the fieldwork is participant observation. '*Participant observation is about stalking culture in the wild*' (Bernard 2011: 258). The quote describes the principle of participant observation—by taking part in the actors' daily routines, the researcher can study actors, their interactions, and their constitutions of reality as they naturally occur (Lamnek 1995: 240).[5] For participant observation, researchers need to go out into the field and take part in the research field's day-to-day life. The idea is that by actively participating in the field for a certain amount of time, researchers can acquire a deep and emic understanding of the people, situations, contexts, or fields that they are studying. Unlike in traditional studies of cultural anthropology, many modern—particularly sociological—fieldworks that employ participant observation do not take place in remote and (from the researcher's point of view) 'strange' cultures. It is in the '*backdrop of commonality that sociological eth-*

---

5   Of course, through their presence and (inter)actions, participant observers already alter the situations in which they are participating.

*nographers attempt to identify differences, i.e. specific features: differences be-
tween them and other types of people, differences of scenes, settings and situations, differences of fields'* (Knoblauch 2005: n.p.).

Using participant observation provides a means to study how different actors practically enact sense-making in situ (Lüders 2013: 390). More specifically, participant observation focuses on how sense-making is put into practice
by different actors in interactions with other human actors, material artefacts, or
local environments. Furthermore, participant observation is the preferred
method when social constructions of reality and processes of negotiation about
the definition of a situation are to be studied under a specific theoretical perspective (Lamnek 1995: 240). More recently, participant observation has acquired relevance in the realm of practice theories. The relevance of participant
observation for practice theoretical studies is based on the assumption that practices are accessible only through observation. *'[I]nterviews and narratives
merely make the accounts of practices accessible instead of the practices themselves'* (Flick 2009: 222). The specific contribution of participant observation is
not only its potential to study the enactments of practices in situ, but more specifically that it allows the study of the bodily activities involved in the enactment
of practices, and the 'witnessing' of the performance of embodied knowledge
(Zahle 2012). This is particularly important not only because participant observation is a direct way to witness practical knowledge, but also because *'individuals are typically unable to state, in general terms, how it is appropriate and/or
effective to act'* (ibid: 51). Practitioners are often not aware of the embodied
knowledge necessary to perform a certain practice.

Participant observation enables studying of not only practical interactions
between different actors, material artefacts, and local environments in the three
projects, but also how these interactions were part of the project's daily and non-
daily practices. Furthermore, participant observation in the projects provided a
means to learn about how actors practically negotiate the project's reality, i.e.
what the projects, and more specifically energy production in these projects, are
about.

Participant observation was used in all three case studies. Throughout the
fieldwork, the level of participation in each project gradually shifted between
being a participating observant and an observing participant (Lamnek 1995:
252; Bernard 2011: 260). The tendency would usually be with being a participating observer, meaning that the observation part dominated my role in the situation. The IBA's building contractor's meetings are an example of a setting in
which I would even define my role as merely an observer. While participants in

these meetings did address me from time to time, this is not because I had obtained a role as a participant, but to ask questions about me and my research or to explain something to me with regard to my role as an observant.

Getting informed consent from all observed people is a potentially difficult task in participant observation (Fluehr-Lobban 2014: 188; DeWald et al. 1998: 272pp.). While informed consent would require all participants in an observed situation to know about the specific character of the situation, it is not always feasible for a participant observer to make his or her role and intentions known to all. Whenever I joined a meeting within any of the projects, I made sure that my presence and identity were announced at the beginning of the event in which 'uninformed' people took part. Nevertheless, it was not always possible to inform people and ask for consent, such as when I took part in the daily routines of the CDT office as well as in large public events in all three cases. This problem of participant observation requires high standards of ethical scientific behaviour. It was ensured that field notes, memos and the likes were treated with the same level of confidentiality as personal data obtained in interviews.

The three projects offered different opportunities for participant observation. In the Scottish case, this method was employed most intensely. Board and staff members allowed me to come to the office of the CDT every day and set up my workspace there. As the office is purposefully designed as a drop-in place for members and partners of the CDT as well as for all other interested people, this enabled a very close encounter with a multitude of daily routines and events and allowed me to engage in informal talks. Additionally, the Trust members allowed me to join more or less all internal (group) meetings, meetings with external actors, and public events. In some cases, the participants of these meetings let me not only listen and take notes but even tape-record the meeting.

As KEBAP does not (yet) have an office space, the opportunities to participate in the project's day-to-day routine were limited. This limitation corresponds with the reality of the project, which did not have a 'material' project place at the time of the fieldwork. Most of the necessary daily tasks were done in members' homes. The project nevertheless offered numerous opportunities for participant observation. I could take part in all internal meetings and also in meetings with external agents. I was also invited to visit all of the project's public events. I took part in at least two project-related meetings and/or events each week throughout the fieldwork. Like in the CDT, the project members in many cases agreed to let me tape-record their meetings. If tape-recording was not possible (because of either physical reasons or dissent from participants), I was allowed to take notes.

The IBA offered the most limited opportunities for participant observation. This was not due to dissent from actors but because of the fact that most of the project work is done by a range of different contractors, who developed certain aspects of the project in their individual working spaces, dispersed all over Germany. Additionally, at the time of the fieldwork, the project planning and design phase had already been finished. I was, however, invited to take part in meetings in which the different contractors working on the building site discussed the proceedings and problems. Unlike the other two projects, the IBA is not an ongoing process. This meant that since the building process had been finished in November 2012, the project did not offer any more opportunities for participant observation.

The opportunities and limitations for participant observation in the different cases largely mirror important realities of the projects. Opportunities for participant observation were largely dependent on 'structural' aspects. These included formal and regular project localities (are there offices or other institutionalized project locations?), the organisation of daily practices (are there people who officially and regularly fulfil certain tasks for the project in publicly accessible places or are tasks fulfilled in a spontaneous way by individuals at any time in their private life?), and the way decisions are made (by one or few individuals in uncommunicated or accessible ways, or publicly by a large number of project members in more or less open discussions). Additionally, the level and regularity of engagements with the public (do the projects aim to be as open for the public as possible, not only in certain events but in their day-to-day activities and especially decisions?) and the duration of the projects (are there continually ongoing new developments or do the processes have a defined start and end?) could be observed. Looking at the possibilities as well as the limitations of participant observation reveals important aspects of the projects and their daily practices.

To complement the semi-structured interviews and participant observation, written documents from the three projects were collected and analysed. Analysis of documents provides a means to study how the reality under study is more or less officially documented (Flick 2009: 255).

Documents from all three case studies have been collected and analysed. These included internal documents like protocols, mails, and notices as well as 'official' documents like annual reports, presentations, leaflets, press releases, and formally published books or journals. These documents were used to access additional information about the projects, particularly background information about the history, developments, figures (i.e. kWh produced, sums needed for a certain investment), and dates. These documents also provide insights into dominant modes of sense-making in the projects at a certain time or within a certain situation. It is important to keep in mind that presentations of sense-making in documents do not mirror an unambiguous or stable representation of a readily defined situation (ibid: 255pp.). Instead, these documents represent a specific version of realities constructed for specific situations (ibid: 259), in which one mode of reality construction has come to dominate a certain situation. Documents like minutes and mails also provided valuable access to ongoing negotiations and conflicts with regard to certain topics, between specific people, or in particular situations. It is important not to take the status quo defined in these documents as established and stable definitions of reality construction in and by the projects. Instead, documents like interview quotes or observation data have to be understood as representations of momentary, fluid, and unstable situations.

2.4 Methods of analysis

Recently, the focus of many debates about methods or methodologies has shifted from data gathering to data analysis (Mey/Mruck 2014: 12). The method of analysis employed in this PhD project is based on 'postmodern' grounded theory as developed by Adele Clarke. Her approach is based on Anselm Strauss' approach to grounded theory. It is fundamental for Adele Clarke's approach that in situational analysis, '*the situation itself broadly conceived is the key unit of analysis*' (Clarke 2005: xxxv; Clarke/Keller 2012: 37). According to this idea and her postmodern aspirations, situational analysis shifts the focus of grounded theory from only human activities to also include non-human, material, and discursive/symbolic elements, as well as the structures and conditions that characterize a situation (Clarke/Keller 2012: 24). Adele Clarke not only asks researchers to be aware of contingency, non-human elements, power, and the relativity of reality, but has developed an integrative '*theory-method package*' (ibid: 37) to social research. This package provides a tool for researchers by which they can analytically grasp methodological ideas and principles.

Most of Adele Clarke's theory-method package—especially the underlying methodological ideas and their theoretical realization—are described in the chapter on theory (Section 3.6). In this section, the focus is solely on her approach to how data should be analysed in order to fulfil the insights of postmodernism, especially the integration of a) ambiguity, complexity, contingency, and instability, b) non-human elements, c) questions of power, and d) a heightened awareness of reality, not as something that is objectively given but as the outcome of sense-making processes.

Because of Clarke's negotiation of objective truth, the focus of situational analysis is on theorizing rather than on the development of substantive or formal theories. As in classical grounded theory, the aim is not to verify or falsify an already existing theory but to develop theoretical middle-range explanations for the specific case. For this purpose, she has developed three different forms of maps that describe a situation in different ways. These maps are not so much heuristic devices as they are analytical tools that help to sensitize the researcher for the complexity, variability and heterogeneity of the situation that he/she is researching. Besides showing complexity and positionality, the three maps are meant to provide a tool for the integration of non-human elements, power, and discourse (Diaz-Bone 2012: n.p.). Instead of replacing the methods of traditional grounded theory, situational analysis is meant to provide further tools that enable the researcher to become aware of relevant complexity issues of a situation. The maps do not substitute but supplement the traditional coding and memo-writing procedure of grounded theory. Clarke repeatedly stresses the importance of memo-writing in particular. Likewise, situational analysis underlines the theoretical sampling approach of grounded theory.

'*Coding means that we attach labels to segments of data that depict what each segment is about. [...] Coding distils data, sorts them, and gives us a handle for making comparisons with other segments of data*' (Charmaz 2014: 4). Through coding, grounded theory aims to force researchers to interact with their data and to start asking analytical questions (ibid: 109). It is through coding that researchers develop conceptual abstractions of their data and start to reintegrate them into theoretical concepts (Holton 2010: 21).

I started with what has been called open (Glaser/Strauss 1973 [1967]), in vivo (Strauss/Corbin 1996), or initial (Charmaz 2014) coding. For in vivo coding, researchers study transcripts or other forms of data line by line. Codes can be attributed to a segment, sentence, expression, or word. As this type of coding has to stick closely to the data, many codes will be comprised of descriptive terms and will summarize the meaning of the content (ibid: 112). While in vivo codes often adopt participants' own expressions, researchers have to be aware that like all qualitative research, in vivo codes are constructs created by the researcher, in which the first order and second order constructions of reality interact. '*We define what we see as significant in the data and describe what we think is happening*' (ibid: 115).

Of course, I had already developed an idea what I wanted to find in my data before starting my first analysis. While early grounded theory approaches prescribed the conducting of initial coding without preconceived ideas or theoretical knowledge, the discussion has developed onward to the idea that '*[to] analyse data we need to use accumulated knowledge, not dispense with it. The issue is not whether to use existing knowledge, but how*' (Dey 1993: 65pp.). Doing a qualitative social research does not mean that grounded theory researchers have to start from tabula rasa. Instead, being part of a scientific community enables researchers to be inspired and informed by scholars who have committed their time and energy on the creation and development of theoretical concepts. While these concepts should never be understood as hard facts, they provide us with a solid foundation and '*sensitizing concepts*' (Charmaz 2014: 117). Making use of sensitizing concepts, however, should not distract researchers from information and insights deriving from the data. '*We cannot analyse the data without ideas, but our ideas must be shaped and tested by the data we are analysing*' (Dey 1993: 7).

To operationalize these principles for in vivo coding, a method suggested by Russell Bernard was used.

> 'This is somewhere between inductive and deductive coding. You have a general idea of what you're after and you know what at least some of the big themes are, but you are still in a discovery mode, so you let new themes emerge from the texts as you go along' (Bernard 2011: 430).

Especially in the beginning of the process, most of the codes, like 'volunteer engagement', or 'accessing funding', were descriptive. In the second round I used theoretical coding to combine my data with the theoretical framework I had created.

During the process of data analysis and recording the results, it became clear that the empirical data did not fit the theoretical framework. Having made this discovery, the initial theoretical framework was dispensed with and the data were coded in vivo again. This process yielded empirically derived insights. In the next step, theoretical approaches that would enable an adequate analysis of these empirical insights were sought. Adele Clarke's situational analysis was found to provide valuable methods for analysis and theoretical approaches. Her proposed method of creating different kinds of maps based on empirical data enabled me to start thinking in a new way of about my data, while also providing a new way to structure and analyse the empirical cases.

Adele Clarke explains the value of her mapping approaches, which might provide a new creative or analytic impetus to researchers who get lost in the process.

> 'Thus these new approaches can address the problem of "analytic paralysis" wherein the researcher has assiduously collected data but does not know where or how to begin analysis. Analytic paralysis is, of course, not supposed to happen in a traditionally pursued grounded theory project [...]. But it does happen, for a wide array of reasons' (Clarke/Friese 2007: 371).

In fact, the idea to create different types of maps did help to overcome the 'analytic paralysis' that occurred after the first theoretical framework was dismissed. Creating maps about each of the case studies made visible all the elements that were relevant to a certain project and their interrelation. This provided a valuable tool to understand the differences and similarities between the projects. Also, these maps showed not only the complexities and heterogeneities existing within each project but also how these were negotiated.

Adele Clarke proposes the creation of three different kinds of maps: situational, social world, and positional maps. Situational maps are meant to show all analytically pertinent human, non-human, material, and symbolic/discursive entities that impact on a situation (see Figure 1). In the first step, the researcher is asked to record all entities coming to his mind on a blank sheet of paper without caring about order or hierarchy. In fact, the main goal of situational maps is to make visible the messiness and complexity of a situation. '*They intentionally work* against *the usual simplifications so characteristic of scientific work*' (Clarke 2003: 559, original emphasis). In order to include their own reflexivity in the map, researchers should be careful to enter not only those elements that seem interesting or important to them but also seemingly unimportant elements mentioned by respondents.

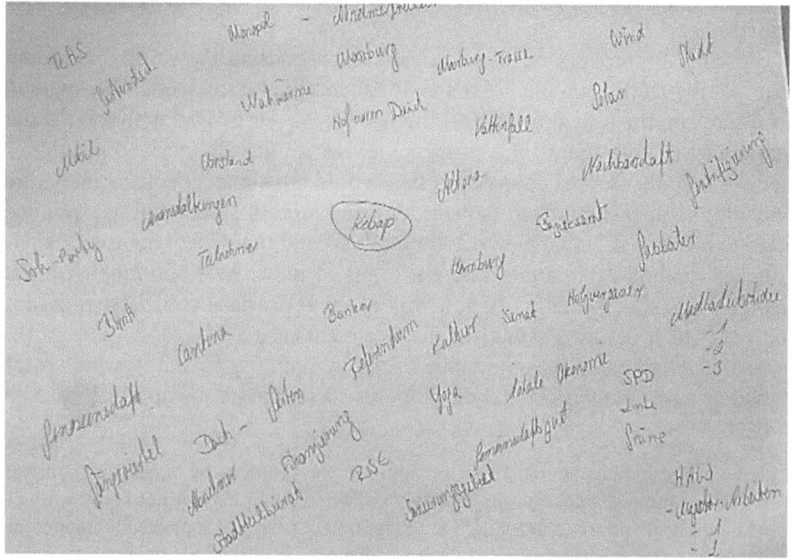

Figure 1: Situational Map KEBAP (Source: AP).

In the second step, situational maps can be reworked in order to analyse relations among the discovered elements (see Figure 2). This step not only shows existing relations and dependencies, but also identifies actors that are attended to and those that are not, as well as the symbolic or discursive elements used. Another way of elaborating the situational map is to order it by means of different categories into which the elements of the first map are placed (human, non-human; local, non-local…). The categories to be used depend on the empirical question. While this categorization might seem as a step away from her original claim to show the instability and messiness of the situation, it has to be said that the aim is not so much to create dualisms as to show the co-construction of seemingly differentiated entities (Strübing 2014: 102). Clarke nevertheless acknowledges the potential 'ordering' function of this analytical step.

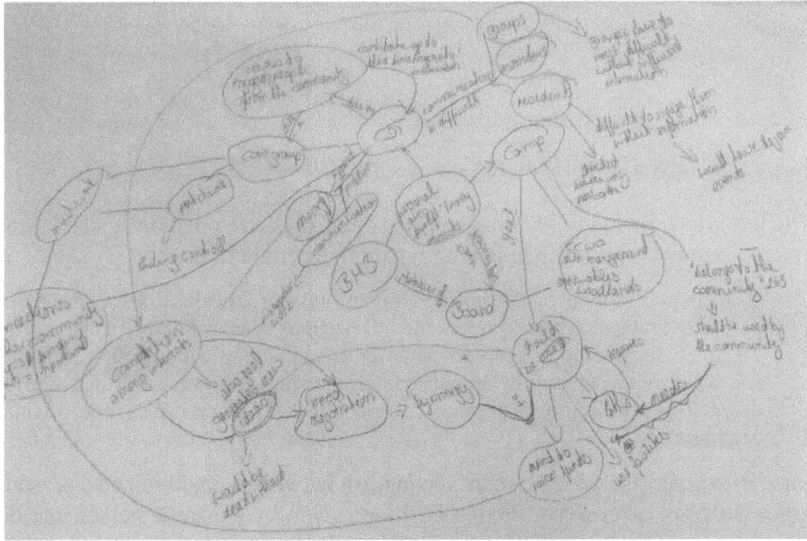

Figure 2: Situational Map CDT - with relations (Source, AP).

Social world/arena maps '*lay out all of the collective actors and the arena(s) of commitment within which they engage in ongoing negotiations*' (Clarke 2003: 559). These maps visualize actor groups and how they relate to relevant issues within a certain situation (see Figure 3). The social world map of a social arena visualizes the involved actors and how they relate to the situation. Researchers should pay attention to all kinds of differences and variations both within as well as among social worlds (Clarke/Keller 2012: 150).

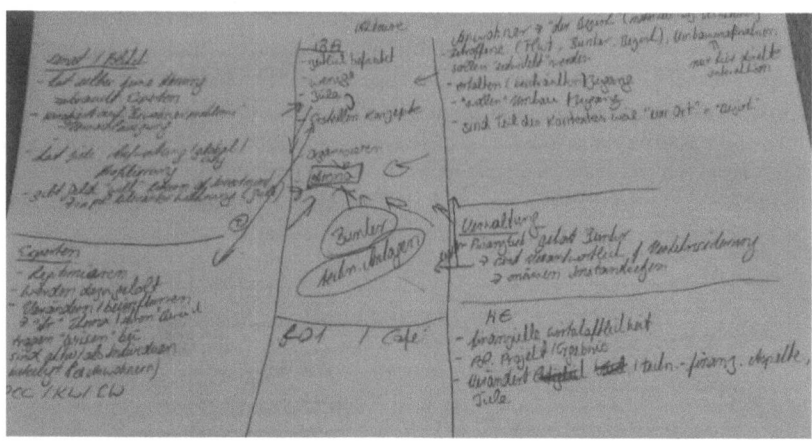

Figure 3: Social World Map of the IBA (Source: AP).

The last type of maps is positional maps, which visualize positions taken or not taken discursively or practically with regard to a certain issue.

> 'Perhaps most significantly, positional maps are *not* articulated with persons or groups but rather seek to represent the full range of positions on particular issues. The maps allow multiple positions and even contradictions within both individuals and collectivities to be fully articulated. Complexities themselves are heterogeneous, and we need improved means of representing them' (Clarke 2003: 560, original emphasis).

Most importantly, positional maps distinguish between a position and a person. Thus, different or even conflicting positions given by the same person can be introduced into the map. This differentiation allows the researcher to show not only the complexities that occur in different groups, but also the ambivalence of human beings (ibid: 556). For example, positional maps have been used to depict the conflict about the meaning of financial aims and activities in the CDT (see Figure 4). During the interview sequences, different positions about the problem become visible. The quotes are ordered in columns, whereby each column represents what was perceived to constitute close positions.

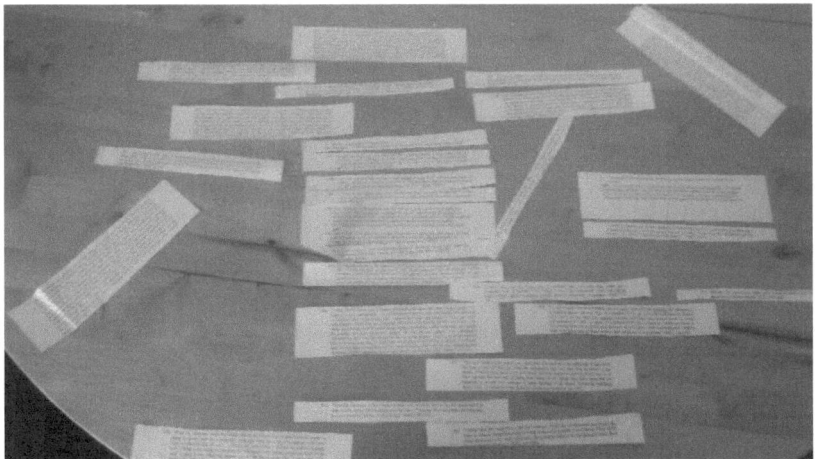

Figure 4: Positional map about financial teleology in the CDT (Source: AP).

The aim of creating the three types of maps is to

> 'do a kind of "social inversion" in making the usually invisible and inchoate social features of a situation more visible: all the key elements in the situation and their interrelations; the social worlds and arenas in which the phenomena of interest are embedded; and the discursive positions taken and not taken by actors (human and nonhuman) on key issues' (ibid: 572).

While analysing the projects with Adele Clarke's method, I started with the construction of situational maps by writing down every element that came to my mind and that I considered to be somehow part of the situation (see Figures 1, 2). Several situational maps were created for every project. After having drawn 'traditional' situational maps, I started to 'play around' with the sources and organisation of situational maps. For example, I coded my data not in the transcript but in the form of a map—trying to figure out which actors mentioned what elements in what order and what kind of relations *they* brought up between different elements. This gave rise to new ways of thinking about the cases and becoming aware of different actors' frames of reference with regard to certain issues. Instead of differentiating data and thus the cases into a range of separated codes, the maps visualized the multitude of links and relations between different elements within one project. The maps also showed the different kinds of elements—human and non-human—that were relevant for different situations in the project. Especially by coding interviews in maps, the actors' heterogeneous and sometimes ambivalent presentations of the projects or aspects thereof came to the fore.

A self-organised workshop on situational analysis with other scholarship holders from the Heinrich-Böll foundation provided an opportunity to have other people draw situational maps of one of my cases. I described one of my cases—the German municipal project IBA—and the participants of the workshop drew a map based on my narrative. Though participants could only include elements that I had mentioned into the maps and would be influenced by my way of describing the project, their maps proved to be highly interesting. More or less all participants had sketched elements or relations that I had not entered into my own maps so far. Taking part in the workshop proved highly valuable in providing insights into my own biases and already existing notions about the projects.

Social world maps did not really work out for me. While I did create social world maps about the projects, they did not provide new insights into my projects. Having used Pierre Bourdieu's concept of social fields and thinking about the conflicts of different actors in these fields, it is possible that I was already looking out for the relevant actor groups in the three projects.

Though I was initially sceptical about the applicability and value of positional maps for my research interest, I found them to be highly valuable. I developed my own way of constructing these maps. Cutting out statements from the transcripts, I positioned and related different statements to each other by arranging and re-arranging them. Proceeding like this allowed me to become much more aware of not only the differences and nuances in actors' statements, about 'sustainability' for example, but also of how different actors, or even the same actors at different times in the interview, would relate 'sustainability' in varying ways to other elements.

Parallel to analysing the cases with Clarke's situational analysis, Theodore Schatzki's concept of social practices provided interesting theoretical insights into the empirical cases. Having developed a theoretical framework meant that theoretical codes could be developed. Theoretical coding was conducted in a manner '*somewhere between inductive and deductive coding*' (Bernard 2011: 430). While being conscious of, and on the lookout for, quotes that would fit codes derived from the theory, theoretical coding also meant searching for data that would contradict the theoretical framework or certain parts of it. While the first theoretical approach had become unsound when trying to combine it with the empirical data, the combination of Adele Clarke's situational analysis and Theodore Schatzki's ontology of social sites not only fit the empirical findings but inspired new insights and the emergence of new analytical ideas from the data.

Recording the results is an important part of the research process. To get results transparent and replicable and to enable intersubjective understanding, original data material like quotes from interviews, documents, and observation notes is extensively used throughout the presentation of the research findings. To avoid using data only as illustrative material (Meyer/Meier zu Verl 2014: 14), data and their interpretation are presented together in the results chapter. Together with exemplification and explanation of the appropriateness of the methodological and methods canon with regard to the research interest and question, this style of presenting findings is meant to realise the qualitative research principles of intersubjective replicability and transparency.

3 Theoretical framework

The theoretical framework presented in this chapter both suits the research interest and corresponds to findings from my empirical research. After dismissing the original theoretical framework, a new framework was developed which was better able to analytically make sense of the empirical findings according to the research interest. With the theoretical framework developed, it has become possible to analyse local renewable energy projects, in each of which renewable energy is associated with different intentions and meanings. While technically the product is the same in all three cases, the product and especially its production are the outcome of very different motivations, meanings, and intentions. It was found that the sense-making of renewable energy (production) is affected by and affects how project members engage with one another, other actors, the local environment, and material artefacts. The theoretical framework has to theoretically grasp and explain this social embeddedness of renewable energy production. I could differentiate four research interests to which the theoretical framework has to correspond. First, the framework needs to provide theoretical tools through which energy production can be analysed as an integrated aspect of the projects' everyday activities. The theory needs to explain the processes of sense-making with regard to energy and related issues in the projects' day-to-day activities. Second, the research interest was not to find discursive or rational explanations but to study how energy is *practically* embedded into contexts. Third, during the fieldwork it became obvious that the meaning of energy and energy production is highly dynamic and contested. Hence, the theoretical framework has to recognize and explain sense-making as a dynamic process. Last but not least, the empirical findings require a theoretical concept that is able to grasp contingencies, ambiguities, and heterogeneities as well as the constantly occurring processes of negotiation occurring in the projects. In a nutshell, the theoretical framework has to theoretically reconstruct the dynamic, continually contested, and negotiated sense-making processes of energy and energy production in the case studies. It has to enable an analysis of energy production not as a homogeneous social phenomenon but as situated social practices.

Practice theories are well suited for providing a theoretical approach to analyse energy and energy production as integrated aspects of the projects' daily activities and to study the practical embeddedness of energy into the different contexts. However, the aim to analyse sense-making as dynamic negotiation processes and to grasp contingencies, ambiguities, and heterogeneities in the projects reveals certain core fallacies of practice theory approaches. Consequently, it was necessary to supplement the chosen practice theory approach with insights from another theoretical concept. The subsequent sections describe practice theories and how they suit the first two research interests. Additionally, an account of the main criticisms of practice theories is given. The approach developed by Theodore Schatzki has already responded to some of these criticisms. This is the reason his approach has been chosen for the analysis, as is explained in more detail in Section 3.5. This section ends with a critique of Schatzki's approach and a summary of the aspects that cannot be 'seen' with his concept. Adele Clarke's situational analysis is then introduced in Section 3.6. As shown in the last section, combining Schatzki's and Clarke's theoretical accounts provides a theoretical framework that is able to recognize and explain all four research interests.

3.1 Practice theories

One of the first things most practice theorists mention is that practice theories comprise an unfixed bundle of approaches, only loosely linked by what has been called family resemblances (Schatzki 2012; likewise, Reckwitz 2004).[6] In fact, critics have stated that practice theories are so loosely related to one another that they cannot be called an integrated theoretical approach or theory as of now. They might never even reach this status as they have yet to show their ability to offer a new perspective or make visible hitherto unseen social phenomena (i.e. Bongaerts 2007). In response to this criticism, practice theorists have emphasised that there are '*good reasons to argue that there is something new in the social-theoretical vocabulary that practice theorists offer. They do form a family of theories which, in certain basic ways, differs from other, classical types of social theory*' (Reckwitz 2002: 244). Furthermore, while practice theories indeed still lack a unified theoretical approach, a common set of core concepts can be identified (Hillebrandt 2014; Feldman/Orlikowski 2011; Rouse 2007).

Fundamentally, practice theorists share the conviction that 'the social' (only) exists in practices. Practices are mostly understood to be collectively shared, embodied, and materially mediated bodily doings and sayings in which different elements (the approaches list understandings, meanings, things and their use, rules, know-how, and intentions) are organised (Schatzki 2012: 13; Reckwitz 2002: 247; Shove/Pantzar 2005). Differences can be noted in the way

---

6   A good overview can be derived from: Reckwitz (2002, 2003) and Brand (2014).

practice theory scholars have interpreted these basic concepts and how they re-late them to one another. Three important points of divergence can be found. First, concepts differ with regard to the level and importance attributed to rou-tinization, and thus the exclusion or inclusion of creativity, ambiguity, and in-stability. Second, approaches vary significantly with regard to whether knowledge is limited to practical forms or if and to what degree theoretical or propositional forms of knowledge are included. The integration of theoretical knowledge significantly influences the ability of the respective approach to in-clude intentionality and conscious reflection. The third main point of differenti-ation pertains to the role and agency attributed to material artefacts.

Before digging deeper into these three lines of conflict and explaining their relevance to this thesis, a short overview of the background and development of practice theory is provided.

3.2 Background and differentiation of practice theories

Practice theories are a specific form of cultural theory (Reckwitz 2002: 244). Cultural theories have to be differentiated as economic-individualistic, norm-oriented, and structuralistic approaches (Reckwitz 2003: 287). *Economic-individualistic* approaches understand human activity as motivated by mostly eco-nomically inspired purpose (profit-maximizing). The economic-individualistic model can be traced back to the ideas of Adam Smith and still exists in rational choice theory, for example. According to the individualistic-economic model, human activities are explained as actions of a rationally behaving (often profit-maximizing) *homo economicus*. Behaviour is differentiated from activity be-cause the latter is guided by teleological individual motivations and the exist-ence and availability of information, which is used to single out the most prom-ising (profit-maximizing) way of acting (Reckwitz 2004: 308pp.). Social order then is '*a product of the combination of single interests*' (Reckwitz 2002: 245).

Norm-oriented approaches aim to explain intersubjective coordination of human activities. Unlike individualistic models, they are interested in the crea-tion and maintenance of social order (Reckwitz 2004: 309). Instead of under-standing the social element as the outcome of individual activities, norm-ori-ented explanations localize the social element at the above-individual level of social norms. These guide individual behaviour (Reckwitz 2003: 287). A fa-mous example is Durkheim, who believed that social facts are categorically dif-ferent from and irreducible to individualistic explanations (Schatzki 2003: 175). Social norms are mainly understood in terms of normative rules. These differ-entiate sensible from senseless or right from wrong behaviour (ibid: 175). Norm-oriented approaches explain human action by reference to collectively shared norms and values, '*i.e. to rules which express a social "ought"; social order is then guaranteed by a normative consensus*' (Reckwitz 2002: 245). Like norm-oriented approaches, *structuralist* theories conceptualize the social element to exist at the level of the non-individual. Instead of collectively shared norms or values, however, they explain social order as an outcome of more or less objectified social structures that are not accessible or changeable for individual actors. A number of scholars taking a structuralistic stance perceive individuals, events, or interactions as merely being bearers of prescribed roles or as taking up positions '*in the structures and institutions that condition, constrain, or form activities*' (Schatzki 2002: 128).

While differing in how they define the 'something' that guides individual behaviour, structuralist or norm-oriented theories conceive of social phenomena as something 'more than' and distinct from individuals. Because of their focus on social facts instead of individual agency, these approaches have been called *societist* ontologies (Schatzki 2002). Societist theories are ontologically different from *individualistic* accounts of the social. Individualist ontologies believe that social reality is created by and through individuals, their constellations, and—in some approaches—their relations. In these approaches, social reality is a labyrinth of (inter-related) individuals (ibid: 125). The social is reduced to the '*sum of the individuals and their actions*' (Røpke 2009: 2491).

In response to the criticism against both societist and individualistic accounts of the social element, *cultural theories* emerged in the 20th century. According to cultural theories, human activities are shaped by collectively shared symbolic patterns of knowledge. Instead of being guided by individual choices or abstract norms, these symbolic patterns create the basic distinctions and schemes. These let actors decide '*which desires are regarded as desirable and which norms are considered to be legitimate; moreover, these cognitive-symbolic structures (of which language is a prominent example) reproduce a social order even in cases in which a normative consensus does not exist*' (Reckwitz 2002: 246). Symbolic patterns of knowledge enable social actors to make sense of the world in a collectively shared manner. Based on this understanding, actors act in ways corresponding to symbolic forms of knowledge.

> 'Social order then does not appear as a product of compliance of mutual normative expectations, but embedded in collective cognitive and symbolic structures, in a "shared knowledge" which enables a socially shared way of ascribing meaning to the world' (ibid: 245pp.).

Practice theories, a distinct form of cultural theories, oppose both individualistic and societal ontologies. They draw on a range of scholars and concepts, among them Wittgenstein, ethnomethodology, and theories of artefacts.[7] Practice theories derive their understanding of rules and rule-following from Wittgenstein. According to Wittgenstein, rules are not self-interpreting. They always need to be interpreted by actors in a given situation. Different ways of behaving can be interpreted to be in either accordance or non-accordance with rules. Providing a formal interpretation of a certain rule would not enhance conformity, because, again, it would function as a rule. From ethnomethodology, practice theories have adopted an interest in how actors practically master their day-to-day life (Hillebrandt 2014: 43). In ethnomethodology, the social aspect can be understood only through the activities of social actors. Through their activities, social actors create the reality in which they live (ibid: 44). Based on ethnomethodology, practice theories also argue against the ahistorical and objectivistic perception of reality.

Pierre Bourdieu and Anthony Giddens are among the first generation of practice theorists (Brand 2011: 178). Until the 1960s, the opposing ontologies of individualism, subjectivism, or constructivism versus societism, objectivism, or structuralism had divided social theory. Bourdieu and Giddens both tried to overcome the cleavage of structure and agency resulting from these two ontologies by creating a theory that would overcome the dichotomy created by these ontological accounts of the social element. While differing markedly with regard to the amount of power ascribed to either the individual (Giddens) or objective conditions (Bourdieu), both developed theories in which the tension between social structures and individuals is mediated by embodied social practices. According to Giddens, practices mediate between structure and agency: *'[t]he basic domain of study of the social sciences, according to the theory of structuration, is neither the experience of the individual actor, nor the existence of any form of societal totality, but social practices ordered across space and time'* (1984: 2).

The different accounts of practice theories oppose individualistic approaches. According to practice theories, social phenomena like knowledge or shared understandings cannot be constructed as possessions of individuals. Instead, socially shared understandings and knowledge are included in practices (Schatzki 2002: 134pp.). The understandings and knowledge included in practices, however, cannot be differentiated from individual actors and their enactment of practices. Knowledge and understandings are continually established, altered, transmitted, or maintained through the actions that compose a certain practice and, consequently cannot be separated from these actions. Individuals performing an action can have their individual *'version'* (ibid: 135) of a practice

---

7   For a detailed account of the theoretical backgrounds of practice theories, see Reckwitz (2003, 2004), Hillebrandt (2014), and Schatzki (1996, 2002).

without necessarily changing the general understanding of that practice. Hence, practice theorists argue that action, one of the key concepts of individualist approaches, '*presupposes something nonindividualist, namely, practices and the understandings they carry*' (ibid.).

With regard to structuralist ontologies, practice theories criticize not only the lack of individual actors and their inability to explain change and dynamic processes but also their tendency to reify that something which they see as constitutive of the social element (ibid: 132). The problem arises because structuralist approaches construct a difference between individuals and this abstract component of the social element which they believe to be the fundamental principle of the social element. While structures determine actors' behaviour, they are not accessible to individual actors. Because of this constructed difference between individuals and the social aspect, individual actions, ideas, believes, and so forth '*cannot be intrinsically* part of *these phenomena—inseparable from them, perhaps, but not an inherent component of the whole societies, social facts, or abstract structures that embed and determine them*' (Schatzki 2003: 180; original emphasis).

Creating a middle ground between structuralist and individualist ontologies, the relation between individuals and the social order is mutual, mediated in actors' everyday practices: '*rather than seeing the social world as external to human agents or as socially constructed by them, this approach sees the social world as brought into being through everyday activity*' (Feldman/Orlikowski 2011: 1241). While they do not dismiss insights from individualistic or structuralistic accounts, practice theorists differentiate their approaches by making the '*practice itself, rather than the individuals who perform them or the social structures that surround them [...] the core unit of analysis*' (Hargreaves 2011: 82).

Since the 1990s, a second wave of practice theories has come into existence. Recent important contributions have come from Andreas Reckwitz (2002, 2003, 2004), Theodore Schatzki (1996, 2002), Frank Hillebrandt (2014), Elizabeth Shove and colleagues (Shove/Pantzar 2005; Shove et al. 2007), and Inge Røpke (2009). Critically distinguishing itself from the first wave, this second generation particularly emphasises the importance of material artefacts and anti-intellectualist forms of embodied knowledge for the constitution of social practices.

In the following section, common features and core concepts of the second wave of practice theories are presented. Based on a critique of some fallacies inherent to a more or less strong degree within these approaches, it is explained why Theodore Schatzki's approach has been chosen in order to analyse the practices of local renewable energy projects.

3.3 The theoretical concept of practice theories

While practices comprise of bodily doings and sayings by individual actors (Schatzki 2002: 72), '*the agents are not the starting point of the analysis, as practices logically and historically precede individuals, implying that practices, so to speak, recruit practitioners*' (Røpke 2009: 2493). Instead of individual actors' activities, underlying norms or structures determining these activities, practices are the basic ontological unit within practice theories (ibid: 2492). Practices are organised bundles of activities—sets of interconnected bodily doings and sayings (Schatzki 2011: 4). As relatively enduring and stable entities, practices are recognizable across time and space (Shove et al. 2007) and are shared by a collective of social actors. In their everyday life, individuals carry out a multitude of practices (Reckwitz 2002: 256).

Definitions of practices commonly include ideas of practices as collectively shared, materially mediated bodily doings and sayings, in which different elements (among others understandings, meaning, rules, know-how) are organised to comprise a practice (Schatzki 2012: 13; Reckwitz 2002: 247). Common features of most practice theories refer to a) practices as bodily doings and sayings, b) the underlying concept of collectively shared symbolic patterns of knowledge, c) an emphasis on practical forms of knowledge, and d) the role of material artefacts. These shared common core concepts are subsequently explained.

Practice theories share a common conception of the importance of the human body and bodily activities. Human agency is foremost understood to be comprised of bodily performances. According to practice theory, human activity is an outcome not of conscious reflection but of practical knowledge and skills needed to carry out a practice. Actors have to know, how and when to perform certain bodily doings or sayings in order to carry out an activity and to be intelligible in their performance of a practice. The body is understood to be '*the locus of agency, affective response, and cultural expression, and the target of power and normalization*' (Rouse 2007: 512). Via the concept of embodied knowledge, practice theories aim to re-integrate practical activities that are necessarily related to body and bodily enactments of the social element into sociological analysis (Brand 2011: 174). Practice theories are interested in the '*skilful performances of bodies*' (Reckwitz 2003: 290).

The emphasis on practical or embodied knowledge includes a critique of intellectualist approaches to human activity. These understand reality to be constructed by propositional forms of knowledge. By basing the social element in the bodily activities of social actors, practice theorists aim to emphasise human interactions as mainly an outcome of unconscious, embodied, and non-representational forms of knowledge. They instead define human practice as outcome of an actor's practical sense (Bourdieu), understanding (Schatzki 2002), or knowledge (Giddens 1984). Practice theorists thereby differ with regard to whether the embodiment of knowledge is seen as a reason for the social stabilization of practices or for their continuing re-interpretation. Proponents of the former view (Bourdieu, Reckwitz) argue for the stability of social practices because by being embodied, social norms and dispositions are not or only barely accessible through conscious reflection. Other practice theorists emphasise the innovative aspect of bodily activities—even those that are generally described as routines (Schatzki, Shove). They argue that no situation or movement can occur twice in exactly the same way. Each repetition of a bodily activity involves an aspect of newness or creativity, which results from the uniqueness of the actors, their experiences and knowledge, and the concrete situation (Schatzki 2002: 74pp.).

Deriving from cultural theories, the concept of practical knowledge employed in practice theories is based on the idea of collectively shared symbolic and cognitive structures of knowledge. As mentioned above, actors make sense of and symbolically organise reality through symbolic patterns of knowledge (Reckwitz 2002: 245). The difference between cultural and practice theories is that the latter believe that knowledge is always situated in practices (Schatzki 1997). Bodily doings and sayings and orders of knowledge are always tied to one another. Practices are based on socially organised ways of interpreting and understanding the world. The shared symbolic patterns of knowledge shape how actors make sense of activities. '*This includes the ideas of what the activities are good for (or why they are considered problematic), the emotions related to the activities, the beliefs, and understandings*' (Røpke 2009: 2492). The symbolical order is realised in practices that support or modify this order through their enactment. The relation between bodily doings and sayings and knowledge is mutual. Without knowledge, actors would not be able to reproduce bodily doings and sayings. At the same time, collective forms of knowledge and understanding are based on visible, bodily activities. Orders of knowledge are thus (re)produced in bodily doings and sayings (Reckwitz 2004: 320). Indeed, practices are defined as basic social phenomenon '*because the understanding/intelligibility articulated within them (perhaps supplemented with normativity) is the basic ordering medium in social life*' (Schatzki 1997: 284).

The practice theorists of the second wave share an awareness of the relevance of material objects (artefacts). By including material artefacts into their concepts, they try to overcome the traditional separation of nature and society that has guided social thought since antiquity. According to this dictum, materiality is a background against which social life proceeds (Schatzki 2010: 126). Modern practice theory approaches argue that social life proceeds in between and amid materiality. Material artefacts are not just interwoven with social life; they are a dimension of it (ibid: 141). Insights from theories, like the actor-network theory which focuses on the interactions of material artefacts and human beings, are integrated into practice theories by combining materiality with social practices. Practice theory scholars emphasise that hardly any practice would be possible without the existence or use of material objects (among others Røpke 2009; Shove et al. 2013, Reckwitz 2003, Schatzki 2002). Not only are material objects used in practices, the material culture is a prerequisite for the creation and recreation of certain ways of knowing and doing (Reckwitz 2004: 322). Like knowledge and symbolic meanings, material aspects play an important role in the creation, modification, and destruction of practices. Knowledge and meaning are embedded in artefacts. Objects are associated with activities, suggesting certain ways of using them. On the one hand, artefacts, because of their durability, stabilize social practices, and support routinized forms of engaging with them. On the other hand, new materials or routines might change or even cease a certain way in which bodily doings and sayings interact with a specific object.

The core concepts—bodily activities, embodied knowledge, and material artefacts—comprise the elements that are organised within practices. In other words, a practice '*is a set of bodily-mental activities held together by material, meaning, and competence*' (Røpke 2009: 2492). Practices, thus, are configurations of heterogeneous elements (ibid.). When performing practices, actors actively integrate the different elements.

> 'The practitioner becomes the carrier of the practice-related beliefs, emotions, and purposes when performing the practice, but these aspects of meaning are seen as "belonging to" the practice rather than emerging from self-contained individuals. Again, this is what makes meaning social' (ibid.).

Actors act as carriers of practices, thereby reproducing social sense-making. However, they are not understood as being merely passive role inhabitants. Instead, practices-as-entities (ibid: 2491) are (re)produced by performances of individual actors. Individuals are active interpreters of practices; they link, integrate, and enact the various elements comprising a practice. Whether performances of practices are to be conceptualized as routinized behaviour or whether the elements of change and creativity within these practices-as-performances (ibid.) are to be emphasised is one of the most contentious issues in modern practice theory and is elaborated in more detail in the next section.

3.4 Critique

Having chosen a practice theoretical approach to analyse the empirical material, I have decided to analyse the projects as outcomes of organised human activities. Practice theories offer a way to focus on those daily practices that occur within energy production projects. Energy is understood and analysed as an outcome of daily social practices. Also, practice theoretical approaches enable a study of energy production as '*practice-as-performance, and performances will always differ between individuals and between social groups*' (Røpke 2009: 2494). Røpke's differentiation of practices-as-entities and practices-as-performances is based on Schatzki's (2002) account of practice theory. He defines practices-as-entities as idealized and abstract forms that are historically and collectively constructed and practices-as-performances as the grounded enactment of practices conducted as and amid everyday contingencies. The concept of practices-as-performances offers a way to understand energy production not as an idealized and homogeneous practice-as-entity, but instead to analyse differences in how energy production is practically done in the three different projects. Last but not least, practice theories offer an approach through which it is possible to theoretically grasp and analyse how sense-making is put into practice within different social contexts.

Nevertheless, when trying to answer the research questions by using only practice theories, two analytical blind spots of practice theories—which have already been criticized by different scholars—become relevant. First, practice theorists have been criticized as being mostly interested in unconscious and embodied knowledge, tending to focus on routinized forms of behaviour. Different scholars have criticized this concentration on routine and stability (Volbers 2015; Bongaerts 2007). It has been argued that because of this focus, practice theories are not able to explain innovation, creativity, and change as important aspects of societies. They would lose sight of the dynamic aspects of societies and human conduct. Critics mostly refer to Bourdieu's concept as an example (i.e. Bogusz 2009). While the degree of stability in Bourdieu's concept remains open for interpretation (Schäfer 2010, 2012), his and other practice theory approaches have been criticized for (over)emphasising the degree of routine and thus stability of practices. Practice theories, however, '*differ extensively over the degree of stability that practices can sustain*' (Rouse 2007: 506). Some scholars have explicitly made dynamic and creative aspects of practices key parts of their concepts. Approaches like those of Schatzki or Shove and colleagues are '*committed to understanding the ongoing dynamics of everyday life. This is again a matter of emphasis: we deal with processes of routinization and normalization, but without supposing that these necessarily result in stabilization or closure*'

(Shove et al. 2007: 11). Other scholars have suggested that different practices are characterized by different degrees of stability and that the question of stability is primarily an empirical one (Rouse 2007: 507).

The empirical information gathered in all three cases suggests that none of them shows high degrees of routinization. Instead, throughout the fieldwork, conditions, actors, relations, and meanings have been constantly changing, implying that the cases are highly dynamic processes. The analysis of the empirical cases consequently requires an approach that not only integrates, but emphasises creativity, change, and innovation. Some practice theory approaches are equipped to explain newly occurring or changing social practices. Practices change when (at least) one of the elements comprising a certain practice is altered (new, changed, abandoned) or when new links between existing or new elements are established and new configurations of elements are created (Shove/Pantzar 2005: 61). Furthermore, practices have been conceptualized to be inherently dynamic as they are enacted by actors who have varying abilities and experiences and interpret specific situations. Even the most routinized or ritualized practice includes a certain degree of change as the bodily activities of actors can never occur in quite the same way twice (Schatzki 2002). By differentiating between practices-as-entities and practices-as-performances, Schatzki provides a differentiated view on practices as collectively shared and durable practices-as-entities on the one hand and the subjective and changing practices-as-performances on the other hand. This differentiation of routinized ideas of practices and their dynamic enactments is one reason for choosing Schatzki's approach to social practices, as it enables the integration of change, instability, and heterogeneity of practice enactments.

Practice theories nevertheless do not provide conceptual tools to theoretically grasp and analyse those specific instances of change and instability that stem from the interaction of different actors when practices are enacted not as individual performances, but as shared and commonly organised projects or tasks. Practice theories cannot explain those instances of creativity and change that result not from individual interpretations but from the negotiation of different actors' ambiguous, heterogeneous, and complex sense-making activities in a specific situation. In other words, they lack tools to analyse practices as outcomes of negotiations in specific contexts. This relates to another criticism of practice theories. These have been blamed for ignoring heterogeneity and ambiguity as important aspects of social practice (Volbers 2015). Critics have argued that practice theories are much too interested in the socially shared aspects of practices. As is obvious in the description of the three cases, ambiguity and heterogeneity are crucial aspects in all projects and their continually changing and unstable sense-making processes with regard to energy and energy production.

Additionally, as already mentioned, it is a major research interest to make energy production visible as a dynamic social process, marked by ambiguities and heterogeneities. Empirical findings and research interests demand a theoretical framework that allows focus on contingencies, ambiguities, and heterogeneities as well as the constantly occurring processes of negotiation occurring in the projects. In fact, heterogeneities and their negotiation play a major role for the creation, change, and abandonment of certain practices in the three projects. In order to analyse them sufficiently, practice theories need to be enhanced with further theoretical concepts. The chosen additional concept has to shift the focus from practices as shared to practices as negotiated phenomena.

Critics have also argued that most practice theory approaches are based on a more or less pronounced theoretical fallacy that results from the differentiation between theoretical and practical knowledge (Volbers 2015; Bongaerts 2007). Deriving from Bourdieu's differentiation between the practical sense of actors and the theoretical rationalization of scientists, most practice theorists adopted the view that practical knowledge and theoretical knowledge are two epistemically different ways of perceiving the world (Volbers 2015). Critics like Volbers (2015) argue that instead of differentiating between theoretical and practical knowledge, as is done in most practice theory approaches, propositional knowledge and intentionality should be integrated into practice theories. He argues that instead of excluding theoretical forms of knowledge from practices, they should be understood as a particular type of practice themselves.

Analysing a context that is as highly influenced by theoretical (scientific) knowledge as energy production underlines the importance of such a (re-)integration. While some practice theories more or less explicitly exclude any form of theoretical knowledge, other practice theorists have included it into their approach. By creating the feature of 'teleoaffectivity', Schatzki's concept makes intentionality and teleological orientation key elements of social practices. He thereby offers a way to include theoretical forms of knowledge. Schatzki's approach, however, is not able to explain the negotiation of *different* intentionalities and thus the different ways in which theoretical knowledge is *made* meaningful in a certain situation.

Scholars have suggested combining practice theories with insights from pragmatism to overcome the analytical gaps or problems of practice theory (Volbers 2015; Schäfer 2012; Brand 2011; Bogusz 2009). While in large parts being based on the same premises, pragmatism differs from practice theories by its concentration on creativity in human conducts and its conceptualization of the relation between practical and theoretical knowledge. Situational analysis—a postmodern form of grounded theory, developed by Adele Clarke—is a theory/methodology bundle based on pragmatism. The concept has been chosen because it is able to offset those fallacies of practice theories that have been identified as relevant with regard to the combination of research interest and empirical findings.

In the subsequent sections, I introduce the concepts of Theodore Schatzki and Adele Clarke. Like all practice theorists, Schatzki argues for a differentiation of theoretical and practical knowledge, and deems practical knowledge— or understanding as he terms it—as one of the key elements of social practices. Schatzki, however, also includes intentionality and thereby theoretical knowledge into his concept of social practices. Furthermore, his approach highlights that practices are always based on interpretations by individual actors and enacted in specific situations and thus are never exactly the same. His approach places practices between routines and innovation. Innovation and change are included as inherent parts of social practices. Combining his approach with Adele Clarke's situational analysis does not fundamentally transform his approach; it merely shifts its focus. The combination furthermore makes instabilities, creativity, complexity, contingency, and their negotiation a focus of the analysis.

3.5 Theodore Schatzki's practice theory approach

The American philosopher Theodore R. Schatzki developed a practice theory in which practices are tied to the social context within which they occur.

> 'The social site is a specific context of human coexistence: the place where, and as part of which, social life inherently occurs. To theorize sociality through the concept of a social site is to hold that the character and transformation of social life are both intrinsically and decisively rooted in the site where it takes place' (Schatzki 2002: xi).

In developing his concept of social sites, Schatzki aims to develop an approach to explain human coexistence. *'By human coexistence, meanwhile, I mean the hanging together (Zusammenhang) of human lives, the togetherness and withness of human beings. Something is social if it pertains to the hanging together of human lives'* (2010: 127; original emphasis). Social sites are the contexts wherein practices are enacted and human coexistence is created, maintained, or altered. Schatzki's practice approach states a continuity of being between social practices and the context as part of which they transpire (2002: 138).

Schatzki's embedding of practices within their contexts is what made me choose his approach for my analysis. His explicit focus on context is in accordance with the methodological principles of qualitative social research explained above. Also, and more importantly, by embedding practices in the context within which they occur, Schatzki enables not an analysis of 'the' practice of energy production (practice-as-entity), but a highly specific analysis of practices-as-performances in and as part of the context within which they occur. His approach calls attention to the specifics of and differences between practices of energy production in different social contexts.

Schatzki's ontology of the social offers a promising approach to understand and conceptualize the empirical findings by understanding practices as contextualized activities. The following sections do not give a comprehensive description of his practice theory. Instead, I introduce and operationalize the aspect of this concept that are relevant to my research interest—the analysis of energy production not as a homogeneous social phenomenon, but as a highly specific and contextualized social practice. According to Schatzki, social sites are '*meshes of practices and orders*' (2002: xii). Hence, I first introduce Schatzki's conceptualization of social orders. In a second step, his concept of social practices is described, before I explain how meshes of orders and practices constitute and are constituted in social sites.

3.5.1 Social orders

According to Schatzki, social order can be understood neither as regularity/pattern or stability nor as interdependence. He argues that empirical findings clearly demonstrate that social order can be maintained despite the existence of irregularity, instability, and multiple forms of dependencies (2002: 17). Instead, his conceptualization of social order derives from the idea that '*how things stand with one entity has to do with how they stand with others*' (ibid: 18). Social order is the '*hanging together*' of things; Schatzki calls it the existence of nexuses (Zusammenhänge). In these nexuses, things 'hang together' even when they are linked haphazardly and arbitrarily. Furthermore, things might still hang together when linkages are contingent and do not adhere to any pattern or schema.

Generally, however, things do not tend to hang together randomly, because arrangements are not 'natural facts' but are created or delineated by human perception. Humans do not tend to randomly perceive things as arrangements. Instead, things are perceived as hanging together in clusters of interrelated determinate stuff (ibid: 1). Deriving from this argument, social orders for Schatzki emerge from how things are laid out and hang together in arrangements

> 'through and amid which social life transpires [...]. All social life is marked by social orders. In such orders, moreover, entities relate, enjoy meaning (and identity), and are positioned with respect to one another. All social life exhibits, as a result, relatedness, meaning, and mutual positioning' (ibid: 38).

Entities are divided into people, artefacts, non-human beings (organisms), and things. Artefacts are items manufactured by humans, while things are what people think of as 'natural' products. The differentiation between the four types of entities is artificial and serves analytical purposes. In fact, the boundaries between the different entities are blurry, as indicated by examples like genetically modified maize or the fact that human beings consist mainly of water and are only able to live because of their symbiosis with milliards of bacteria. Generally, it can be argued that nowadays most (natural) things are influenced by human activities at least in some ways. Schatzki agrees with Latour and others who have successfully challenged the idea that the social is a result of human activities alone. Instead, '*because human activity is beholden to the milieus of nonhumans amid which it proceeds, understanding specific practices always involves apprehending material configurations*' (Schatzki 2005a: 12).

Within arrangements, social orders are constituted by the ways in which entities are *related* to one another, the *positions* they take up towards one another, and the identities and *meanings* they possess (2005b: 51) (see Figure 5). Existing *relations* between entities of an agglomeration are social insofar as they contribute to the coexistence of human beings (2002: 38). Schatzki identifies four types of relations but does not limit the range of possible types to these four. According to his concept, entities can be related to one another through causal relations. Causal relations exist if one entity's actions make something happen or if one entity's actions or conditions lead to another entity performing an action (ibid: 41). While the first form only brings about a state, the second type leads to a reaction. This, however, does not mean that one entity's actions can bring about a predefined reaction. Responses to a certain action are contingent and could always also have been different. Nevertheless, causal relations bring about changes in the flow of action (ibid: 42). Spatial relations exist because entities are physical. Being material, they are related to one another in physical space. Most objects not only take up place but are related to one another spatially via activities, which also have to take place somewhere. Entities can furthermore be related to one another through intentionality if one of them performs an activity towards, or has thoughts, beliefs, intentions, or emotions about the other (ibid: 44). Prefiguration relates to '*how the world channels forthcoming activity*' (ibid.). Prefiguration is not limited to activities. Instead, entities can facilitate, restrict or enable actions of other entities.

Entities that are related to one another by causality, spatiality, intentionality, or prefiguration form an arrangement. '*This means that entities form arrangements by virtue of taking up or occupying relations of these sorts (usually a combination of them)*' (ibid: 46). Defining arrangements via their relations includes an understanding of these aggregates as not being marked out by 'naturally existing' boundaries (ibid.). Demarcating boundaries instead are relative to perception, ideas or interests. As will be elaborated later, distinct arrangements are furthermore created and distinguished through practices. In this thesis, demarcation of boundaries is based on the self-representations of actors, for example when certain human actors such as project members or material entities like energy production technologies are understood to be part of a project's energy production. On the other hand, demarcation is also shaped by my research interest and perspectives. According to Schatzki's definition, the fact that these borders only exist in the perception of actors (participants and researcher) does not mean that the setting of boundaries is an absolutely relative or random activity. Instead, demarcation is an outcome of sense-making practices by participants and me, and—according to what has been said in the methodology chapter—is related to objectified constructions of reality. With regard to the three case studies, relations between different entities are not understood to be naturally given. Instead, they are studied as phenomena which require analysis.

The *position* of an entity within an arrangement is partly created by its relations to other entities in an arrangement. '*An entity's position in a given arrangement is its location in that arrangement's plexus of relations*' (ibid: 53). The position of an entity is not only an outcome of spatial aspects like distance, direction, or inclusion, but also by its causal, intentional and prefigurational relations to other entities. An entity's position is also related to the meanings it attains within a certain context. With regard to the analysis of the case studies, local residents for example are positioned within a project's social order because they are spatially (close proximity to the production sites) and intentionally (activities of the project are shaped to cause minimal negative impacts on them) related to the project's energy production.

Meaning and identity of an entity are related to the relations and positions entities take up in the arrangements of things. Meaning relates to '*what something is*' while identity demarcates '*who someone is*' (ibid: 47). Identity is distinguished from meaning by the fact that only entities with an understanding of their own meaning possess identity. While Schatzki acknowledges that his conception of meaning/identity comes close to contemporary notions of 'roles', he underlines that, unlike in most role concepts, the constitution of meaning through prerogatives and obligations associated with a certain role, in his view, is not a one-way process; meaning also constitutes obligations. Meaning and identity are invariably multiple (ibid: 53). Most, if not all, entities are bearers of many different meanings and have different identities. Furthermore, the identity of a person or the meaning of an entity might change not only according to the various arrangements in which it is included, but also within one arrangement.

Identities and meanings are dependent on the ever-changing positions and relations of the bearer of a meaning or an identity and the understanding of other
entities from the aggregate. Being labile in nature, meanings and identities
change in predictable or unpredictable ways depending on contexts, circumstances, and events (ibid: 54). Within this thesis, the meanings of certain elements, like certain energy production technologies, are analysed. Solar panels
for example might not only be meaningful as items of energy production, but
might also function to signal a project's aspirations to the outside world. Taking
into account the existence of multiple meanings ensures the integration of complexity and ambivalence into the analysis.

Entities of an order are

related to one another via	positioned towards one another	have meaning/identity for one another
causality: if one entity's actions or conditions make something happen or lead to another entity performing an action	as outcome of an element's relations	Meaning: what something means within an arrangement
spatiality: most objects are related to one another spatially (also via activitities)		Identity: if this something has an understanding of itself
intentionality: if one element performs an activity towards, or has throughts, beliefs, intentions, or emotions about the other		
prefiguration: Entities might facilitate, restrict, or enable actions of other elements		

Figure 5: Constitution of social orders (Source: AP).

Schatzki's approach of social order is particularly valuable for the analysis of
the three case studies. At the same time, however, his approach has certain fallacies with regard to my research interest. His approach is valuable because

> '[t]o proclaim the interrelated meanings and identities of arranged items a key com-
> ponent of social order is to declare being central to order. It is to acknowledge, first,
> that there are no arrangements that are not arrangements of somethings and, second,
> that social somethings, perhaps somethings in general, are somethings as parts of
> arrangements' (Schatzki 2005b: 51).

Put in a more prosaic way, a major advantage of Schatzki's approach is that it acknowledges the contextuality of the meanings, positions, and relations of certain elements. The social orders of energy and energy production included in the projects hence are understood not as naturally occurring phenomena, but as resulting from the particular ways in which different entities take up positions, relate to one another, and have or are given meaning within a certain social context. At the same time energy and energy production are analysed with regard to how they shape the context within which they occur.

The problematic aspect of his account with regard to the research interest is that, while it acknowledges the fluidity of meanings and identities, Schatzki's approach is not able to explain instability as an inherent aspect of social orders. Even more importantly, Schatzki's approach does not provide analytical tools to see and understand relatedness, positions, and meanings as outcome of continuously ongoing negotiations between certain ways of sense-making.

This gap in his conceptualization of social orders is partly offset by the fact that he sees social orders as always related to social practices. The meaning of energy and energy production is not just an outcome of their (more or less stable) positions and relations, but is constituted in practices.

> 'Understandings […] are carried in social practices and expressed in the doings and sayings that compose practices. In particular, what something is understood to be in a given practice is expressed by those of the practice's doings and sayings that are directed toward it. Meaning, consequently, is carried by and established in social practices' (2002: 58).

As social order is manifested in the arrangements of entities and as these entities derive their meaning through social practices, orders do not exist by themselves. Instead, they are established, demarcated, maintained, altered, or destroyed through human practice. More precisely: '*social order is instituted within practices*' (2005b: 53). The following section introduces Schatzki's account of social practices and how they constitute and are constituted by social orders.

3.5.2 Social practices

Schatzki defines practices as those human activities that create social worlds. Practices are entities of '*spatially-temporally dispersed, open-ended sets of doings and sayings, which are organised by common understandings, "teleoaffective structures" (ends and tasks), and sets of rules*' (Schatzki 2005b: 58). Understandings, teleoaffective structures, and rules are the elements by which bodily doings and sayings are organised into social practices.

According to Schatzki, practices are spatiotemporally dispersed '*because each of them takes place somewhere in objective space at some point in, or over some duration of, objective time*' (2012: 15). Practices are open-ended as they are not composed of any specific number or pattern of activities. They consist of bodily doings and sayings, whereby sayings are a sub-form of bodily doings (doing-saying). Bodily doings and sayings are '*actions that people directly perform bodily and not by way of doing something else*' (2005b: 56). The fact that practices are embodied already implies a fundamental relation between practices and arrangements of entities: practices are enacted by human bodies (entities), generally in interaction with other human bodies, often involving various non-human entities (2012: 16).

Activities are organised into social practices through four principles (see Figure 6). While actions are individual bodily doings and sayings, the *organisation* of a practice cannot be ascribed to any specific individual. An action belongs to a practice when it is organised by an array of practical rules, understandings, teleoaffective structures, and general understandings. '*This array is distinct from, and differentially incorporated into, the minds of participants*' (2003: 192).

Practical understandings involve knowledge of what constitutes a certain activity and how to carry out this activity (2012: 16). Practical understandings include the necessary knowledge about how to perform a certain bodily doing or saying required to carry out a certain action. They also help to single out and perform, (but do not determine) what makes sense in a given situation. With regard to bodily activities, practical understandings provide ideas about what activities are appropriate or possible. While being performed by individuals, practical understandings are social in two senses: '*(1) in that multiple people carry on the practices involved and possess versions of the understandings carried therein and (2) in that both the intelligibilities and the practices carrying them are [...] "out there" in public space accessible in principle to anyone*' (2003: 184). Practical understanding, therefore, is one way in which the activities of individuals are organised through socially shared forms of knowledge. While social practices cannot be performed without practical knowledge, Schatzki emphasises that practical understandings do not determine practices. Within the figure of practical understandings, Schatzki integrates practice theory's concept of embodied knowledge into his approach. While some scholars tend to associate this concept with high levels of routinization and stability (Section 3.3), for Schatzki even the most routinized or ritualized practice includes a certain degree of change, as the bodily activities of actors never occur in quite the same way twice (2002). Each enactment of a bodily doing or saying slightly differs from all previous instances in which these activities were performed. Because of the involved instances of interpretation and creativity, practical understandings guide and shape social practices but never determine them. Practical understandings are relevant to many practices within the projects. Practical knowledge is not only required but also created through a project's activities. The projects not only necessitate and provide theoretical knowledge about energy production, but also enable participants and residents to (bodily) experience energy production as a day-to-day activity.

Rules are explicit formulations that school or enjoin people in particular actions. Examples include statute laws, 'rules of thumb', and explicit normative enjoinings onto actors (ibid: 185). Rules are social in the sense that they are '*ubiquitous in human life: humans are always formulating or producing them*' (2012: 16). Because they are ubiquitous in human life, rules play a role in nearly every practice. Like practical understandings, rules are important for the organisation of practices but do not directly or singularly determine them. Here, Schatzki draws on Wittgenstein's idea of rules and rule interpretation. While rules shape an actor's activities to a certain degree, both applicability and meaning of a rule are always interpreted by actors. Rules influence many aspects of the projects. Not only is energy production a highly regulated activity, but related aspects like building permissions also significantly influence the enactment of social practices in the three projects. All projects have also created rules for themselves. As the analysis will show, neither external nor internal rules remain unquestioned in the projects. The relevance of external political incentive programmes or of the internal aim to be sustainable and what they mean for social practices within the projects are constantly questioned.

The most important elements in the organisation of practices are *teleoaffective structures*. The notion combines the aspects of 'teleology' as orientation towards certain ends, and affectivity as the emotions and moods that can or should be expressed or felt when carrying out a certain practice. A teleoaffective structure is '*a set of teleological hierarchies (end-project-activity combinations) that are enjoined or acceptable in a given practice*' (2012: 16). Generally, people enact bodily doings and sayings to realise a certain end with emotional connotations. The social aspects of teleoaffective structures arise from the socially given acceptability of ends and expressed emotions. Teleoaffectivities are crucial when analysing the contextuality of energy production practices in the project. As shown in a later section, energy production is associated with very different ends and ideas in each project. Energy might be produced to realise the aim of community development, sustainability, resistance against the conventional energy system, or the creation of a lighthouse project.

General understandings are abstract senses about the values, worth, nature, or place of things expressed in people's doings and sayings. General understandings relate to the felt 'rightness', 'beauty', or 'nobility' of a bodily doing or saying and are not about its end. While Schatzki differentiates general understandings from teleoaffectivities, the analysis of the empirical findings highlights the interrelatedness of the two elements. Schatzki's conceptualization of general understandings gives the impression that general understandings already exist 'primordially'. When analysing general understandings related to energy production practices, it becomes apparent, however, that a certain activity is only understood to be right or wrong with regard to the teleologies inspiring the activity. While producing energy with a woodchip burner might be understood to be a 'right' practice when the aim is to produce energy in as decentralized a manner as possible, it becomes a 'wrong' activity when trying to produce energy as cheaply as possible.

Activities are organised into practices by

teleoaffectivities	a) teleology: orientation towards certain ends, b) affectivity: the emotions and moods that can or should be expressed or felt when carrying out a practice
general understandings	abstract senses about the values, worth, nature, or place of things expressed in peoples' doings and sayings
practical understandings	(embodied) knowledge about how to perform a certain doing or saying required to carry out a certain action
rules	formulations that school or enjoin people in particular actions

Figure 6: Organising principles of social practices (Source: AP).

While Schatzki does not offer a theoretical explanation of how the four organising principles are internalized, he describes how principles come to influence individual actions. According to him, rules, understandings, and teleoaffective structures influence the practices that *make sense* for a person to perform in a certain situation (2005b: 55).

Schatzki's differentiation of the four organising principles is one reason his approach has been chosen for the analysis. They offer a way to reasonably dismantle the practices of energy production in the three case studies. In particular, the concept of teleoaffective structures offers an interesting conceptual tool to analytically realise the research interest and question. By analysing the understandings, rules, and teleoaffectivities, the seemingly homogeneous activity of energy production can be shown to vary decisively in terms of all four principles across the three projects. Furthermore, using these principles in the analysis provides theoretical figures to explain ambiguity, complexity, and heterogeneity between different ways of sense-making within the projects.

Schatzki's concepts are operationalized for the analysis through the four organising principles. Practices related to energy and energy production in the three projects are analysed with regard to what and how teleoaffectivities, understandings, and rules have been shaping them. Analysing energy-related practices in terms of their organising principles focuses on the ways by which different (ambiguous, heterogeneous, complex) intentions, motivations, aims, and forms of knowledge shape the production of energy. The elements are mainly studied by analysing interviews, documents, and notes from the participant observation.

Analysing practices with Schatzki's approach means to understand that these are not composed of regularized or homogenized patterns of actions. Instead, irregularity and ad hoc or even unique actions are elements of a practice (2005b: 57). The activities comprising a practice are not stable. Instead, doings and sayings, as well as the '*understandings, rules, and teleoaffectivities that organise them, can change over time in response to contingent events*' (ibid: 61). Even more importantly, human activity is indeterminate. '*What I mean is that nothing regarding teleology or motivation can determine or fix, prior to activity, what a person does or why. It is only with the occurrence of activity that what a person does and why become determinate*' (2011: 5). This indeterminateness of practices is closely related to Schatzki's conception of time. Instead of the ordered succession of '*objective time*' (2013: 35) in his concept the past, present, and future are always included in an activity. Ideas and desires for the future influence actions and ends, while the past influences the experiences and hence the sense-making of participants. The past, however, is itself indeterminate. It '*is not fixed or laid down prior to a person acting*' (2012: 19). How past experiences are perceived, interpreted, and applied in a certain situation is only determined at the moment a decision is taken and the resulting activity takes place. Included in this conceptualization of indeterminateness is a conceptual inclusion of individual agency and creativity. What happens is not determined by practice-order bundles but always includes elements of agency. Individuals make sense of a situation and act upon this sense-making, thereby bringing their individual future and past into it. This conceptualization allows for a (limited) inclusion of contingency, heterogeneity and complexity. Different individuals may make sense of the same situation differently or attach varying meanings to it.

Instead of elaborating upon the aspect of ambiguity and contingency, Schatzki is more interested in how these complexities are interwoven in a certain situation. According to him, the practices of different actors are interwoven '*through the existence of common, shared, and orchestrated elements*' (ibid: 20). Practices are *common* when participants act towards the same ends, purposes, or motivations or at the same places/paths and do so because this is enjoined in the normative organisation of the practice. Activities are *shared* when they do so not because this is enjoined but because it is still acceptable. *Orchestration* of activities occurs when elements of one person's practice are dependent upon elements of another person's practice (ibid: 20).

While describing forms of interwoven practices, Schatzki's approach is not interested in explaining *how* practices are interwoven or orchestrated with one another. Moreover, the idea of conflict or negotiation is missing in the concept of interwoven practices. How are elements orchestrated when these elements do not match easily or are even opposed to one another—such as the different intentions motivating different participants' engagement and attitudes towards renewable energy and energy production? Excluding conflict and negotiation implies a homogenization of included activities and/or actors. Ambivalence and complexities are excluded from these theoretical figures. To offset these shortcomings of Schatzki's approach, it is necessary to combine his practice theory with another theoretical concept in order to analytically grasp this thesis's research interest and questions. Before introducing Adele Clarke's situational analysis, however, the third important dimension of Schatzki's theoretical approach is explained.

3.5.3 Social sites

Within the concept of the social site, Schatzki brings together orders and practices by embedding both in the context of the social site. The notion of 'context' already carries in it the idea of something being embedded into something else. Schatzki refines and theorises contexts into the concept of the site. '*Sites, in general, are* where *things exist and events happen*' (Schatzki 2002: 63, original emphasis). The 'where' does not only signify a location in physical space, but also embraces social meanings that are captured in the term context. In the most general sense, a context is '*the sum total of everything other than itself that determines these*' (ibid: 60). Contexts not only surround or immerse the elements of which they are the context; they also have powers of determination. They determine the meaning and significance of, confer value to, and prefigure, i.e. constrain and enable, what occurs in them (ibid: 62).

Social sites are a specific type of context. They embed different types of entities and phenomena. These entities and phenomena not only occur in a certain context but are intrinsically part of the site. '*[A] context is a* site *when at least some of the entities that occur in it are inherently components of it. That is to say, for something to be or to occur in a site context is for it to be or to occur* as *a constituent part of its context*' (ibid: 65; original emphasis).

I understand energy production and all the elements organised in it (practices and entities as well as their relations, positions, and meanings) to be both an outcome of specific social contexts and inherently part of the context amid which they occur. With regard to the analysis, this means that I do not only analyse how different aspects of energy production are shaped by the context, but instead also examine how these aspects influence the social site of which they are inherently part.

The elements making up a social site are those entities forming the social orders of a site and the practices that constitute and give meaning to these arrangements. These practices and arrangements of entities are inevitably bound up with one another in social sites. On the one hand, doings and sayings are carried out by bodily entities (humans) and are normally directed at other human entities, often involving other material entities. Most material entities, on the other hand, would not exist without, or are at least affected by, human practices. Practices effect, give meaning to, and are inseparable from arrangements. Schatzki distinguishes two forms of determination by which practices establish social orders: '*Social orders are established within the sway of social practices because practices mould the forms of determination that are responsible for them*' (2005b: 53pp.). Also, practices determine orders through the institution of meaning (ibid: 55). Conversely, social orders channel, prefigure, facilitate, and are essential to practices (2012: 16). Mostly, however, orders and practices are co-constitutive. Schatzki identifies five types of relations between practices and arrangements (see Figure 7).

Causal relations: activities create, alter, and destroy arrangements or parts of arrangements while arrangements of entities '*befall*' (ibid.) the participants of a practice and induce activities. '*The causality involved is not of the bringing about sort, but instead of the leads to variety: both the properties of material entities and the events that occur to them lead people to perform actions and practices to take certain courses*' (2010: 139). Causal relations also exist between entities, like between biomass boilers, pipes, and houses. These relations might then have effects on the practices that are carried out amidst them.

Prefiguration is '*the difference that the past makes to the nascent future*' (2012: 16). Prefiguration is not limited to how practices or arrangements might have enabling or constraining effects. Instead, prefiguration makes certain actions easier, more complicated, promising, relevant, difficult, or more lasting. Prefiguration can be conceptualized as a qualification of possible paths (2010: 140).

Constitution is the way in which arrangements constitute practices when they are '*essential to or pervasively involved with practices*' (2012: 17). The term 'essential' relates to practices that could not be carried out in the absence of certain entities or arrangements. Pervasive elements are those elements that shape a practice but that could be substituted by other entities. Constitution also relates to an arrangement that would not exist without a specific practice. Most arrangements would not exist or would exist in a very different form without the practices that constitute them and that are carried out amidst them. Practices and arrangements are co-constitutive (2010: 140).

Intentionality refers to how practices and arrangements are related to participants' thoughts and imaginings about the respective other when using/doing an arrangement/practice.

Intelligibility emphasises the importance of arrangements and practices having meaning or being intelligible as such and such (ibid.). Entities are intelligible as what they are and what they are meant for or how they can be used in practices. Intelligibility, thus, is instituted in practices carried out by humans (2012: 141).

Orders and practices are related to one another via:

causal relations	Activities create, alter, and destroy arrangements or parts or arrangements while arrangements of entities 'befall' (ibid.) the participants of a practice and induce activities
prefiguration	Orders/practices make certain actions/orders easier, more complicated, promising, relevant, difficult, or more lasting
constitution	A) 'essential' constituting: orders/practices that could not exist/be carried out in the absence of certain entities or arrangements or practices; b) pervasive constituting: elements that shape an order/a practice but that could be substituted by other entities/practices
intentionality	Practices and orders are related by participants' thoughts and imaginings about the respective other when using/doing an arrangement/practice
intelligibility	Arrangements and practice having meaning or being intelligible as such and such

Figure 7: Types of relations between orders and practices (Source: AP).

Practices are also connected with other practices and orders are connected with other orders. Practices, for example, overlap when one action belongs to different practices. More importantly, different practices overlap when they are subject to the same organisational elements, like rules, ends, or projects (ibid: 151). Finally, practices overlap when they are related to one another through intentionality. Intentionality also relates orders to one another. Different orders might overlap through connections. Furthermore, orders are connected to one another through causal, spatial, and prefigurational relations between any of their components (2002: 156). By means of these relations, overlaps, determinations, and facilitations, *'practices and orders form an immense, shifting, and transmogrifying mesh in which they overlap, interweave, cohere, conflict, diverge, scatter, and enable as well as constrain each other. Such is the nature of the social site'* (ibid: 156pp.). The conceptualization of sites as meshes of orders and practices allows an analysis of social phenomena in which different heterogeneous, inconsistent, and even seemingly contradictory entities and practices contribute to the situation (see Figure 8).

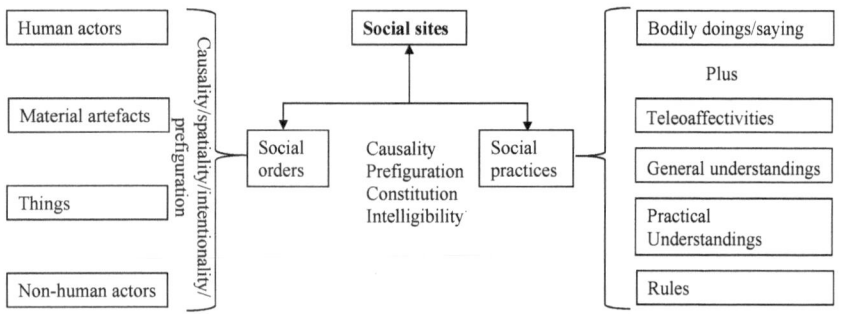

Figure 8: Social sites (Source: AP).

Analysing the three case studies with the concept of social site *'directs attention to how practices and arrangements causally relate, how arrangements prefigure practices, [and] how practices and arrangements constitute one another'* (2010: 146). The concept of social sites enables the analysis and theoretical explanation of how different elements within the three case studies constitute and are constituted by varying teleoaffectivities and understandings. An additional value of Schatzki's approach lies in his emphasis on the contingency and instability of social sites.

> '[P]ractices and orders are not just contingently but also incompletely and precariously packaged into bundles. Indeed, the very notion of a "mesh" of practices and orders is meant to suggest that activities and arrangements form a great evolving horizontal web of interweaving practices amid interconnected orders. Whatever consolidations of practices and orders occur in the mesh are contingent, regularly disrupted and always perforated by the moving rhizomes of dispersed practices that lace through the social site' (2002: 154).

This conceptualization of social sites as continually changing and shifting meshes is congruent with the methodological reflections and empirical findings that prefigured construction of the theoretical framework. Using the concept of social sites, attention can be directed to complexities and instabilities of practices of energy production and how these practices relate to the specific context. Nevertheless as is repeatedly pointed out in the last sections, Schatzki's approach alone does not suffice to theoretically analyse and explain some key aspects of the research interest and questions. The main pitfall of Schatzki's concept is its inability to explain the constitution, modification, and destruction of social practices. While presupposing the existence of social sense-making in social practices, Schatzki's concept is unable to explain how the already ambiguous and heterogeneous understandings and teleoaffectivities of one social actor are orchestrated or shared with the practice elements of a multitude of various other actors (and entities) making up and enacting the social site.

Adele Clarke's situational analysis provides an approach by which social sites are understood as *negotiated* and *contested* meshes of orders and practices. Clarke's conceptualization of Anselm Strauss's theory provides a way to analyse ongoing negotiations. Additionally, through her conceptualization of negotiations, she also provides an approach that further enhances the role of instability, ambiguity, unpredictability, and context-specific creativity within the theoretical framework. By combining Schatzki's approach with situational analysis, I relocate the focus of the analysis from the reproductive and routinized aspects of embodied practices onto their negotiated and situated character.

The following section introduces Adele Clarke's situational analysis and then explains how her concept can be combined with Schatzki's approach in a meaningful way to analyse renewable energy production as context-specific and negotiated social practices of different actors.

3.6 Adele Clarke's situational analysis

Adele Clarke, a former pupil of Anselm Strauss, has a long experience in feminist theory and aims to develop an approach for qualitative social research that integrates theoretical and methodological positions into a coherent '*theory-method package*' (Clarke/Keller 2012: 37). In the following sections, I describe Clarke's general ideas before introducing the theoretical concepts she uses to '*resituate [...] grounded theory around the postmodern turn*' (Clarke 2003: 553). The last part of this section explains how Clarke's concepts are combined with Schatzki's practice theory approach in order to theoretically realise the research interest in a way that takes the methodological premises and empirical findings seriously.

3.6.1   General underpinnings and conceptualizations

According to Adele Clarke, situational analysis is an effort to regenerate and update grounded theory such that it includes important insights from postmodernism. While she is convinced that parts of grounded theory have always been postmodern, she contends that, in other ways, it is still deeply rooted in modern and positivistic ideas (Clarke/Keller 2012: 26). Modernism, as Clarke sees it, is defined by its search for stability, universality, generalization, simplification, regularity, and the likes. These fundamental propositions lead to a research that, in her eyes, has long tried to erase 'context', variation, and complexity. While postmodernism is not a unified programme or approach, its proponents have shifted attention to issues like instabilities, localities, partialities, complications, fragmentations, contradictions, and complexity (Clarke 2003: 555). According to these approaches, reality is not only generally constructed; a multitude of realities are constructed by people with different and changing perceptions while they are situated in different and changing situations or positions.

Adele Clarke identifies four remaining positivistic reflexes of grounded theory. The first is its inadequate inclusion of contingency, instability, and fragmentation. The second reflex is an implicit assumption that 'reality' is out there, waiting for the researcher to find it. Third, grounded theory does not integrate non-human elements distinctively into its analysis. Finally, it has hitherto neglected questions of power. Clarke wants to overcome the first two modern reflexes of grounded theory by revitalizing pragmatistic assumptions underlying grounded theory, '*which are to be activated through methodological and theoretical repositioning and appropriate practical measures*' (Strübing 2014: 101; translation AP). Her first criticism concerns the integration of complexities:

> 'the methodological implications of the postmodern primarily require taking situatedness, variations, complicatedness, differences of all kinds, and positionality/relationality very serious in all their complexities, multiplicities, instabilities, and contradictions' (Clarke 2003: 553).

Clarke does not only want to *increase* grounded theory's acceptance of diversity, heterogeneity, instability, and contingency, but aims to make it *focus* on these issues. This aim requires both an extension of grounded theory's method canon, and a modification of some of its basic theoretical concepts.

Deriving from postmodern understandings about the complexity of social life, Clarke's second criticism concerns the implicit assumption underlying grounded theory, which claims that one (unified) reality exists out there for the researcher to find. This perspective on the empirical process lacks reflexive consideration of how research 'objects' are being constructed in (social) research. To overcome this fallacy of grounded theory, Clarke extends it with postmodern understandings of knowledge and truth. Postmodernism has challenged positivistic ideas of knowledge and truth, which had their background in enlightenments' enthusiastic celebration of rationality, humanism, and the existence of objective truth. Contesting these ideas, postmodernism emphasises that there is no 'objective truth' whatsoever. Instead, knowledge and truth are socially produced realities (Berger/Luckmann 1991 [1966]). Research can never be objective; it is an expression of the ideas, interests and positions of the researcher, the situation, and the context of the research. Constructivist approaches have challenged not only the 'objectivity' of social sciences, but also the idea that natural sciences yield objective truth (Knorr-Cetina 1999). Instead of being related to objective truth, knowledge only exists in the form of '*situated knowledges*' (Haraway 1988) which are produced by historically and locally positioned groups of people. Knowledge has come to be understood in terms of power and domination rather than as an expression of truth (ibid: 576).

These postmodern conceptions of reality, truth, and knowledge have led to severe methodical challenges for social scientists. Some scholars have even proclaimed a '*crisis of representation and legitimation*' (Clarke 2003: 553) of social research. Adherents of this view have generally questioned whether research of 'the social' is still possible when complexity and perspective are assumed as inherent aspects of social reality. More specifically, the paucity of means and methods to integrate complexity and ambiguity in social research has been severely questioned. Faced with the complications of attending to complexity, heterogeneity, and contingency, some postmodernist research methods like autoethnography or interpretive and biographical approaches have abandoned all efforts to reconstruct the social constitution of reality and instead now focus on representing individual '*voices*' (ibid: 556). While she is supportive of these approaches, Clarke is more interested in researching '*the social*' in social life (Clarke/Keller 2012: 31). She argues that researching 'the social' '*is not impossible after the postmodern turn but quite different*' (Clarke 2003: 555). Methods and methodology of qualitative social research have to be developed or enhanced to sufficiently enable the integration and representation of the complexities of the empirical world into research and analysis.

According to Clarke, grounded theory, as developed by Strauss and Strauss/Corbin, provides an aptly basic concept to overcome the methodological challenges posed by postmodernism. According to her, grounded theory has always integrated a basic methodological awareness of situatedness and context while remaining interested in analysing social situations and the social element, instead of individual voices (Clarke 2003: 556). In particular, Strauss's consideration of Georg Herbert Mead's concept of symbolic interactionism has provided grounded theory with the '*capacity to be distinctly perspectival in ways fully compatible with what are now understood as situated knowledges'* (ibid: 555; likewise, Clarke/Keller 2012: 29). Using an approach in which reality is understood to be constructed in social interactions, however, is not sufficient to push grounded theory entirely around the postmodern turn. To develop the inherent potential of grounded theory, Clarke draws on pragmatistic ontologies and epistemologies.

In pragmatism, the modern belief in the existence of objective truth is tempered with the conviction that truth is constantly produced through human interaction with others and the (material) world around them. Reality is constantly in the making. Humans have to cope with situations for which they can find only temporary solutions, as no solution can constantly and perfectly fit all encountered situations (Gronow 2012: 16; Strübing 2008: xx). The outcome of these situations poses a 'new' truth about reality, on the basis of which human interaction proceeds. The created reality is perceived as 'objective' reality by the acting individuals. It may potentially create new problems that have to be solved experimentally. Consequently, reality is always created in practice and in interaction with the entities making up a certain situation. It is context-bound, unstable, and always temporary (Volbers 2015: 207).

Based on these insights, Clarke abandons grounded theory's focus on 'basic social processes', i.e. its focus on the basic human activities constituting social phenomena. Instead, she emphasises the importance of researching more than the *'knowing individual'* (Clarke/Keller 2012: 30) or focusing on action alone. According to her, the focus needs to be *'fully on the situation of inquiry, broadly conceived'* (Clarke 2003: 556). While traditional grounded theory focused on human action and only saw contextual elements as conditioning this action, in *'situational analysis, the situation itself is the key unit of analysis'* (ibid: 559). Such a perspective has to include different human, non-human, material, and discursive/symbolic elements, as well as the structures and conditions that characterize a situation (Clarke/Keller 2012: 24). Understanding how elements and their relations shape situations is the focus of research. Situational analysis motivates researchers to analyse empirical situations, which are understood to be complex, contingent, ambiguous, heterogeneous, and only temporarily stabilized. To theoretically grasp how situations are constituted, Clarke's situational analysis supplements the traditional social process metaphor of grounded theory.

Clarke's emphasis on not only integrating complexity and instability but making them a focus of research provided an important inspiration for how to approach my empirical material. Clarke's approach has encouraged an understanding of ambivalences and complexities in the empirical material as valuable findings. It is, however, her conceptualization of Anselm Strauss's social

worlds, arenas, and negotiations of orders that provided the theoretical tools necessary to grasp and explain these findings. Even more importantly—with regard
to the research interest and question—the concept of situational analysis provides an approach that allows a theoretical explanation of energy and energy
production, not despite, but because of complexity, ambivalence, and the likes.
Consequently, the concepts of social worlds, arenas, and negotiation of orders
are explained in more detail.

3.6.2 Social worlds, arenas, and negotiation of orders

In developing his concept of social worlds and arenas, Strauss critically differentiated his ideas from functionalistic theory, which '*sees a profession largely
as a relatively homogeneous community whose members share identity, values,
definitions of role, and interests*' (Bucher/Strauss 1961: 325). In his research,
mainly based in the medical sector, Strauss showed that professions are neither
homogenous nor static but the result of continuously ongoing processes of interaction, in which possible professional standards, relations, and methods are
constantly negotiated (Strübing 2005: 171). Structures of the profession are likewise continually produced, reproduced, and modified in interactive processes of
negotiation (ibid: 172). Through interaction, not only with each other but also
with already 'structuralized' processes, actors form new constellations and patterns of interaction, in which they are themselves embedded via their '*organised
identities*'. In these negotiations, groups constantly emerge, develop, fall apart,
re-integrate, re-organise, break up, and form new constellations (ibid.).

Adele Clarke has integrated Strauss's concepts of social worlds, social arenas, and negotiation of order into her postmodern approach to grounded theory.
Social worlds are created through fluid, constantly ongoing negotiations within
'*groups with certain commitments to certain activities, sharing resources of
many kinds to achieve their goals, and building shared ideologies about how to
go about their business*' (Clarke 1991: 131). The activities performed in a social
world are derived from its (negotiated) substantive area or core activity. Strauss
conceives of the activities as being localized and involving the employment of
material things and forms of organisation.

> 'In each social world, at least one primary *activity* (along with related clusters of
> activity) is strikingly evident. [...] There are *sites* where activities occur; hence
> space and a shaped landscape are relevant. *Technology* (inherited or innovative
> modes of carrying out the social world's activities) is always involved. Most worlds
> evolve quite complex technologies. In social worlds at their outset, there may be
> only temporary divisions of labour, but once under way, *organisations* inevitably

evolve to further one aspect or another of the world's activities' (Strauss 1978: 122, original emphasis).

By constructing his concept of social worlds around activities, Strauss defines social worlds as processes (Strübing 2005: 178). The process character is further strengthened by conceptualizing relevant sites, technologies, and organisations as (sub-)processes in which actors are involved. By emphasising process instead of stability, Strauss conceptualizes conflict, negotiation, and power as inherent parts of social worlds. Social worlds are contested processes whose core or re-lated activities and legitimate forms of commitment are (re)established in nego-tiations. Social worlds can only be temporarily stabilized through collective agreement.

Being a pragmatist, Strauss emphasises that social worlds do not exist as such; they only become visible in their practical consequences. An important aspect of the analytical differentiation of a social world is that actors relate to one another practically by means of symbols. Individuals relate to one another using objects and activities that are deemed relevant. Social worlds are further differentiated from one another by the '*shared commitment*' (Clarke 1991: 131) of their members. Participation in a social world is established and maintained through a sufficient degree of self-commitment and engagement with regard to a negotiated core activity (Strübing 2005: 180). Being an outcome of activity, membership itself is dynamic and only temporary; those who no longer contrib-ute to the activities of a social world are no longer part of it. Active participation is not a question of either/or, but a continuum of more or less '*authenticity*' stemming from the degree of commitment shown and the level of legitimacy given to the activities of a person (Strauss 1978: 123).

Social worlds are not distinctive entities and do not require exclusive mem-bership (Strübing 2005: 181). People participate in multiple social worlds, even at the same time. Likewise, different social worlds intersect and overlap through human activity. Because of these intersections and the gradual nature of partic-ipation, the boundaries of social worlds are fluid. Intersections and fluidity es-tablish a dynamic interaction of social worlds with their environments (ibid: 183). Because participants do not all interact in the same way with environments and because resources are not shared equally by all participants, different per-spectives are established, which contribute to the continuous development of different segments within social worlds. Strauss calls these segments social sub-worlds (ibid.). Intersection and segmentation further enhance the dynamic char-acter of social worlds as they '*imply that we are confronting a universe marked by tremendous fluidity; it won't and can't stand still*' (Strauss 1978: 123).

Different social worlds and sub-worlds can be analytically differentiated in each of the case studies. As defined by commitment and active engagement, important social worlds in the CDT and KEBAP, for example, are working groups and the board. Each working group is organised around a shared interest and core activity, like planning and/or running the energy production installations, gardening, or heritage. Social worlds constitute the social spaces in which energy production is being planned and carried out in the three projects (CDT, KEBAP, and IBA). Both in the IBA and the two other case studies, social subworlds related to energy and energy production constantly exist, overlap, and interact in many different ways with one another but also with other social (sub-)worlds. Interaction between multiple and fluid social worlds implies that different ideas, interests, and activities have to be negotiated. In order to analytically grasp these spaces of (conflictive) negotiation between different social worlds, Strauss developed the concept of social arenas. While social worlds conceptualize the co-production of community, cohesion and similarity, social arenas analytically grapple with negotiation and solution of problems caused by dissimilar or even opposing perspectives (Strübing 2005: 189).

Social arenas are sites in which participants of different social worlds negotiate issues, perspectives, and problems. Arenas thus comprise of the '*interaction by social worlds around issues—where actions concerning those are being debated, fought out, negotiated, manipulated, and even coerced within and among the social worlds*' (Strauss 1993: 226). Negotiation starts with the definition of the problem that is at issue in a certain arena. Problem definition is not a one-time activity but a constant process of defining, redefining, and rejecting definitions. This process accompanies the development of problem solutions. Types of negotiation, for both problem definition and solution, range from bargaining to coercion, and include relatively harmonious as well as immensely conflictive processes. Clarke defines an arena as '*a field of action and interaction among a potentially wide variety of collective entities. It [...] includes all collective actors [...] committed to acting within it*' (1991: 131). The notion of field, like the term arena, is problematic insofar as it seems to imply that social arenas are social locations wherein actors interact directly and openly with one another. Direct confrontation, however, is by no means the only form of interaction. In fact, negotiations often occur in indirect form, for example when unsatisfied employees passively protest by working 'by the book', thus consciously reducing the efficiency of work processes (ibid.).

Just as individuals participate in more than one social world, social worlds are related to more than one social arena (Clarke/Keller 2012: 149pp.). Different social worlds participating in an arena have to relate themselves or become related to the problems discussed in that arena. As Strauss phrases it, social worlds and the problems have to match: '*Matching is an active process, carried out by the participants. They select and reject issues, and reshape them in accordance with their own image and aims'* (Strauss 1993: 229). To support their claims, different social worlds can combine their efforts by creating (temporary) coalitions. Social arenas, like social worlds, might intersect and overlap with each other.

IBA, CDT, and KEBAP are all comprised of multiple social arenas. Different social worlds, like working groups and board members (CDT, KEBAP) or employees of the IBA and Hamburg Energie, negotiate and bargain about their ideas, understandings, or perspectives in such arenas. It is in these social arenas that the meaning of energy within the respective project is created in negotiation processes of social sense-making. As shown in the analysis, these processes can cause severe conflicts. Additionally, different social worlds in the projects fight over a number of different resources (money, places, volunteers). The solutions found in these processes are always only momentary and ambivalent, as participating or affected groups interpret them in very different ways, making negotiation a constant activity in each project.

With the concept of *negotiated orders* Strauss conceptualizes the idea that in every interaction, actors have to deal with pre-given versions of an already negotiated order. Strauss introduces the idea of already existing conditions or 'structures', among which actions unfold. In Strauss' concept, structures are relatively enduring and stable elements of a certain situation which turn into the shaping conditions of this situation, henceforth framing activities (Strübing 2005: 191). It is important to note that structures do not determine actions. While actors have to deal with the existing conditions, structures do not pre-define how actors deal with them (ibid.). Actors (unconsciously) negotiate these existing orders while perceiving, interpreting, and acting upon these conditions. In the concept of negotiated orders, reality is created in practice and in interaction with the conditions that make up a certain situation. Likewise deriving from pragmatism is Strauss conceptualization of social worlds, arenas, and negotiated orders as always only temporary and unstable understandings, agreements, rules or contracts (ibid: 193).

Orders in the three projects might be comprised of (formal) rules and laws, or enduring conditions resulting from the material artefacts or the local environments that are made meaningful in the projects. Strauss's concept of negotiated orders is highly relevant with regard to research interest and question. Throughout the analysis, structures and conditions are understood not as determining a situation, but as being negotiated and thus socially made sense of within a certain context. In these contexts, different ideas and perspectives of participating social worlds also have to be negotiated with one another and the social conditions.

The important contribution of Adele Clarke's situational analysis is that it enables researchers to analyse a phenomenon by focusing on meaningful groups and acting collectives while starting from the individual (Clarke/Keller 2012: 147). For this purpose, Strauss's concept of social worlds and arenas is important. Social worlds are where individuals become social beings. Social worlds, arenas, and negotiated orders are important tools for Clarke's intention to tackle the methodological challenges posed by postmodernism, especially to understand and elaborate *'what has been meant by "the social" in social life— before, during, and after the postmodern turn'* (Clarke 2003: 557).

As described above, Clarke has criticized grounded theory for its lack of integration of non-human elements and social power. In order to overcome these two fallacies, Clarke combines grounded theory with insights from actor-network theory. By combining grounded theory with actor-network theory, Clarke fulfils the postmodern demands of integrating non-human actors into social research. This demand is based on the premise that humans do not exist or act in isolation, but that society and nature mutually create each other. Humans and non-humans depend on each other (Clarke/Keller 2012: 37). Because it does not focus on human properties alone, Clarke sees Strauss's version of grounded theory as fundamentally capable of integrating non-human elements (ibid.). Focusing on situations instead of action, however, enhances grounded theory's ability to include human and non-human elements into the analysis. Thus, in her conceptualization, the analysis of social worlds, arenas, and negotiation of orders requires the researcher to consider the situative engagements of actors with the material artefacts co-constituting the situation.

While Clarke's integration of material artefacts does suit my research interest, the theoretical framework developed for this thesis is based on the way Theodore Schatzki includes material artefacts and other entities. His concept offers a more pronounced way of analytically engaging with these elements by not only including entities into the analysis, but also considering their positions, relations, and meanings. At the same time, Schatzki's concept has to be refitted with Clarke's concept of negotiated orders. The analysis shifts the focus from a pure consideration of entities and their positions, relations, and meanings, to study how these (material) conditions are continuously *made* meaningful and negotiated in complex and heterogeneous ways within the three projects.

The example of the material artefacts shows that it is not only Schatzki's approach that benefits from its combination with situational analysis. Rather, the two theoretical concepts offer different perspectives that can be combined productively. The next section details how and to what effect the two approaches are combined.

3.7 Combining situational analysis and social sites

Clarke's situational analysis is roughly based on the same ontological premises as Schatzki's theoretical concept of social sites. Both theoretical approaches are based on postmodern and post-pragmatic ideas. Schatzki might not emphasise his postmodern aspirations as openly as Adele Clarke, but it is clear that his approach is also based on postmodern insights, as shown in his explanation of the value of his approaches:

> 'I believe that one noteworthy outcome of writing histories and analysing contemporary phenomena with these experientially resonant concepts is that history and the contemporary world seem less systematic or ordered and more labyrinthine and contingent than they do when described and analysed through the conceptual armature of many other theories' (Schatzki 2010: 146).

The quote makes it explicit that for Schatzki, like for Clarke, contingency and complexity are important aspects of reality. His approach to practice theory pays attention to these aspects.

Another important ontological postmodern stance of both scholars lies in the ways in which they understand the relevance of materiality to social life. Both Schatzki and Clarke believe that society is inherently tied to material entities and arrangements. Most importantly, both approaches emphasise the situatedness of social life. 'The social', according to both scholars, is created and based in the context in which it is physically and socially located. Clarke repeatedly argues that research that only focuses social action is not sufficient if it does not pay attention to the situation within which activities unfold. *'Action is not enough. Our analytic focus needs to be fully on the situation of inquiry broadly conceived'* (Clarke 2003: 557). Schatzki formulates that the social is created within its context. His concept of social sites defines sites/situations as meshes of orders and practices. He combines and intertwines action and situation by elaborating how these two constantly (re)generate and modify each other.

For both scholars, 'the social' only exists in specific situations or contexts. For Clarke, the situation consists of all those elements that are part of the situation.

> *'The conditions **of the** situation are included **in** the situation.* There is nothing like "context." The conditional elements of the situation have to be specified in the analysis, *because they are constitutive for it,* and not only surround or frame or contribute to the situation. They *are* the situation' (Clarke/Keller 2012: 112; original emphasis, translation AP).

This definition of situations is compatible with Schatzki's concept of social sites. His concept also states that most, if not all, entities comprising a social site are *inherently* part of that site. Schatzki's social sites, like Clarke's situations, are made up of the elements that occur in them and thus cannot be separated from them. Likewise, Schatzki and Clarke both state that sites or situations are made up of different entities and their mutual relations (Clarke/Keller 2012: 25; Schatzki 2002). According to both theorists, all entities that take part in a situation influence and constitute all other entities in some ways (Clarke/Keller 2012: 114). While being ontologically combinable, one question remains to be answered in this section: What is the theoretical benefit of combining the two approaches?

This thesis aims to analyse energy and energy production as fundamentally social phenomena. More specifically, my research interest is to study energy and energy production as the outcome of highly specific and contextualized social processes. Apart from this research interest, four requirements have been carved out in the beginning of this chapter, which the theoretical concepts have to answer in order to live up to the methodology and the empirical findings. First, the framework needs to provide theoretical tools with which energy production can be analysed as an integrated aspect of the project's *everyday activities*. Second, the theoretical framework has to be able to theoretically explain how the respective context *practically* influences the interactions of project members with one another as well as with external actors, local environments, and materialities with regard to renewable energies.

I have already argued that these first two requirements can be sufficiently fulfilled by using practice theory approaches. Like all practice theory approaches, Schatzki's concept focuses on everyday activities, and thus the way in which socially shared patterns of knowledge are practically enacted. By using Schatzki's approach, energy production is studied as a practice-as-performance, and hence as a practice that—while drawing on socially shared symbolic patterns of knowledge—is always context-specific. Practices in his concept are bodily doings and saying that are organised by the elements of teleoaffectivities, understandings, and rules. Employing these elements for the analysis of the cases shows how the different actors' ways of sense-making crucially shape the interpretation and enactment of energy and energy production in the projects. Energy and energy production can be shown to be influenced by very different teleoaffectivities, understandings, and rules. Based on Schatzki's combination of practices and social orders in social sites, it is further possible to not only analyse energy and energy production as the outcome of actors' practices, but also take into consideration how these practices are embedded in their social context. This conceptualization not only emphasises the contextuality of energy production, it also provides tools by which non-human elements, their relations, positions, and meanings can be included into the analysis. The concept of social practices as evolving amid social orders provides, for example a way to analyse how seemingly similar material objects (the bunkers, the technologies) might in fact be employed differently not only in each project but also by different actors in one project.

It has been argued that while practice theory provides tools for the analysis of the case studies, it is unable to meet the last two requirements deriving from the empirical findings and methodological principles. With regard to these requirements, the theoretical framework has to recognize and explain sense-making as a dynamic process. Last but not least, the empirical data demands a theoretical concept that can grasp contingencies, ambiguities, and heterogeneities as well as the constantly occurring processes of negotiation occurring in the projects. Adele Clarke's postmodern version of grounded theory is able to enhance Schatzki's practice theory approach with regard to fulfilling these requirements. In particular, Clarke's re-introduction of Anselm Strauss's theoretical figures of social worlds, social arenas, and negotiated orders is a valuable extension to Schatzki's concept. These direct the focus on ongoing negotiations and contestations occurring in the three social sites (see Figure 9). Practices of energy production can be analysed as evolving out of constantly ongoing negotiations of different actors' understandings and teleoaffectivities and of the interactions of these actors with the social orders creating the respective situation. Using Clarke's approach, Schatzki's practice theory can be extended to not only include but focus on ambivalence, complexity, instability, and contingency and to thus come to understand energy production as an outcome of constantly changing dynamic and contextual processes.

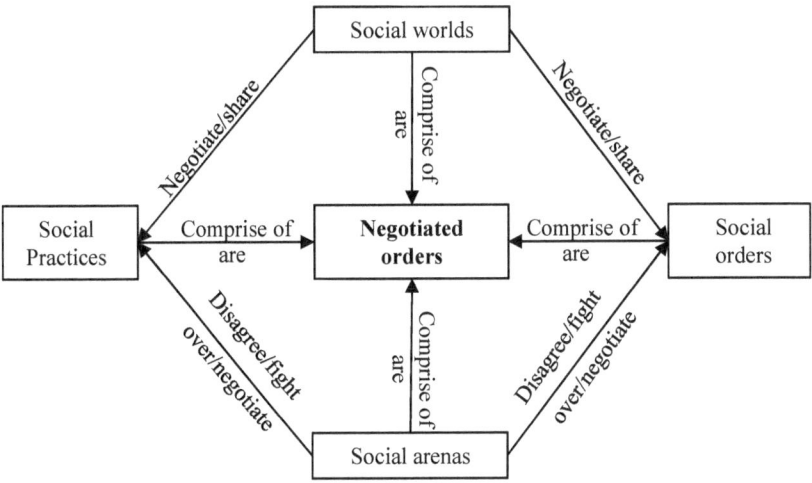

Figure 9: Combining Clarke's and Schatzki's concepts (Source: AP).

The concept of *social worlds* offers a way to analyse how social practices and orders are socially shared as an unstable outcome of negotiations of actors interested in the same types of core activities. Scrutinizing Schatzki's concepts of social practices and orders with the concept of social worlds directs attention to how social practices and orders are negotiated and shared in social worlds. Employing the concept of *social arenas* draws attention to how social orders and practices are contested, fought over, and negotiated in the projects. The idea of social arenas emphasises that practices and orders are neither stable entities nor are shared homogeneously among different actor groups but are constantly contested, fought over, and (re)negotiated. Taken together, social worlds and social arenas modify Schatzki's practice theory in the sense that negotiation, heterogeneity, and complexity are not merely implicit aspects but are made the focus of the research. Clarke's conceptualization of *negotiated orders* enhances Schatzki's concept of social orders by drawing attention to how social orders or the relations, positions, and meanings of their elements are negotiated by different social worlds. Understanding social orders as negotiated orders conceptualizes the relations, positions, and meanings that shape a certain arrangement as a temporary and unstable outcome of ongoing negotiations.

The framework enables the analysis and explanation of energy and energy production as the outcome of contextualized dynamic social processes. These processes consist of daily social practices that are shaped by constantly negotiated and contested forms of social sense-making. The framework is furthermore in accordance with the empirical findings, which suggest that dynamic, negotiation, and ambivalence define the projects and should thus not only be included but rather be made a focus of the analysis.

In the following chapters, the theoretical framework is used to analyse the three case studies. The three projects are understood to be social sites. Instead of understanding them as meshes of orders and practices, they are here envisioned as being made up of social worlds and social arenas, within which different social practices and social orders are created, shared, negotiated, and contested.

Social worlds are constellations of human actors and other elements that share (at least) one social practice as a common core activity. They share certain teleoaffectivities as well as general and practical understandings and rules. Social worlds share certain ways of relating to and acting amidst social orders. In fact, social worlds are defined as sharing the same ways of positioning, relating to, and ascribing meaning to certain material artefacts, human beings, things, and non-human entities. Both social practices and social orders are only temporarily stabilized in social worlds and are marked by complexity, ambiguity, and heterogeneity.

Social arenas are social spaces within the social site of the respective project, which evolve around the interactions, negotiations, and contestations of certain social practices or specific teleoaffectivities, understandings, or rules. The negotiations occurring within social arenas might also concern the existence or the legitimacy of social orders. Elements that occur in more than one social order might mean very different things for different social worlds or be related in different or even contradictory ways to other elements within one social world.

The structure of the analysis is derived from the elements of social practices and social orders. By structuring the analysis according to teleoaffectivities, general understandings, practical understandings, and rules, the projects are represented as outcomes of social practices. Within the sections, the focus, however, is on how social practices are based on the negotiation and contestation of their organising elements and thus comprise dynamic and unstable processes. The second part of the analysis is structured around the elements of social orders—human actors and material artefacts comprise their own sub-sections while things and non-human beings are clubbed together in a sub-section called 'local environment'. Analysis of these sub-sections concentrates on how relations, positions, and meanings are shaped by and shape sense-making of different social actors.

4 Sense-Making of renewable energy production in the three cases

Within this chapter, the case studies are analysed, using the theoretical framework developed above. Each case study is preceded by a description of the social and political context within which it evolved. This description consists of a short overview of main characteristics of the place. This is followed by a brief explanation of main historical and material aspects of the military site within which energy is (meant to be) produced. Subsequently, the historical background of the projects—up to their official starting point—is explained. Relevant political events which are an important part of the respective project's background are then described. The last subsection recites the development of the respective project up to the time of the fieldwork.

4.1 Describing the CDT

The village and camp of Comrie are situated in central Scotland, about 70 miles north-west of Edinburgh and 50 miles north-east of Glasgow at the start of the Highlands. About 2,000 people live in Comrie's parish boundary area (including the households in the surrounding glens). The village is a historical conservation village that is situated in a national scenic area and has received a number of awards for its beauty.

Apart from its scenic location and award-winning beauty, Comrie is relatively close to the urban areas of Edinburgh and Glasgow. The combination of these factors has motivated many affluent people to move into the village. This trend took off when modes of transport and communication were sufficiently enhanced to allow for quick commute and options for home-office-work constellations. Besides these commuters, a large percentage of new residents are retired professionals who worked 'down south', i.e. in London, Edinburgh, or Glasgow.

4.1.1   Cultybraggan Camp[8]

Cultybraggan Camp lies about one mile from the village of Comrie. In 1941, 100 Nissen huts were erected on its 80 acres to host up to 4,000 German PoWs and their Polish guards. PoWs and soldiers lived in the Nissen huts. As can be seen in Figure 10, these are semi-cylindrical buildings, mainly made of corrugated iron. The front and back are made of different materials—usually wood and brick stones.

Figure 10: Nissen huts in Cultybraggan Camp along the sports area (Source AP).

In 1949, the last PoWs were released from the camp. It was subsequently taken over by the army to be used as a training camp for British soldiers. Many of the huts were torn down to make space for the new requirements. A sports area and a rifle range were built. A 10-acre hill ground was purchased nearby by the Ministry of Defence (MoD) for training manoeuvres.

In 2001, the MoD decided to get rid of the now derelict camp. The last training session at the camp took place in October 2001 and the camp was prepared to go on the market. Mainly due to administrative reasons, it took the MoD more than five years to actually put the camp on the market. This delay was

---

8   All information in this chapter derives from Comrie's oral history group publication 'Camp 21' and from interviews with members of this group. For further information, please refer to the publication.

fortunate for the local community, which only that year had set up the necessary structure to organise a community buy-out.

4.1.2   Prologue to the CDT

The new residents have raised Comrie's level of qualification and income above the national average. The influx of highly affluent people led to an increase in property prices (well above the surrounding villages and towns), creating a dynamic in which only increasingly affluent people are able to afford living in the village, causing property prices to rise ever higher. Due to this situation, young people in particular increasingly face difficulties when trying to rent or buy a flat or house in the village. The high prices are also an obstacle for small businesses. This impedes young entrepreneurs from starting a business in the village and also affects some long-established businesses.

An often-mentioned reason for choosing Comrie as a place of settlement is the village's well-known community spirit.

B7:   'I moved to the village with my family 18 years ago. That was a very clear decision to live in the sort of community that Comrie is' (CDT, Board Member 7: 14–17).

Besides often being mentioned in interviews, this spirit is also manifested in the fact that there are about 65 active groups (in a community of about 2,000 inhabitants), including groups in the sectors of sport and leisure, volunteering and charity, ecology and nature, art and culture, heritage and history, and geology and other natural sciences.

In March 2005, the Westray Development Trust[9] (WDT) gave a presentation about its activities in Comrie. The presentation was initiated by an inhabitant of Comrie who knew about the activities of the WDT through his professional engagement.

---

9   Development Trusts are community owned and managed organisations that *'aim to achieve the sustainable regeneration of a community or address a range of economic, social, environmental and cultural issues within a community'* (www.dtascot.org.uk 2014). Development Trusts are legal and organisational structures, enabling communities to find and pursue common goals, to apply for funding, and to interact officially (as well as informally) with other organisational institutions. Generally, Development Trusts aim to generate income for the community. All income generated thereby has to be reinvested for the public benefit. The umbrella organisation for Development Trusts in Scotland is the 'Development Trust Association'. The DTA was founded as a charity in 2003 and had 206 full and provisional members and 42 associated members in January 2014 (ibid.).

B7:   'In the process of doing my work there, I made some very good friends, and they
       were passing through Comrie, and I said, "if I get a group of people together,
       would you just speak to them about what you've been doing?" So I stuck some
       posters up in the village, and about 40 people turned up at the White Church hall
       [...]. And they listened to the Westray one and they decided that they would like
       to try and do something like that in Comrie. So that resulted in the formation of
       the Comrie Development Group [...]. Over the period of a year, we did a lot of
       development work in the village, contacted all 56 of the local groups and did all
       that sort of stuff. Then we held a big community meeting to look at the sort of
       priorities that Comrie might address if they were to set up a trust. And that com-
       munity meeting was attended by 150 folks in the White Church. Out of that, eve-
       rybody said, yeah, we'd like to do this and we'd like to set up a trust. Then it just
       started growing arms and legs' (CDT, Board Member 7: 43–56).

Inspired by the WDT presentation, a group of people first came together in May
2005. During the next year, contacts with the existing village groups were established and a public community meeting, organised in October 2005, was attended by about 150 people. In this and subsequent meetings in 2006, a vision
for a development trust was formed which '*promotes all aspects of local economic development*' and creates '*affordable assets to live, work, and play*' for
the community (CDT 2008; CDT 2012).

   In March 2006 an interim Board was set up that managed the Comrie Development Group and was responsible for forming a Trust—creating and registering the legal organisation, communicating and registering with the DTA, and
managing all the associated issues. In June 2006 everything was sorted and the
Development Group launched a weekend festival with a range of activities and
events in July, just before the Comrie Development Trust was formally constituted as a company limited by guarantee.

   One of the first activities of the CDT in August 2006 was to register an
interest in Cultybraggan Camp. The desire to purchase the camp for the community was inspired by knowledge from one of the group members who worked
for the MoD department that was running and organising the camp, which was
being used as a training camp at that time.

B2:   'And it was, in my tenure at the time that we decided to close—the army Ministry
       of Defence decided—to close Comrie. As a result, I put the proposal through in
       2001 to close it. It was subsequently closed—I think the final training done there
       was October 2001. In the meantime, I had actually bought a house in Comrie,
       eventually to retire up there. At the same time, the Comrie [...] Development
       Group had formed with the intention of setting up a trust to do good deeds for the
       village and all the rest of it. [...] it [the camp, AP] was going to go on the market.
       I, because of the job I was doing with the MoD, knew when that was going to
       happen. It had been raised at the Comrie Development Group, which I think at
       that stage had just become the Comrie Development Trust. What was likely to

happen to it—and I was able to tell them—that it was likely to be developed, to be bought by developers, with the intention of probably building large expensive houses of which we had quite a few in Comrie already. As it was, the village thought that wasn't a terribly good idea and wasn't what they would want that facility to do, although they weren't quite sure what they wanted it to do. Anyhow, at the end of the day, it was decided that they'd do a community right-to-buy, they put in an application for a community right-to-buy' (CDT, Board Member 2: 6–28).

Both old villagers and the new residents were against the development of the camp into housing blocks. Besides fearing a further increase in estate prices, many residents felt that the camp was an important aspect of the village's identity. Thus the idea of doing a community buy-out of Cultybraggan Camp was met with enthusiasm from most residents, many of whom were members of the newly founded CDT. After the MoD officially indicated the sale of Cultybraggan Camp in February 2007, the CDT reacted immediately by organising and holding a community ballot in May. The background structure that enabled the CDT to purchase Cultybraggan Camp is the *Scottish Land Reform Act 2003,* the creation of which is closely related to the history of Scottish landownership and the devolution of the Scottish Parliament in 2001. Both background aspects are described briefly in the following section.

4.1.3 Scottish landownership

The question of landownership and the laws and historic events related to it are tightly interwoven with Scottish and/or British history as well as Scottish identity. Most dramatic social and political events in Scottish history are bound up with questions of land and landownership (Chenevix-Trench/Philip 2001: 139). Scottish landownership is a showcase of the distribution of power in the United Kingdom. Indeed, '*Scotland's history is, to a large degree, a history of landed power. Until the 19th century landed power was synonymous with political power'* (Wightman 1999: 15).

For a long time, landownership in Scotland was exclusive. Until 2003, only 1200 landowners held two-third of Scotland's land (Bryden/Geisler 2007: 28; Wightman 1999). Until the Land Reform Act of the Scottish Parliament in 2003, Scotland was the '*last country in the developed world to have a feudal system of tenure and [...] the most concentrated pattern of private landownership anywhere in Europe'* (Wightman 1999: 13, see likewise 1996: 5; Callander 1998: 1,7).

While hereditary rights were reformed during the 18th and 19th century throughout Europe as an important step towards the abolition of feudalism

(Wightman 1999: 38), in Scotland these reforms only happened near the end of the 20th and beginning of the 21st century. The legal and political process that finally abolished the feudal system of landownership started in 1980 and went through a political process lasting more than 20 years before the Abolition of Feudal Tenure etc. (Scotland) Act was passed by Westminster Parliament in 2000 and came into legal existence on 28 November 2004.

Following a referendum in 1997, the Scottish Parliament was convened by the Parliament of the United Kingdom through the Scotland Act 1998. The Scottish Parliament is a devolved parliament whose legislative powers are derived from the British Parliament and only cover those areas that have not been specified for the Parliament of the United Kingdom.

For many decades, the benefits of being part of Great Britain and thus part of a major economic and political force in the world, as well as a sense of common identity, outweighed any desire for autonomy in Scotland. This changed in the 20th century. '*In the final quarter of the twentieth century, a social-political agenda had emerged in Scotland at odds with the one in England. People in Scotland saw themselves as more socialist, liberal, and less British-national than people in England*' (McCrone 2017: 46). The growing awareness of a Scottish identity manifested itself for example in a greater divergence in the voting behaviours of Scottish and English voters. Because of the relatively small number of Scottish voters, the composition of Westminster was decided by English voters. While this had always been the case, it became a problem when Scotland started to vote differently from England. This circumstance was increasingly perceived as a '*democratic deficit*' (McCrone 2005: 13). The changing voting patterns were accompanied by the emergence and growing significance of the Scottish National Party (SNP). The SNP's main aim has been, and still is, a complete Scottish independence.[10]

Devolution of the Scottish Parliament began with the Blair government, newly elected in 1997. The government held a successful referendum that led to the Scotland Act 1998, whereby legislatorial powers were devolved from the British to the Scottish Parliament. The first election was held on 6 May 1999 and the first session of the newly established Parliament took place. The creation of the Scottish Parliament was a prerequisite for the land reform. First, the number of Scottish bills to be dealt with in Westminster often prevented decisions that related to 'only' Scottish issues (Watson 2001: 24pp.). Second, many members in the House of Lords owned land in Scotland and thus it was not to their

---

10   Their activities not only contributed to devolution but also led to a formal independence referendum which was being held in September 2014. Within this referendum, however, most Scottish people voted for remaining within the United Kingdom.

interest in changing landownership (Bryden/Geisler 2007: 28; Wightman 1999: 11).

For the newly devolved Scottish Parliament, Land Reform was one of the most pressing issues. In fact, it had been one of the main pre-election promises of the Labour Party. Already in February 2001 '*the new Scottish Executive in Edinburgh issued the Consultation Paper on Land Reform that led to a Draft Land Reform Bill. Two years later, the new Parliament passed land reform legislation*' (Bryden/Geisler 2007: 28). The Land Reform Act 2003 is devoted to making land accessible to the public either by means of general access rights (Part 1) or, more specifically (Part 2), by giving communities the pre-emptive right to buy the land on which they live and to which they have a '*substantial connection*' (§ 38, 1 b, Passage I). Local communities (defined by parish boundaries) are given pre-emptive right to buy (parts of) the feudal estates in which they live. The procedure starts with the willingness of the landowner to sell land. If a community has registered interest in buying land, a ballot has to be organised, in which at least 50% of the adult inhabitants of the community have to take part, of whom at least 75% have to approve of the buy-out. Furthermore, in order to register an interest in land, a community body with an elected board of directors—who are members of the community—has to be set up. The Land Reform Act also regulates the purposes for which any possible profits deriving from the landownership might be used. The established company

> 'is empowered to own and manage land and other property. Its assets, and any income derived from them, do not belong to the members and may not be distributed (except, for example, as reasonable payment for services or goods). It must manage these assets for the long-term benefit of the geographically defined community for whose sake it was formed' (Hoffman 2013: 291pp).

Most of the communities registering an interest in land decide to own and manage the land through a development trust. Besides being an adequate form of organisation for the purpose, the high percentages of communities that decide upon a development trust when buying land are due to decisions made by the first community buy-outs.

4.1.4 Running the CDT

In accordance with the requirements of the Land Reform Act, the CDT in May 2007 held a community ballot in which 72% of the residents took part (1,404 people). Of these, 97% voted in favour of the community buy-out. The Comrie community buy-out thereby became the most inclusive ballot under the Scottish

Land Reform Act. The high percentage of support for the Trust has been maintained throughout the years. Of Comrie's 2,000 inhabitants, more than 600 are members of the CDT.

In the same month in which the community ballot took place, the CDT secured a grant to undertake a feasibility study to evaluate possible future developments of the camp. Having reached a bargain with the MoD and now in possession of a feasibility study, including a business plan, sustainability strategy, and a legal structure report (CDT 2012), the CDT was able to apply for funding to purchase Cultybraggan Camp in September 2007. The purchase was funded by a £350,000 loan from the Tudors Trust and refinanced through a commercial loan from an ethical green bank (CDT 2011).

Another important step was taken in August 2008 when the CDT, after a successful application for funding, was assigned one of the four national exemplar projects of the then newly established Climate Challenge Fund (CCF) (CDT 2008: 10). The CDT's application relied heavily on the sustainable development of Cultybraggan Camp. More important for the exemplar project status, however, was the second strand of the Trust's activities—the 'Climate Challenge' project which was designed to mirror and deliver the Scottish government's 'Climate Change Delivery Plan'[11] at the local level. The Delivery Plan covers the issues of electricity, heat, transport, waste, rural land use, and behaviour change (CDT 2012). In Comrie, these issues were tackled through street-to-street energy audits, support and advice for private installation of renewable energy sources, the creation of allotments, a community orchard, sports and recreation facilities, educational programmes, and further activities relating to human activities that are directly or indirectly related to energy consumption and/or production (CDT 2008). In December 2008, CCF funding enabled the CDT to open an office on one of the village's main roads and to employ two part-time staff members—a Climate Challenge Manager and a Climate Challenge Support Worker/Office Manager—who were responsible for organising the street-to-street energy audit and the insulation programme. Additionally, six part-time staff members were recruited to carry out the audit (CDT 2011: 2).

In February 2010, two full-time staff positions were funded by the CCF to support the general development of the Trust. A Delivery Plan Manager and a Delivery Plan Assistant were accompanied by four part-time Delivery Plan Advisors. In March 2011, a second application to the CCF was unsuccessful. Because of Comrie's status as an exemplary project of the CCF, the CDT managed

---

11   The Scottish governments' Climate Change Delivery Plan sets out how Scotland will meet the targets of the Climate Change (Scotland) Bill to reduce carbon emissions by 42% by 2020 and achieve an 80% reduction in greenhouse gas emissions by 2050 (The Scottish Government 2009).

to get some additional funding in order to keep going till the next application round:

CCF: 'That [Climate Challenge Project] came to an end March 2011. They [the CDT, AP] were unsuccessful in an application for a project that would've taken them from 2011 to 2012. They spoke to the Scottish government and said, "Look, you've started a project with us that's really quite exciting. If you don't continue this in some way, we're going to end up having to stop some of the work." And they got some additional funding to help them with revenue costs for staff, because they had other loans and all the rest of it; […] to allow people […] to continue what they have started under the CCF project, because the government didn't really want to see things just come to a halt. Then, in 2012, they applied and were successful in receiving funds for the current project' (Climate Challenge Fund, Staff Member: 43–53).

In order to be able to keep the main staff positions (Delivery Plan Manager and Assistant) these had to be changed from full-time to part-time positions and to be supported by other funding sources. Additionally, it was possible to employ a third part-time worker who was responsible for office management of the CDT. After a successful application for funding in 2012, a fourth staff position was created as a joint venture between CCF and another Scottish government programme. The position of Energy Advisor is financed partly by the CCF and partly via the Scottish government's Green New Deal scheme. All four jobs were secured by funding till May 2015. After that, either the CDT would have had to achieve financial self-sufficiency and thus be able to pay staff from its income, or new funding streams would be required.

In the first four years (2007–2010), concentration and resources were split more or less evenly between the two strands (Cultybraggan Camp and the Climate Challenge) of the CDT. In the community and idea exchange meetings that led to the creation of the CDT, the community had outlined a vision and aims for the Trust. The vision, which stated that the CDT should develop the community in a sustainable way, was qualified by three general aims—community, economy, and environment. After the community buy-out, the vision and aims were, in three major community meetings, broken down into certain activities related to the aims and vision (see Figure 11). To improve the quality of life in the community (community aim), the Trust should support community groups and projects in the village and the camp. To generate local economic activity, create jobs, and achieve long-term financial sustainability (commercial aim), the Trust would set out to initiate income-generating projects, develop activities that keep the money in the community, increase opportunities for the community to invest in the future of the village, and initiate income-generating activities at the camp. To reduce Comrie's environmental footprint and develop its ability to adapt to climate change (environmental aim), the CDT would deliver the Climate Challenge projects in the village and develop the camp on the principles of sustainability (CDT 2012).

Vision	Aims	Activities
Improve Comrie's sustainability and resilience	**Community** Improve quality of life in the community	**Community** Support community groups and projects in the village and the camp
	Economy Generate local economic activity; create jobs and achieve long-term financial self-sufficiency	**Economy** Increase income generating projects that keep the £ local; increase opportunities for investment; increase income generating activities at the camp
	Ecology Reduce Comrie's environmental footprint and develop its ability to adapt to climate change	**Ecology** Deliver the Climate Challenge projects in the village and develop the camp on the principles of sustainability

Figure 11: CDT's Vision, Aims, and Activities (Source AP).

The vision and the CCF funding have significantly shaped the evolution of activities of the CDT. The third important factor was the engagement of the members of the community. Being a volunteer organisation, most of the work in and for the CDT relies on the private engagement of its members. For the first four years, this interplay between vision, funding, and activities was more or less self-enhancing, especially in the ecological realm. The aims of the Trust specified the Climate Challenge as a main part of the Trusts' activities. The initial focus on climate change and sustainability was strongly supported by the board of the Trust at that time.

B7:   'The people who formed the first board of the Trust, who really stuck together for that first four or five years, the personalities involved in that had quite strong visions. So they helped to mould—also, they are very articulate young professional people whose passions were shared for sustainability and community empowerment. But also they knew how to speak that language. And I sort of include myself in that. There were people who were chief execs at voluntary organisations and quite senior within environmental groups. So I'm sure that the leadership of the Trust put a certain angle on that stuff as well' (CDT, Board Member 7: 153–160).

Getting the CCF funding further intensified the environmental focus. This was due to the fact that the money received from the CCF was restricted to the projects it had been obtained for, i.e. the Climate Challenge and the sustainable development of the camp. Having two Climate Challenge Delivery staff members who spent their entire workforce on developing the sustainability aim of the Trust was a further advantage compared to the other aims of the Trust, which were handled by volunteers.

In accordance with the vision and aims, a detailed vision for Cultybraggan Camp was developed in further community meetings. The camp was divided into five zones—a food quadrant (including allotments and a community orchard), a sport and recreation quadrant, a commercial quadrant (for local businesses), an eco-hub (an area for ecological activities, projects, and enterprises), and a future and heritage centre (CDT 2008; CDT 2011; CDT 2013). The first activities undertaken by the CDT *for* the camp involved getting planning permission and funding to do the necessary essential infrastructural work (drainage, water, electricity, waste treatment) and the allotments. The first group activity taking place *at* the camp was in April 2009, when the area of what had once been PoW Block 1 was transformed into 34 allotments. As of today, the allotments are one of the most active groups within the CDT and at the camp. About 40 families have (or share) plots to grow vegetables and fruit. Over the years, the camp became the venue of different groups, activities, and events. Apart from the allotments, a further important activity is the community orchard. Fruit trees and berry hedges are cultivated and pruned by the volunteer orchard group. In September 2009, refurbishment of nine Nissen huts (creating 12 workspace units) was completed and the huts were let out to local businesses or entrepreneurs. These include joiners, builders, garages, gardening contractors, an equine laundry, and a company providing arts and crafts workshops. Two of the huts have been sold, while the rest are let out on long-term rent.

Besides these examples of success, there are other projects that could not be realised during the duration of the fieldwork, due to lack of either funding or determined volunteers. An example is the sports quadrant. An application to the Scottish Sports Association was unsuccessful, after which nobody took the responsibility and workload to improve the applications or to look for alternative funding sources. Another major disappointment was the unsuccessful application for funding to build a Heritage and Futures Centre at the camp. The centre had been planned to combine a local history exhibition about the camp and the village with an exhibition showcasing the Trust's aims and endeavours for a low-carbon future. A feasibility study of the project had been received but funding for the realization of the project proved to be inaccessible.

Over the years, the number of activities taking place at the camp has been constantly increasing. A very short overview of the most important ones shows the variety of events and activities. In June 2009, the CDT *'hosted a hugely successful Climate Change Conference for Communities on behalf of The Scottish Government'* (CDT 2011: 2), which was attended by over 120 delegates. Also successful was the 'Climate Challenge Youth Conference', attended by more than 100 children of all ages. A Waste Night and the launching of a foundation at the villages' fortnight were two other major events for the CDT in 2009. The biggest event in 2010 was a music festival—the CultyQuake. In 2011, the first wedding was celebrated at the camp. Both the Scottish National Orienteering Championships and the Original Mountain Marathon started from there. In 2012, the camp participated in the regional 'doors open day'. In 2013, the Heritage Group organised many events at the camp. A highly successful open doors day with a WWII re-enactment group and exhibition attracted large crowds. The Heritage Group also organised guided tours through the camp.

With regard to the production of renewable energy at the camp, energy efficiency and renewable energy feasibility studies supported by Community Energy Scotland (CES), the Energy Savings Trust, and other public funding bodies (CDT 2013: 13) were undertaken in 2010 and 2011. In August 2010, a feasibility study came up with recommendations for energy efficiency measures and recommended a district heating biomass scheme to provide primary heating and hot water. Installation of photovoltaic panels and a hydroelectric scheme for producing renewable *electricity* on site was proposed by the report. Both installations were viable to claim the newly introduced (1 April 2010) FITs of the UK government. The applications for funding and negotiations with the possible contractors took a while, but the first 10KW photovoltaic system was installed at the camp in December 2011. The income from the sale of one of the commercial huts was used to finance the project. After its installation the system produces about 10% of the electricity used in the camp. The idea was that the realization of the hydro scheme would account for the rest. Over the years, however, all available options around the camp were tested and had to be abandoned for various reasons. A site near the village had to be given up because of complicated and unresolvable questions of ownership. Other sites turned out to either not have enough flow or to be inhabited by salmons, making them unfeasible for a hydro scheme. A site at a derelict fish farm nearby would have been technically feasible but proved to be financially unviable.

Due to various characteristics of the locality, wind energy (which is very common in the area around the village) is also not an option.

S3:   'Wind is—I don't know what the term is, because there's so much sheltering from
        the hills, it's not great for wind turbines unless you actually fit it up on the hill.

And the hill ground goes up so far. It's also a site of national scenic beauty so it would be very difficult to get plan permission for it, for wind' (CDT, Staff Member 3: 117–121).

An additional (supposed) obstacle further complicating this mixture of physical and administrative circumstances is public opinion. Generally, all interview partners agreed that they were relieved about the fact that wind energy was out of the question for 'objective' reasons (i.e. the physical reasons). This meant that they would not have to face the discussions and conflicts that would have been aroused within the community by the issue of a potential wind farm. At the time of the fieldwork, the Trust was looking into possibilities of extending the PV system. Two different possible (commercial) contractors did feasibility studies. Different sites and systems (both technical and financial) have been proposed.

With regard to the production of renewable *heat,* the feasibility studies done in 2010/2011 all agreed that a biomass boiler was the best option for renewable heat production in the camp. The results of the studies, however, varied with regard to the best system size, as it was unclear at that time how many users would be connected to the system. In the middle of 2011, the Energy Savings Trust announced the Scottish District Heating Loan Scheme (SDHL). As the SDHL was only available in 2011 and 2012, with no certainty of being available beyond 2012, the CDT decided to take a risk and apply for the scheme. The application was based on those potential customers which were already on site and on users who were supposed to move into the camp. The main risk was that the system revenue would not fully cover the capital cost until a certain capacity level was reached, despite the Renewable Heat Incentive (RHI) payments.[12] Nevertheless, the Trust decided to install the biomass boiler system, arguing that this decision was in accordance with the CDT's vision and aims. '*On balance the Trust decided that the opportunity to fulfil one of its' key criteria for sustainability, minimizing the use of fossil fuels, was too good to miss and the application was submitted*' (CDT 2013: 14). In November 2011, the Energy Savings Trust announced that Comrie was one of the first nine projects that were given loans from the SDHL scheme. After a tender process, the installation of the biomass boiler and the first part of the District Heating System were completed in May 2012. After the first few months, however, it became obvious that the system was inefficient.

In 2012, the Trust was successful with another major funding application to the CCF. In the third round of the CCF, which ran till the middle of 2015, the

---

12   The British Government's non-domestic Renewable Heat Incentive was launched in November 2011 to provide payments to industry, businesses, and public sector organisations. 'The RHI pays participants of the scheme that generate and use renewable energy to heat their buildings' (Gov.UK).

CDT got funding for the 'more than the usual suspects' project—which aims to engage other than the 'usual suspects' in an energy efficiency challenge—for organising the Youth Climate Challenge Conference and providing advice and help to other newly established community groups working on climate change. While these projects are still in line with the original values and aims of the Trust, the community became more and more unsatisfied with the work of the CDT. Different aspects contributed to the feeling of dissatisfaction.

The Trust was concentrating a great deal on developing the camp, fulfilling the funding requirements, and working towards financial self-sufficiency. Many people became increasingly frustrated with what they saw as stagnation or concentration on the wrong topics and issues. The problem arose partly from the fact that many people do not have much knowledge of how organisations like a development trust work, and specifically are not aware that the Trust is liable not only to the residents of Comrie. Planning permissions and other administrative and legal procedures and restrictions also bear upon the Trust and the realization of its projects, as do funding schemes and money lenders. Funding schemes in particular impose restrictions upon the Trust. Most funding comes from 'restricted funds'. This means that the money has been awarded for certain projects or purposes and must not be spend on anything apart from these. Such restrictions on accessed money are not common knowledge among many people. They do not understand why—despite the large sums that have been accessed through different funding applications—the Trust still does not have enough money to realise many of the once-planned projects. Many people have come to believe that the Trust must have 'lost' much of the money it has accessed through funding (informal talks with residents).

At the end of 2011, members of the Trust, especially at the board level, felt that the conflicts were more than the 'normal' level of interpersonal conflicts and issues over topics that always arise when a number of actors come together to pursue their individual interests within one project. The board members individually and together decided to push for change. The General Annual Meeting in December 2011 led to major changes in the Trust, its board, and its activities. In the meeting, five members of the old board stood down and new board members were elected by the congregated members. Also this meeting was dedicated to collecting and discussing the members' ideas and desires for the future development of the Trust. During this and subsequent meetings, it became obvious that the Trust needed to concentrate more on the community.

The creation of the Heritage Group in 2012[13] further shifted interests and activities in the Trust. The group was born out of genuine interest in the heritage of

---

13   Its predecessor is the oral history group that researched the history of the camp. This group, however, dispersed after publishing a booklet about the history of Cultybraggan Camp.

the general area, the village, and Cultybraggan Camp. Within one year, the Heritage Group had become the most active group in terms of the number of participants as well as the number and attendances of events. With the money raised in these events and through a successful application to the Heritage Lottery Fund, the group succeeded in raising the capital needed to refurbish one of the camps' huts into a heritage centre. As seen in the analysis, the ideas and interests of the Heritage Group often disagree with other activities, like the production of renewable energy. The main reason for these conflicts is that while the Heritage Group aims to preserve the camp, most other groups need to change aspects of the camp in order to realise their interests.

4.2 Analysis of the CDT 'a vehicle to serve the community'[14]

4.2.1 Teleoaffective structures: Serving the community

The basic idea of creating a community development trust was introduced to Comrie during a talk given by the representatives of an already existing development trust. The event was attended by about 40 community members, who not only liked the idea of a development trust but afterwards engaged in activities to start a trust themselves. In the first step, a community development group was set up, which contacted other groups and inhabitants of the village to ask about the ideas and interests that they would want to see fulfilled by a development trust.

B2:   'At the same time, the Comrie Development Group had formed with the intention of setting up a trust to do good deeds for the village and all the rest of it' (CDT, Board Member 2: 12–13).

The quote describes the activities of the development group, thereby explaining the idea underlying these activities. While the '*intention of setting up a trust to do good deeds for the village*' had motivated the activities of the development group, what these '*good deeds*' would be was not settled and in fact was meant to be decided by the residents—and future members—themselves. This differentiation indicates the existence of two teleologies. The key aim for the CDT from the beginning has been to serve the community—to do good deeds. The second teleology, which can be seen not only in the quote but also in the practice of the development group, which contacted the existing groups in the community and asked them what a development trust should do in order to best serve

---

14   Staff Member 3 in an informal talk.

their needs, exemplifies a participatory teleology. It is not the future trust that decides its own activities but all members of the community. Creating or enhancing benefits for the community and being a participatory process are in fact the two core teleologies guiding and determining the activities of the CDT. Most interviewees talk about these concepts when they are asked what the Trust is about.

S1:   'I think that is essentially what we want to do: We want to facilitate things that are happening in Comrie—just help people make stuff happen' (CDT, Staff Member 1: 67–68).

B7:   'The CDT? What it's about is what this community wants to make it about. That's what the community development trust is about. It's about the idea that this community has a vehicle, has a way of people making things happen that they want to happen' (CDT, Board Member 7: 137–140).

S2:   'I describe it as a group of people who have come together for the good of the community to try and do community-led projects. [...] It provides an organised collection of people, a collective to then apply for funding to carry out the projects as defined by the community' (CDT, Staff Member 2: 315–318).

B5:   'Well, you should know that what the Trust is trying to achieve: It's trying to achieve various things to make the village a better place to live in for everybody' (CDT, Board Member 5: 7–8).

In all these quotes, the activities and projects of the CDT are explained as not being defined by any specific group. Instead, the Trust is described as a facilitator, a vehicle, or a collective that enables the inhabitants of Comrie to realise their interests and ideas. The common way of defining the Trust by these two teleologies and their combination indicates that these have come to be negotiated orders of the CDT.

As negotiated orders these teleologies structure proceedings within the social arenas of the Trust. In the following chapters, it is argued that most conflicts and negotiations draw on and thereby reproduce these negotiated orders of realising a community interest and being supported by the community members. Also these negotiated orders, their interpretations, and the conflicts surrounding them, affect the production of renewable energy in the CDT. Energy production is negotiated in two social arenas. First, energy production is an element in the social arena within which the sustainability agenda is negotiated. Second, energy production is negotiated and has to legitimize itself with regard to its ability to contribute to the community's well-being and is thus always part of the CDT's central social arena. This chapter first analyses how energy and energy production are negotiated with regard to the CDT's sustainability agenda. In the second step, the arguments through which the legitimacy of energy production in the Trust's main social arena is contested are scrutinized.

Arguing that energy and energy production practices are made sense of within the Trusts' sustainability agenda presupposes that energy production is made sense of as a sustainable practice. Consequently, the following section examines how energy production is embedded into the CDT's sustainability agenda. To show the complexity and heterogeneity of the sustainability context within which energy production is embedded, the variety of activities deriving from and related to the sustainability context of the Trust are described.

In the community and idea exchange meetings leading to the creation of the CDT, the community outlined a vision and certain aims for the Trust. The vision stated that the CDT should work for the sustainable development of the community. The vision was split into three aims: enhancement of the community, economy, and environment of the village. After the Trust purchased Cultybraggan Camp, these aims were further specified into certain activities. To reduce Comrie's environmental footprint and develop its ability to adapt to climate change (environmental aim), it was agreed that the CDT would deliver the Climate Challenge projects in the village and develop the camp on the principles of sustainability. Besides a number of activities dedicated to reducing the use of energy and increasing energy awareness in the village, the production of renewable heat was, from the very beginning, part of the CDT's strategy to realise its environmental sustainability vision.

I: 'Do you know how the original idea to use renewable energy—how this was introduced to the camp?'
S3:   'I think it goes right back to the beginnings of the Trust. One of the aims and objectives of the Trust is to go forward in an environmentally sustainable way' (CDT, Staff Member 3: 131–134).

The interviewee relates renewable energy production to the sustainability aim of the CDT. He describes energy production both as an outcome of the Trust's sustainability motivation and as an important instrument for realising this intention. He makes sense of energy and the production of renewable energy at the camp with regard to the environmental sustainability agenda that had been taken on by the CDT. In the quote, energy production is thus related to sustainability and not an isolated activity within the context of the Trust. Understanding energy production as part of the broader aim of environmental sustainability implies that energy production is not a singular or isolated element in the context of the CDT. Instead, making sense of energy production as a sustainable practice relates it to a range of other activities in the Trust, which are also made sense of as being environmentally sustainable.

S2:   'My vision is more integrated to actually doing more environmental stuff and not just the energy, not just renewables, because there's so much to be had, and people just you know reducing their demand—reducing their energy usage, which would be a good way of doing that' (CDT, Staff Member 2: 198–201).

The interviewee expresses her aim to broaden the range of activities deriving from the CDT's environmental agenda. Energy production is by her embedded into the Trusts' environmental sustainability teleology, from which it derives its meaning. The quote also shows that energy production is not an end in itself, but that energy production is made sense of as a social practice that contributes to realising the teleology of environmental sustainability. Energy production is not understood as a stand-alone issue or an isolated social practice. Instead, it is described as an integrated element within environmental sustainable practices.

The production of renewable energy is most prominently related to the issues of carbon consumption and efficiency. Many sustainable activities of the CDT target carbon consumption and efficiency in the village.

I: 'So when you speak about renewables, you speak a lot about saving money?
B5:   Well, that's the whole idea of the thing. Ehm, why burn heating of any descriptions, whether it be coal fires or central heating or anything. Why let it burn and fly out through your roof if you can <u>contain</u> it! So you don't spend as much money? But it's not only from a money point of view, it's from an energy point of view. You're <u>wasting</u> energy. And we can't go on and on and on and on wasting energy or we'll have no energy left' (CDT, Board Member 5: 273–279).

When asked about the Trust's renewable energy production, this interview partner relates it to its potential for saving money. On being further inquired about this kind of sense-making, he directly relates energy production to energy efficiency measures. By arguing that '*we can't go on and on and on and on wasting energy or we'll have no energy left*' the interviewee makes sense of carbon-savings with regard to sustainability considerations. Renewable energy production is thereby related to energy efficiency and consumption measures. According to his argument, energy production does not make sense as a stand-alone issue. Renewable energy is instead given meaning within the teleology of carbon savings, of which it is defined to be one integrated aspect. The production of renewable energy is thereby *intentionally related* to (Schatzki 2010: 140) energy-saving and efficiency practices, as all these activities serve the purpose of reducing the wastage of energy. Likewise, it appears from the quote that renewable energy production is not only given sense with regard to environmental—more specifically carbon-saving—issues. Energy production, in this quote, is also related to a financial teleology of money-saving. Like in the environmental teleology, producing or using renewable energies only becomes a practice within the financial teleology—according to the interview partner—if it is combined with other elements in meaningful ways. In stating '*Why let it burn and* *fly out through your roof if you can <u>contain</u> it!'*, the interviewee relates renewable energy (*'it'*) to insulation measures. Within the financial teleology, using heat from any source does not make sense without also using energy efficiency measures. In a nutshell, the quote describes producing or using renewable heat as integrated elements of environmentally and economically sustainable social practices.

The quote above indicates that members of the Trust might be motivated by environmental and/or economic considerations. There are, however, a range of other aims to which the production of renewable energy is related. Resilience provides an additional meaningful concept for some members of the CDT.

> B4: 'I think, well, the purpose of the Trust is to build a resilient community [...]. Step-by-step it's trying to reduce that reliance on the big energy companies. By producing our own and getting the people in the village to produce their own renewable electricity. It's also the allotments trying to reduce your reliance on the big supermarkets and growing more locally. [...]. So yeah, turn it to all those things to make it more self-reliant rather than being affected by all these external shocks not instigated by ourselves but by government. [...] If you look at the overarching aim of the Trust, it's to build a resilient community. Which is all those things, just to make it <u>less</u> reliant on external sources of energy, food, transport, et cetera' (CDT, Board Member 4: 279–290).

The production of renewable energy in this quote is intentionally related to other activities that are also given sense with regard to the teleology of resilience. When it is made sense of as a resilient practice, renewable energy is related to practices like gardening or local economy measures. Energy production within this argument is a practice through which the community of Comrie can become less reliant on '*external*' influences, which are seen as more prone to shocks and crises. Like with sustainability, putting energy production into the context of resilience integrates the production of renewable energies with 'green' and 'local' practices. Thereby, the production of renewable energy is again seen as not an end in itself but an integrated activity organised as a social practice by the teleology of resilience.

While the quotes are limited with regard to the actors and voices presented, it appears that actors in the CDT differ with regard to how they make sense of energy and renewable energy production. Actors might relate renewable energy to not only different teleologies but also different combinations of various teleologies. For all actors, however, renewable energy production is neither an isolated activity nor does it in itself realise a certain teleology or combination of teleologies. Instead, within the respective organising teleology, the activity of energy production is related to other activities. How and to what other activities energy production is related, derives from the respective (combination of) tele-ologies. The relation between energy production and other activities within the different teleologies does not result from any naturally existing relationship be-tween the different elements; instead, relations are re-created in the CDT's mem-bers' social sense-making. Neither the concepts of resilience and sustainability, nor their application to the production of renewable energy have been created by the members of the CDT. Both are socially already existing forms of sense-making. Nevertheless, the activities, the teleologies, and their specific combina-tions re-create these concepts in ways specific to the context of the social site.

By making sense of renewable energy production with regard to different teleologies, the activity of renewable energy production is organised as a social practice in flexible and fluid ways. Flexibility and fluidity imply that the activity itself and its relations to different teleologies are negotiable and contestable. Particularly, the sustainability agenda has increasingly become contested in the CDT. The central role of sustainability within the vision and aims of the Trust has been questioned by different social worlds within the CDT (this aspect is elaborated in Section 4.2.5). In fact, the negotiations, contestations, and conflicts about sustainability had been one of the most virulent social arenas in the CDT at the time of the fieldwork. Different groups increasingly challenge the rele-vance and legitimacy of sustainability. In 2012, sustainability became so prob-lematic as a general concept that the board felt the need to revise its strategy.

S1:   'Since it's started, the Trust has always been about sustainable development, but last year there was a big change. [...] we went through a strategy review. What is the CDT, what is the role of the CDT, what does it do, why does it exist, and so on. I think essentially, we still do the same things, we still stand for the same things, but we just see it differently. People didn't like sustainability. They thought it was a bit jargony. So they were like "What does that even mean?" And so we went through the whole thing. [...] What we actually came up with was a defini-tion of sustainability, but we didn't want to call it that. Its jargon, you know. So we now say that the aim of the Trust is to promote the long-term well-being of the community, rather than the sustainable development of the community' (CDT, Staff Member 1: 41–42, 47–56).

One of the criticisms mentioned in the quote relates to the inapproachability and *'jargony'* character of the term sustainability. The critique challenges not the practice of energy production, but the concept through which this practice is made intelligible to the members of the Trust. Likewise, it appears from the quote that the teleology itself has not been altered in the review process. Instead, what the process *'actually came up with was a definition of sustainability, but we didn't want to call it that'*. The process confirmed the aims and the activities subsumed under the term sustainability. What had been changed in the negotiations is the terminology through which the aims and activities are made intelligible.

In a later sequence of the interview, the interviewee explains how the renegotiated terminology affects the carbon-reduction strategy of the Trust.

S1:   'So rather than saying, "We have a carbon challenge, and we challenge you to reduce your carbon footprint" and all that kind of thing. It's like, well actually you know my thing always is go where peoples' interests are. And then or as SM4 says, sell them ice cream and then sell them broccoli. ((laughing)) Advertise ice cream and then sell them broccoli when they arrive. Because I'm like, yeah we just advertise the things that people are interested in, and once they're there we'll have something that is getting some sort of message across. You know and I think that's the way that the Trust is moving' (CDT, Staff Member 1: 158–165).

Within the carbon reduction strategies of the CDT, sustainability is no longer mentioned. Instead of trying to make sustainability intelligible to its members, the Trust is now trying to realise its sustainability teleology without using the controversial term or related concepts. Because the strategy revision has altered neither the teleoaffective structure nor the practices, but only the ways in which these are made intelligible to the actors involved, board and staff are able to still pursue the 'old agenda'. The evolving strategy of reframing these activities can be described as passively not making sustainability an issue anymore. This strategy serves to momentarily stabilize the context of the CDT, as it distracts attention from the contested aspects while the practices themselves are still carried out.

Besides the passive strategy of not mentioning sustainability, another way through which the activities of the CDT are negotiated in practice is by actively reframing the activities of the CDT. Thereby the activities are made sense of in ways compatible with the teleoaffective structures of different social worlds within the Trust.

S1:   'Yeah. How is it different? We're not targeting individual homes so much. I think we still have environmental objectives. They're just less explicit. How has it changed? Because we've still got SM2 post, which is about helping people individually and their homes register their energy consumption. I think the focus our energy advisor has tried to take is rather than saying "How can I help you reduce your carbon footprint?", she's been saying, "How can I help you spend less money on your bills?". So I suppose the focus is more about money and less about carbon. Although for us it's about carbon, we're not necessarily selling it that way' (CDT, Staff Member 1: 107–113).

After describing the passive strategy as making environmental objectives '*less explicit*', the interviewee explains how another staff member has actively aligned her practices with the interests and ideas of the community members. She has started to make her job intelligible to the community members by making sense of it with regard to financial instead of carbon savings. Neither the activities themselves nor the energy advisor's teleology has changed. What has changed is how she negotiates these activities in interactions with other actors. The activities are made meaningful to them by embedding them into members' daily ways of sense-making. Instead of organising energy-saving or production practices in reference to the socially contested concept of sustainability, the CDT now relates energy-related practices to forms of social sense-making that are more meaningful in the day-to-day context of the 'target population'. The interviewee repeatedly emphasises that the board and staff of the CDT have not relinquished their environmental or sustainability aim. Within the social world of the board and staff, carbon-saving is presented to be a meaningful core activity still. For this social world, sustainability was confirmed through the strategy revision process. What has changed is the way how certain activities are negotiated in interactions with different social worlds in the CDT.

Within the social world of the board and staff, the production of renewable energy is not solely related to environmental sustainability. As described in the quote, energy-related issues have come to be reframed from mainly environmentally friendly practices to practices that improve the financial resources of the community members. This reframing of energy-related issues is necessary within the context of the CDT as renewable energy production is not an end in itself but has always been one of many elements through which the Trust aims to serve the interests of the community. Because these ideas and interests are constantly changing, the Trust needs to either constantly alter the range of activities pursued, or negotiate the ways in which these activities can be made sense of in ways meaningful to its members. With regard to energy production, the board and staff have decided to reframe their energy-related practice from being environmentally friendly to being economically viable practices.

Besides being embedded and challenged in negotiations about the Trust's sustainability agenda, energy and energy production are also directly contested with regard to their ability to live up the CDT's core teleoaffectivity of serving the community. The most contested issue relates to the ability of energy production to generate financial income for the Trust. Generally, the teleology that communities engage in the production of renewable energy to generate a financial income that can be fed back for the public benefit of the community is based on socially shared forms of sense-making. This form of social sense-making is, for example, institutionalized in British and Scottish energy policy and related organisations like CCF or CES. While being based on socially shared forms of

sense-making, these patterns of knowledge are nevertheless interpreted and negotiated in the context of the CDT. It is with regard to the particular context of
the Trust that energy production as an income strategy has come to be severely
contested by different social worlds in the CDT. As shown, the conflicts about
the financial viability of renewable energy production are closely related to financial debts of the Trust, as it needs to repay three large loans. These debts
have led to the creation of a social arena within which the financial situation and
teleologies are argued about by different social worlds. By being negotiated in
this social arena, renewable energy is actively related to a range of other elements that make up the social site of the CDT.

In order to purchase the camp and the biomass boiler, the CDT has taken
loans from different institutions.

S4:   'Roughly speaking, the Trust owes 363,000–365,000 pounds to those three [Tri-
odos Bank, Social Investment Scotland, and Tudors Trust for buying the camp
and repairing its basic infrastructure, AP]. And on top of that, it owes 180,000
[pounds] to the boiler. I think financially, it's one of those things that it's something that if the Trust wouldn't have the loans to repay, life would be much easier
for everybody' (CDT, Staff Member 4: 703–706).

In this quote, the woodchip boiler is described not as part of the CDT's sustainability, but of its financial debt context. Buying Cultybraggan Camp and repairing its basic infrastructure (fresh and waste water, electricity, metering, and traffic safety measures) have encumbered the Trust with financial debts. The woodchip boiler is part of this financial context because of the amount of upfront
investment necessary to purchase it. To make this investment, the CDT had to
take another loan, increasing the debts of the Trust. The inclusion of energy production in the financial debt context became more pronounced and a constant
issue when the biomass boiler system turned out to be inefficient and—instead
of generating financial income—forces the CDT to pay extra for running the
system.

Relating the production of renewable energy to the CDT's debt context effects the perception of not only the existing but also possible future energy-related practices in the Trust.

S2:   'We've got 10 kilowatts of solar panels there [at the camp, AP] just now. We're
going to increase that to 30 kilowatts at some point. But we need to make sure that
the problems we've had with the district heating system—the biomass boiler is
being much less efficient and costing a lot of money—those problems need to be
ironed out before we put any money into anything else' (CDT, Staff Member 2:
135–139).

It appears from this quote that it is not only the specific technology of the biomass boiler that is related to the financial problems. Instead, energy production

practices and technologies are generally put under financial reserve. Within the specific context of the Trust, the production of energy has changed its meaning from being an activity that generates income to one which has to be financially supported through other activities.

Having to repay the Trusts' debts has come to be a core activity for the social world of the board and staff members.

B6:   'So all of that is out there; the potential is there. But we have to convert that to reality and actually identify money-making projects. So that to me is our next task. We have the mortgage, and that has to be repaid' (CDT, Board Member 6: 142–144).

Being in debt forces the members of the Trusts' board and staff to make the repayment of these debts a priority. Most other activities of the Trust have thereby come to be scrutinized and evaluated by this social world with regard to their financial aspects. In this quote, the main task of the board and staff in fact is explained to be not the community members' ideas and interests but the identification and realization of money-making projects. This change in the board and staff members' perceptions and activities has led to a major conflict within the CDT. For the board, activities and projects should generate financial income in order to enable the Trust to repay the loans for the camp and the boiler instead of the boiler generating income in order to realise other activities. While most other social worlds in the Trust principally acknowledge the significance of the mortgage and debts, they nevertheless severely oppose an evaluation of their respective core activities within this teleology.

GL: 'I can see why there are conflicts. The CDT board, quite rightly, are—their aim is to maintain the camp and to get rid of the loan, to pay off the loan. After the community buy-out there was obviously a big chunk of money that's been that needs to be paid back, and this board has tried to work out, in their opinion, the best way of doing that. Some of the things they have done and are planning to do are very contrary to what heritage is all about' (CDT, Group Leader: 156–161).

In this case, it is a group leader from the Heritage Group who complains about the board's teleologies and activities. She generally acknowledges the rationality of the board's activities in relation to this social world's teleoaffective structure. In fact, according to her, the board has '*quite rightly*' made it their core activity. Instead, she criticizes the effects this has had with regard to the core activities of the social world of the Heritage Group. This mismatch exists not only between the board and the Heritage Group, but also between the board and nearly all other social worlds in the CDT. Many groups and activities were not realised because they would have required money, which the board felt the Trust could not afford. Also, different activities in which the board has engaged have direct negative influences on different group activities. Most prominently, the CDT has sold some parts of the camp ground. This has led to a loss of potential activity space for different groups. More importantly, selling camp ground conflicts with the original idea of the Trust—to purchase and maintain Cultybraggan Camp for the community. Within this specific context, selling camp ground is a highly contestable activity. With regard to the social practices and orders of the members' social worlds, the teleologies and activities of the board and staff appear to be directly destructive. Consequently, the members of other social worlds in the CDT contest and try to resist the financial teleology and its effects on their social worlds. Because of its negative influence on the financial situation and the effects this situation directly has on other activities in the CDT, the upfront investment and the inefficiency of the biomass boiler have contributed to and are influenced by other social practices of the social site in specific ways. The biomass boiler and its financial effects, consequently, have not only affected the social practice of renewable energy production, but also had effects on many other social practices within the social site of the CDT.

The financial teleology is also questioned among the members of the social world of the board and staff. Some members of the board and staff are critical of activities motivated by the financial teleology.

S1:   'What Cultybraggan Camp should aim to do is have big events! External events to bring in money into the community. Although SM4 has a bit of a bugbear with those kinds of ideas because he said, "I thought we were trying to reduce our carbon footprint here?". What we're doing by boosting the tourism is we're asking people to travel to the area. I don't know what the answer is, but he has a point. Definitely has a point' (CDT, Staff Member 1: 248–252).

Using the example of big events, the interviewee describes potential conflicts between activities resulting from the financial teleology and activities within the environmental sustainability teleology. While big events like the CultyQuake music festival generate financial income for the CDT, these events are often not compatible with the ideas of environmental sustainability. The financial teleology in this case does not directly prevent or impede upon activities of other social worlds within the CDT. Instead, activities resulting from the financial teleology might unintentionally have negative effects on other social worlds in the Trust.

The last two quotes exemplify how different social worlds in the CDT have to negotiate their respective teleoaffective structures. The problem specifically arises when different activities and/or their underlying teleoaffectivities cannot be reconciled with one another. Even if the context has been momentarily stabilized, new elements or relations might change the situation at any point. As exemplified in the financial teleology of the board, the teleoaffective structure of one social world might change due to other elements of the context—like purchasing and inefficiently running the biomass boiler system. The new teleoaffective structure of one social world might cause new or effect existing negotiations.

Both interviewees in their quotes have accepted that financial considerations are a dominating teleology within the social site of the CDT. The dominance of the financial teleology results from both the severe potential risks associated with not servicing the debts and the relative dominance of the board. The power of both the board and the financial teleology, nevertheless, is constantly questioned and has to be negotiated within the Trust. While the more particular teleologies of sustainability and heritage do not provide powerful enough arguments to question the financial teleology, the Trusts' core intention of serving the community is a potent argument in these negotiations.

S4:   'I think what has happened is that for Cultybraggan Camp, I think the development has forced the Trust to look at things financially of what is and isn't possible sort of thing. They worry about the money a lot. And I think that this, for a couple years or so, has gotten it away from the community elements. The difficult financial elements of Cultybraggan Camp, I find there's a big focus on that, and there's been the funding around the Climate Challenge funding, which has sort of focused around the Climate Challenge agenda. So it's sort of been money and find money and climate change that's been the main things. I think in the past years I think there has been this recognition actually we do need to re-engage with people and work out how to engage' (CDT, Staff Member 4: 383–395).

The quote summarizes the ambiguities and conflicts that have evolved from the negotiation of the CDT's teleoaffective structure. The financial teleology—presented as an externally induced necessity—conflicts with other aspects of the Trust's teleoaffective structure. Because the core aim of the CDT is to serve the ideas and interests of the community, both the general dominance of the financial teleology and the specific ways in which the board is pursuing this teleology can be challenged through the community argument. According to the quote, the community element had become somewhat subdued by the pressing financial necessities in the last few years but recently has gained back its power. In this quote, the community aim is presented as being in opposition to not only the financial but also the climate change-related teleologies and activities of the Trust. The climate change agenda is related to the financial agenda as it has been 'the funding around the Climate Challenge funding, which has sort of focused around the Climate Challenge agenda'. In this quote, the climate change teleology is described as resulting not so much from the community interests as from the availability of financial funding. While, according to the quote, financial considerations and the climate change teleology have dominated the Trust's activities in the past, more recently '*there has been this recognition actually we do need to re-engage with people and work out how to engage*'. Re-engaging with the people and thereby with the community elements is set in opposition to the financially motivated climate change teleology in the quote. In fact, within the negotiations leading to the strategy revision process described above, one of the main arguments against the CDT's sustainability agenda has been the financial motivation of this agenda. Members complained that it did not derive from the interests of the community—the concept of sustainability not even being intelligible to many members—but, as explained in the quote above, was mainly motivated by the availability of grants and loans (this aspect is further elaborated upon in Section 4.2.4).

This presentation of the sustainability agenda as not being legitimated by the community is contested by the members of the board and staff.

B7: 'Actually a lot of the time what I'm doing is reminding people about the strategy for the Trust and the camp that came out of this really good community engagement process. [...] And I'm not reminding people because I don't think it should change—because I think the whole nature of this is that it should change—but I think it should change by thoughtful engagement, not by personal priorities or current kind of pet hobby horse things' (CDT, Board Member 7: 187–191).

In this quote, the sustainability strategy for the Trust and camp are presented as resulting from '*this really good community engagement process*'. For the interview partner, it is not the sustainability but the 'new'—particularly the heritage—social worlds that have not been legitimated by the community, but instead derive from '*personal priorities and current kind of pet hobby horse things*'. Both sides make use of the CDT's aim to be a participatory project. The board member in the quote above argues that the sustainability agenda has been legitimated by a formal community engagement process. The members of the opposing new social worlds in return try to prove their legitimacy through the number of people who are interested in and engage in their activities—and thus through informal forms of consent. The example shows that all social worlds engaging in the social arena interpret and make use of the community argument with regard to their own teleologies.

The board members not only legitimate the financial teleology and its resulting activities by means of the formal community engagement process, but also employ informal community arguments to legitimate their activities.

B6: 'My biggest concern at the moment is that if we're going to be sustainable, we have to be able to wash our face: in other words, pay the mortgage and pay the overhead. At the moment, I have concerns about how we're going to do that. Because unless we can generate more improved income and regular income at a particular level, then it's going to be very difficult I think to service the community' (CDT, Board Member 6: 44–49).

While the interviewee underlines the importance of the financial aims and activities, this aim is presented as being motivated by a community teleology itself. The Trust needs to focus on financial activities to survive in order to be able to further service the community. According to this argument, financial activities are not inspired by financial teleologies but instead are a requirement of the context within which the Trust is embedded. The main teleology in this quote is to keep the CDT alive because otherwise '*it's going to be very difficult I think to service the community*'. By this statement, the speaker declares himself as being in accordance with the teleoaffective structure of the CDT, as his activities are motivated by the desire to secure the Trust as a community asset. Financial ends are not defined as an aim for profit or income in itself but as an activity that enables the Trust to service the needs and interests of the community.

In order to assert the financial teleology in the social arena of the different group interests, the board members employ different arguments. Besides emphasising the inevitable character of the financial teleology, they legitimate this teleology in reference to the CDT's community aim by arguing that the Trust cannot serve the community if it does not service the financial necessities first. This last argument implies a temporal hierarchy of teleologies. Financial considerations have created a hierarchical ordering of practices in time.

B2:   'Once we can pay off the debt, then I think we can have another exercise of saying to the village, "Look, we've paid off the debt. We're potentially going to come into some income stream. Now, what do you want to do?"' (CDT, Board Member 2: 267–269).

The board member argues that '*once*' the financial necessities are fulfilled, the CDT will be able to get involved in activities, tasks and, projects requested by actor groups from the community. By introducing the element of temporal order into the teleological hierarchy, temporality is made a relevant aspect in the CDT's teleoaffective structure. Certain teleologies or related practices are not generally denied but are postponed into the future. In this quote, the argument does not focus on the type of activities that can be realised—those which produce financial income—but on the chronological order of activities.

While—as argued above—energy production has come to be associated with the Trust's debts, it is still also given meaning as a potential income-generating practice. In order to legitimate their activities and aims, the Renewables Working Group relates these to the financial teleology. The interplay of costs associated with energy production and its projected financial turnout are exemplified in the Renewables Working Group's communication about their activities and plans.

'The Renewables Working Group have been investigating the potential to increase solar power at the camp. Some ideas being explored are:
-    Increase array from 10 kW to 30kW. This would cost in the region of £27,000, with a payback of 5.5 years and a 20-year income of £98,624
-    Install a 10 kW array on the site of the bunker. Costs and income to be confirmed.
-    Install up to 150 kW array on a sloped roof constructed on the rifle range site. This would require a roof/frame to be constructed at additional cost but the installation itself would cost around £170,000 with a payback period of 12 years and the income over 25 years is estimated at £1,140,367.
-    Ground mounted arrays at various points around the camp and also on the sewage treatment site' (webpage Renewables Working Group, 2015).

Throughout this presentation, renewable energy production is explained in terms of money. Contributions to carbon reductions, wellbeing, or sustainability of the community are not mentioned. On the one hand, energy production is presented as being causally related to finance because of the necessary financial investments. On the other hand, the outcomes of energy production practices are also presented and measured in relation to their potential financial income generation. This way of presenting energy production can be understood in relation to the hierarchy of teleologies in the context of the CDT. As financial requirements have come to be increasingly important, working groups and their activities are also increasingly explained and legitimated in terms of these teleologies. The presentation positions energy production in the Trust's teleoaffective structure by relating it to the financial teleology.

Being a community organisation, any financial income generated has to be fed back into the community. Within the CDT, income will either be used to pay back the debt or to enable further community actions. When describing the amount of money that could be earned via the production of renewable energy, the Renewables Working Group presents the described options in terms of how much potential income they would generate for the community. As it also transpires from an earlier quote (Board Member 2: 267–269), generating financial income is not presented as an aim in itself but is always related to the aim of serving the community. Generating money through the production of energy within the context of the CDT thus is given meaning as a way to enable the community to realise its ideas and interests.

The production of renewable energy in the CDT is not an isolated practice; it is embedded into the social site through the teleoaffectivities underlying and

motivating it. Neither energy production nor its sense-making are stable or un-
contested. Instead, renewable energy production is given meaning and negotiated within the CDT in many ways. Also, it is not an outcome of one single
rationality or teleoaffectivity but is embedded into different teleologies and
made sense of with regard to different socially shared forms of knowledge. Both
the different teleologies as well as energy production as a viable element with
regard to a certain teleology are contested within the context of the CDT. Energy
production, as a viable element of the CDT's teleoaffective structure, has to be
constantly negotiated among the different social worlds of the Trust. At the same
time, energy production has to be positioned within negotiations of different teleologies in the CDT's social arena. In order to remain meaningful within these
conflicts, renewable energy production needs to be related to the arguments relevant to the conflict. As shown in this section, energy production in the CDT
derives its meaning from the teleologies of sustainability, resilience, and financial income generation. At the same time, these teleologies, as well as energy
production itself, always have to legitimate themselves with regard to the community aim. Renewable energy production in the CDT is thus related to the teleology of community in manifold ways. It is shaped and legitimated by the ideas
and interests of the community, in reference to which it constantly has to legitimate itself.

4.2.2   General understandings: sustainability or a healthy lifestyle?

According to Theodore Schatzki's theory of social practices, general understandings are abstract senses about the values, worth, nature, or place of things
expressed in people's doings and sayings. Thus, general understandings relate
to the 'rightness', 'beauty', or 'nobility' of a bodily doing or saying. While teleoaffectivities are related to the ends associated with a certain practice, general
understandings focus on abstract senses associated with the activity itself. General understandings, defined as abstract senses about the value and worth of a
practice itself are influenced by the teleologies within which energy production
is embedded in the CDT. Individual people's understandings of renewable energy and its relation to the aims and intentions to the CDT, however, are also
part of the social site.

Looking at an already presented quote again, it can be shown that the general understanding of renewable energy within the CDT is closely related to an
individual's ideas and interests.

I: 'So when you speak about renewables, you speak a lot about saving money?

B5:   Well, that's the whole idea of the thing. Why burn heating of any descriptions, whether it be coal fires or central heating or anything. Why let it burn and fly out through your roof if you can <u>contain</u> it! So you don't spend as much money? But it's not only from a money point of view, it's from an energy point of view. You're <u>wasting</u> energy. And we can't go on and on and on and on wasting energy or we'll have no energy left' (CDT, Board Member 5: 273–279).

In this quote, energy production is valued with regard to the financial and sustainability teleologies. The interviewee first relates the production of renewable energy to the aim of saving money. Saving money is *'the whole idea of the thing'*. In his next two sentences, he broadens his argument by including not only heat from renewable sources but any kind of heat. It transpires, that for him energy production from sources *'of any descriptions'* does not have a worth or value in itself so much as it does as part of an integrated set of energy-related practices. Producing or using renewable energy is only a valuable practice if it is combined with energy efficiency, carbon-saving activities, and facilities that together help to stop *'wasting energy'*. Only when integrated in such a way, the production of renewable energies will lead to financial savings and carbon reductions. With regard to both financial and environmental considerations, for this interviewee the production of renewable energy is generally understood as a valid practice not in itself but only in its combination with other practices and material arrangements. The interviewee's general understanding of energy production shows his sense-making of energy production as one integrated practice within (economically and environmentally) sustainable tasks and projects.

The interviewee's understanding of renewable energy is based on socially shared forms of sense-making in the CDT and beyond. As explained, the production of renewable energy in the Trust is only one aspect of the CDT's sustainability agenda. The 'integrated' understanding of renewable energy can be observed in the daily activities of the CDT. The Trust's aim to decrease the village's energy demand has led to the creation of different jobs like the Energy Advisor or Energy Awareness Risers.

I: 'Would you mind telling me more about the energy awareness?
GL: Well, that was basically to go around to people's houses and see how they're getting on with the way they use their energy and if they were struggling financially. Just to make them aware of all the different options they could have. […] I was basically there to support them to save energy, save money, help the environment, the whole, thing' (CDT, Group Leader: 29–31, 38–40).

The energy awareness officers were employed in the early phase of the CDT. They were mainly responsible for carrying out the street-to-street energy audit, which was a main project in the Carbon Challenge. In the social site of the CDT, these activities were intentionally related to the production of renewable energy, as both kinds of activities are made sense of with regard to the Trust's aims of sustainability and financial savings for the community. Again, the production of renewable energy in the CDT is made sense of with regard to teleologies like financial or environmental sustainability. The general understanding of renewable energy in the CDT derives from this embeddedness.

B4: 'I think reducing everything! ((laughing)) would be ideal. People should use—well, A, people should—in terms of their household—get everything as insulated et cetera as possible. We've all become accustomed to having houses stupidly hot. Don't need houses that hot. It's ridiculous. Put a jumper on. Cut your fuel bills. It's better than living in a stuffy house anyway. I don't know why people insist on having these really hot houses. [...] and even when people do insulate and everything, they think, "Oh, I can still have it that warm". And they don't actually do any energy efficiency savings kind of thing. So there's that whole rebound thing: "Oh, I'm saving here, so I'll just use more here". I think people need to use less energy, need to use more renewable energy, need to cut down what we consume (CDT, Board Member 4: 268–274).

In this quote, heating a house without engaging in energy efficiency measures is generally understood as an environmentally and financially non-viable practice. In fact, heating houses—no matter what energy sources are being used—without cutting down on consumption leads to '*stupidly hot*' houses, a situation which is here described as '*ridiculous*'. Instead of heating houses so much, this interviewee—who is one of the board members most active in the planning and organisation of renewable energy production in the CDT—proposes that residents should '*put a jumper on*'. Cutting down on consumption here prevails over the production of renewable energy. Renewable energy is understood as non-viable if it is not embedded into carbon reduction strategies. By implication, it can be argued that the production of renewable energies in the context of the CDT is understood as a sensible or viable practice only if it is integrated into energy reduction and efficiency measures. As described above, this general understanding of renewable energy as an integrated activity is realised in the CDT's daily activities, in which awareness rising campaigns, steps for consumption reduction, and energy efficiency measures are carried out along with the production of renewable energy.

It is an important aspect of the ambiguities and complexities in the CDT that not all members share this general understanding about renewable energy production as part of an integrated sustainability approach.

I: 'A big proportion of the development activities have been devoted to climate change or to environmental issues. What do you think about that in general?

B1:   There seem to be two different abstract concepts. The climate change, Carbon Challenge, energy reduction. I have difficulty associating it with what is happening at Cultybraggan Camp. I know that we're doing a district heating system, but to me that is just a big, beautiful machine @that you can work with@ It's not terribly efficient, and I don't think we've saved any carbon through using it. Because, really, we've had problems in getting people to make use of it. It's rather expensive. Even if you take the time and trouble to connect up with the system, the fuel costs are only marginally better than gas or whatever. So I think a lot of thought has been put into these issues, which are, I suppose, politically motivated. But there's not much sense' (CDT, Board Member 1: 106–116).

In the end of this sequence, the interviewee argues that there is '*not much sense*' in trying to save money through the biomass system because '*the fuel costs are only marginally better*'. Using the biomass boiler, in his general understanding with regard to the financial potential, appears to be a non-viable practice. His estimation of the potential carbon-savings relates the biomass boiler to the different people and their activities at the camp. Within the specific arrangement of elements (elaborated upon in more detail in Section 4.2.6) in Cultybraggan Camp, the biomass boiler has become inefficient. It appears that the general understanding of the practice of producing heat through a biomass boiler is closely related not only to the teleologies motivating the practice, but also to the specific context within which this practice is enacted.

Asked about climate change and environmental issues, the interviewee mentions the district heating system right at the beginning of this quote. The topic had not come up before in the interview. Instead he actively re-creates the connection between climate change and renewable energy production in this quote by stating that he has difficulty connecting the two. It can be argued that by connecting the two—supposedly unconnected—issues, he is deliberately differentiating his position from the practices of sense-making of climate change and energy production within the CDT. The reason for the interviewee's denial that energy production makes sense with regard to climate change/sustainability can be explained when combining the quote with a later sequence from the interview. In this sequence, he answers whether he personally believes in climate change.

B1:   'No, not at all. I don't think we have much evidence to say that any kind of global warming is not simply a cyclical weather event over a number of years. That maybe in 10 years or 20 years, we'll be saying, "Oh, we're going back to the next ice age. We're going to be globally frozen." I don't think that's well established at all.

I: Do you think there are other reasons it might be important or interesting to have these modern techniques? Are there other reasons for them?

B1:   Oh yes, I think so, I think it's healthier. I think rather than thinking about the fu-
        ture, thinking about the present. That there's a lot of benefits in having a healthier
        lifestyle. It seems to be better to burn underline(wood)than to burn oil, because you're not
        putting so much black @tiny things@ into the atmosphere and the air is nicer to
        breathe, so that seems to have an impact on the present day' (CDT, Board Member
        1: 172–182).

The interviewee represents those actors within the CDT who are critical of the
Trust's green agenda. While being in favour of the woodchip boiler especially
as a technological artefact and '*a big, beautiful machine*', the use of these artefacts for him is not legitimated by climate change or sustainability. His argument
instead is based on individual present-day health issues. Health, he argues, can
be improved by using renewable energy technologies, because burning wood
produces fewer fine particulates than burning fossil fuels.[15] In relation to the
teleology of preserving the health of the community members, energy production is generally understood as an appropriate practice by him. It is within this
teleology that renewable energy production is understood to be '*healthier*' and
thereby '*better*'. While rejecting the worth or value of this practice when evaluated from a climate change or economic perspective, his positive general understanding of renewable energy practices results from a '*healthier lifestyle*' teleology. While his estimation of the economic value of the practice is based on
the specific context to which this evaluation relates, his general understanding
of the practice is based on his individual ideas and convictions.

He thereby deviates from the ideas and convictions of most other actors in
the CDT. These last two quotes exemplify the conflicts and ambiguities to which
renewable energy production has been subjected in the CDT. Summarizing the
described conflicts about renewable energy technologies, sustainability, and climate change, the CDT integrates actors who are critical of sustainability, because they doubt climate change as the fundamental idea underlying the Trust's
carbon-reduction activities. The quotes indicate that it is questionable whether
acceptance of the production of renewable energy could be stabilized within the
context of the CDT if it would be organised only with regard to sustainability.
Within the CDT, however, production of renewable energy is embedded into a
context in which it is first and foremost defined and valued in its ability to contribute to the wellbeing of the community. Analysing renewable energy production within the CDT as embedded in a community context enables a different
reading of the two last quotes (BM1: 106–116, 172–182). In the second quote
(172–182), it appears that somebody who does not believe in climate change can

---

15   While he criticizes the reliability of evidence on climate change, he is not able to provide more
      evidence for his claim than that it '*it seems to be better*' except that he thinks it is healthier.

still be supportive of renewable energy production within the CDT if he understands the Trust to be an organisation working for the—here bodily—well-being of the community. Renewable energy production is legitimated not because it reduces carbon emissions with regard to a future scenario but because it improves the present-day health of the community members. The point of reference through which energy production is legitimized and from which its general understanding derives is the present-day community, not the future environment.

The CDT's conception of sustainability and its energy production projects are not only motivated by environmental concerns. Instead, environmental concerns are always interwoven with and have to be negotiated with other teleoaffectivities. With regard to the different teleologies, the general understandings of renewable energy become ambiguous. The general understandings of renewable energy production in the CDT are related to and shaped by underlying ideas, believes, and teleologies. Different actors, even within the same social world of the CDT, might have heterogeneous or even opposing understandings of renewable energy production with regard to certain forms of knowledge or teleologies. Production of renewable energy in the CDT is subject to ambiguous general understandings by different actors. Ambiguity results not only from different teleologies but also from heterogeneous general understandings within the social site. The general understanding of energy production in the CDT is thereby embedded in a context in which community, environmental, and economic aims inform the practice of renewable energy production.

4.2.3 Practical understandings: knowing the community

Practical understandings inform the practice of renewable energy production in the CDT in different ways. Knowledge and skills are important resources in planning and realising activities related to energy production and consumption. Knowledge becomes important when it is necessary to mobilize financial resources through funding applications, for example.

S2:   'Actually, I think lots of people don't even know about the energy savings trusts, whereas to me, it seems like, "How could you not know?" Even my husband, a few months ago, I said, "Who would you go to for energy information?" He was like "((keeps silent for some seconds))"' (CDT, Staff Member 2: 188–190).

The quote exemplifies how resource mobilization strategies through funding application are based on specific types of knowledge. Applying for funding requires knowledge about the relevant organisations and institutions. The quote indicates that, for this actor, knowing about these institutions and organisations has become so natural that for her, '*it seems like, "How could you not know?"*'. The interviewee has been engaged in renewable energy for large parts of her private and professional life. Knowing about relevant institutions has become so much embodied by her that she does not consciously consider having this type of knowledge in her normal day to day life, but uses it more or less unconsciously.

The next quote underlines the importance of knowledge and skills for the realization of renewable energies in the CDT.

B7:   'the people who formed the first board of the Trust, who really stuck together for that first four or five years, the personalities involved in that had quite strong visions. So they helped to mould—also, they are very articulate young professional people whose passions were shared for sustainability and community empowerment. But also they knew how to speak that language. And I sort of include myself in that as well. There were people who were chief execs at voluntary organisations and quite senior within environmental groups. So I'm sure that the leadership of the Trust put a certain angle on that stuff as well. And then finally, of course, what happens is that when you're accessing money, you tell a story to help you access that money. It's only practical to do so' (CDT, Board Member 7: 161–170).

The quote describes the personal requirements involved in the mobilization of financial resources through funding and how the personal competencies of certain actors have become part of the social site. The interviewee describes the people who formed the first board, people who had a strong '*vision*' and '*passion*' for sustainability. Using Schatzki's vocabulary, they were driven by strong teleoaffectivities towards sustainability and community empowerment. The quote describes the practical understandings necessary or at least helpful for the practice of applying for funding. The members of the first board were '*articulate young professional people*' who '*knew how to speak that language*'. They not only knew what stories to tell when applying for funding, but were capable of telling these stories in the appropriate style. The interviewee relates the actors' having these skills and knowledge to them having being involved in sustainable or renewable energy activities in their private and/or professional lives. The skills resulting from these involvements are an outcome of both consciously accumulating theoretical forms of knowledge and practically and unconsciously learning how to do certain things, especially '*how to speak that language*'. The actors and their knowledge became an inherent part of the CDT not only because their engagement enabled many of the activities pursued by the Trust, but also because by bringing their teleoaffectivities, knowledge, and skills into the social site, they '*put a certain angle*' to the aims and activities in the CDT. The quote thus also serves to show the importance of certain actors for the constitution of social orders (see Section 4.2.5)

Expertise or theoretical forms of knowledge are relevant when planning, organising, and realising renewable energy projects. When the necessary type of knowledge could not be mobilized from among its members, the Trust accessed professional knowledge from external actors. With regard to renewable energy production, the CDT has commissioned a number of feasibility studies about different technologies and their usability, potentials, and risks. While these forms of professional expertise are required, they are also criticized by most actors. Not only the members of the CDT, but even those institutions which officially require the existence of professional expertise are ambiguous in how they see professional expertise.

CCF: 'Where we see consultants writing applications, you lose the community voice, and then it's hard to understand, "Does the community really want this?". Or is this just a consultant writing this up, saying "This is what we need to do in this community?". We want it to be done by the community for the community, not something that is done to the community. [...].

I: Is it maybe becoming too professional, sometimes, or...

CCF: We have no problem with consultants working with projects and helping them, but it needs to be they are working <u>for</u> the project and not dictating what should happen' (Climate Challenge Fund, Staff Member: 251–259).

This quote describes the ambiguity towards expert knowledge. On the one hand, expert knowledge is seen as essential for the successful planning and realization of community projects. On the other hand, when a project is based on too much expert knowledge, it is understood to '*lose the community voice*'. The quote suggests that only the members of a community know what the community '*really want[s]*'. This type of knowledge, according to the quote, is not something that can be consciously learned by consultant; it only stems from actually living in a certain community. The statement differentiates between the theoretical knowledge of consultants and the practical knowledge that constitutes the community voice. Experts are only allowed and able to contribute theoretical knowledge to community projects.

In the daily activities of the CDT, professional knowledge is constantly incorporated in and negotiated with experience-based knowledge of the community members. The decisions about certain technologies in the CDT, for example, are all based on professional feasibility studies. The recommendations of the professional experts, however, are not followed word for word, but are negotiated with regard to other elements from the social site. In the discussions and activities of the Renewables Working Group, the professional recommendations are integrated into the context of the CDT.

S2:   'Community Energy Scotland believes it's not really viable. It doesn't make <u>finan-cial</u> sense. It doesn't make sense in the terms of the amount of effort and resources and <u>time</u> that would be spent on it to get it up and running as well. So basically, hydro at the Drummond fish farm site is not an option' (CDT, Staff Member 2 at Renewables Energy Group Meeting: 15–17).

The quote exemplifies how professional recommendations are embedded into the specific context. With regard to a potential site for a hydro scheme, the expert from CES estimated that the site would not be financially viable. According to the interviewee's estimation, the site also does not make sense in '*terms of the amount of effort and resources and <u>time</u> that would be spent on it to get it up and running as well*'. The staff member implicitly argues that the project is not 'interesting' enough for enough members of the CDT to justify the effort and time needed to realise the project. This estimation is based on practical knowledge about the interest and willingness of the CDT members to engage in such an activity. The quote exemplifies how seemingly objective entities, like time and resources, which are necessary to realise a certain project are subjectively made sense of with regard to (other elements from) a certain context. While prefiguring the activities within (certain social worlds of) the CDT—by numbering the resources necessary for the realization of a certain project—expert knowledge does not determine these activities, as it is always embedded into a specific context.

Wind turbines are an example of how theoretical knowledge is negotiated with practical forms of knowledge specific to a certain social site.

S1:   'Oh look, I was just going to invoke the 'W' word—wind farms, wind mills. […] That's a dirty word around here' (CDT, Staff Member 1: 503–504).

The professional recommendation to not install wind turbines is based on specific environmental aspects of Comrie and Cultybraggan Camp (Section 4.2.6). While explaining why wind is not a viable option in Comrie, all interviewees, however, combine this experts' advice with the social conflict potential of installing wind turbines. Knowing about the conflict potential of wind turbines is not the outcome of consciously acquired knowledge; it results from the actors' embodied experiences of living and engaging in and with the community. The combination of the professional recommendations and practical understandings about the community resulted in a rejection of wind energy as a viable option for Comrie. In fact, the degree of discontent about wind energy even dominates the expert knowledge. It is the main reason the community has not yet commissioned a feasibility study evaluating the potential for wind energy at locations other than the camp. Knowing about this conflict and its severity for the involved actors was a meaningful argument in the negotiations about potential sources for renewable energy production in the CDT. The conflict never took place explicitly in the CDT. Instead, the decision of the CDT to not further pursue the option of wind turbines was based solely on the actors' practical knowledge about its conflictive potential, which resulted from different people already having had discussions to that effect with members from the community.

The CDT is not only based on both theoretical and practical knowledge, but also creates both types of understanding among its members. In part, the production of knowledge and skills results from deliberate training efforts promoted or provided by the CDT. Good examples are the Energy Awareness Officers. The four officers got training for the job, which enabled them to educate the local residents about energy consumption and efficiency.

GL: 'They came here and did some training to let us know how to go about, what questions to ask, what to be aware of, and just the general how to be an advisor. But we also went to one of the energy efficient houses that they showed just to have a look at different options and things like that. So it was nice to have a bit of training. It meant you got to get to know the other people that were doing it as well' (CDT, Group Leader: 64–68).

The interviewee describes how she got trained to be an Energy Awareness Officer. The training was provided in order to enable members of the community to carry out the street-to-street energy audit as part of the Carbon Challenge. The Energy Awareness Officers acquired knowledge about climate change, carbon savings, and energy efficiency measures to pass it on to other community members. Providing the training can be seen as a deliberate effort to build up necessary theoretical knowledge in order to realise the CDT's sustainability aims.

Creating practical knowledge about sustainability within the community is also an important aim. People should act sustainably not only because they have been trained to do so. Instead a common desire expressed in the interviews is to make sustainability a normal part of residents' day-to-day life.

B2:   'I think it's [sustainability, AP] so important that we should take it for granted that it underpins everything we do. At the moment we're making big issues about it; you hear about it, […]. I think it is much more important than that. It should be important enough that we don't discuss it. Does that make sense? It should just be second nature that everything we do is sustainable, is what I call "least action" in other words, you do things in the simplest way without disturbing things. That should just be second nature. The fact that we say you know, the fact that we have to write sustainability into our programmes and plans I believe is wrong. I think it just naturally should be there. But that's an educational process. So that's where I think we could do a lot just by doing it by example' (CDT, Board Member 2: 175–184).

The interviewee does not name any particular consciously planned or organised activity or strategy through which sustainability should be supported by the CDT. Instead, for him, sustainability should inform everything done by the Trust and its members. Statements like '*It should be second nature*' or '*it just naturally should be there*' indicate that, for him, sustainability—while it is based on theoretical knowledge—should become embodied knowledge. Within the necessary '*educational process*', the Trust could play an important role '*just by doing it by example*'. In setting an example, the CDT would not so much supply theoretical knowledge to its members at it would transfer practical knowledge about how things are done in a sustainable way. The production and usage of renewable energy is one way by which the Trust sets a practical example. While programmes like the street-to-street energy audit mainly transfer theoretical knowledge about energy savings and efficiency measures, practically producing and using renewable energy for the camp shows people the potential and practicality of locally produced renewable energies.

Furthermore, the Trust provides people with the opportunity to become actively engaged in planning and running renewable energy installations. The main responsible body for renewable energies in the CDT is the Renewables Working Group. Its members keep themselves informed about potentially available technologies, appoint professional institutions to create feasibility studies, discuss the results of the feasibility studies, search for and get in touch with potential partners, and find out about and negotiate terms with companies producing, selling, and installing the chosen technologies. Also, the members of the working group have been trained in running and maintaining the solar panels and the woodchip boiler systems. If problems occur—as with the biomass boiler system—the members of the working group are also the ones who identify the problem, discuss it, and try to implement solutions. In the case of the biomass boiler, this meant finding out why the system is running inefficiently, and then contacting and negotiating with the subcontractor who built and installed the technological equipment. Next, the members applied for and received funding (some months after the conclusion of my fieldwork), enabling the CDT to extend the system by connecting more huts to it. In the process of engaging in all these different activities, the members of the working group not only acquire and use a wealth of theoretical knowledge about energy technologies, but also accumulate, necessitate, and make use of practical forms of knowledge. They not only theoretically know how to maintain and run the system but have also acquired the practical skills to do so. Likewise, the members of the working groups are themselves responsible for writing funding applications. As explained earlier, they thereby learn about existing institutions and how a successful funding application should be written.

Besides the undeniable importance and relevance of theoretical knowledge, practical forms of knowledge significantly shape and are shaped by energy-related practices in the CDT. Knowing about potential and existing conflicts or problems in the community is one example of how practical forms of knowledge have informed the production of renewable energy in the Trust. Additionally, it is the members themselves who organise and run the production of renewable energies, thereby acquiring and putting to use theoretical and practical knowledge about different systems. The kind of practical knowledge acquired and how it is used closely relates to the context of the CDT. It is related to the social order, and especially to the arrangement of actors and material artefacts in the CDT. Thus, that the members of the Renewables Working Group learn how to apply for, plan, run, and maintain renewable energy systems is related to the fact that the CDT is a volunteer organisation (Section 4.2.5). None of the funding that pays for the four staff members has been applied for and thus cannot be devoted to running or maintaining the energy infrastructure. Consequently, nearly all the work put into the production of renewable energy in the CDT is done by volunteers, none of whom have had any prior experience in planning, realising, or running renewable energy technologies. Having to organise the production of renewable energy almost completely based on volunteer engagement also means that the chosen technologies need to be manageable by non-professionals with limited amount of time and resources (Section 4.2.6). Installing easy-to-maintain technologies conversely means that it is possible for non-professionals to run and maintain the systems. Within the specific context of the CDT, it is necessary that members of the Trust acquire and put into use both theoretical and practical knowledge about the planning, realization, organisation, running, and maintenance of the renewable energy technologies.

4.2.4   Rules: Making use of the Land Reform Act and the RHI

Two sets of governmental rules are important for the context of the CDT's energy practices. The first relates to Scottish landownership and the second to the (related) structures of British and Scottish energy policies. The context within which landownership and community energy are based in Scotland has already been described in Section 4.1.3. This section is interested in investigating how evolving rule sets have shaped the context and are made meaningful within the context of the CDT.

The background and process of modern Scottish landownership, especially the Land Reform Act 2003, have been an essential aspect of the creation of the CDT. The Act prefigures the CDT and its practices in the sense that it has enabled the inhabitants of Comrie to purchase and own the camp. Purchasing the camp again was one of the key intentions for establishing the CDT.

S1:   'There was a big push to up the membership when the opportunity to buy Cul-
       tybraggan Camp came up. The idea of buying Cultybraggan Camp and stopping
       somebody going around and building houses there was popular across the board.
       So I think we do have a lot of members who are members because of that and for
       no other reason' (CDT, Staff Member 1: 324–328).

The quote shows that the possibility of doing a community buyout on the legal
ground of the Land Reform Act was meaningful in the context of Comrie because of the affections and teleologies of Comrie's residents. The Land Reform
Act was made meaningful in the context of Comrie because the rules organising
community buyouts were meaningful as an instrument enabling the Trust to realise the inhabitant's teleoaffectivities. Only because it was meaningful with regard to the community's interests, the Land Reform Act came to prefigure the
community buyout. The rules provided by the Land Reform Act came to shape
the activities of the Trust because they were relatable and meaningful in the
context of the CDT.

The British and Scottish government's energy policy is directly related to
the production of renewable energy in the CDT. The most relevant set of rules
deriving from this policy concerns the regimentation of community energy projects. Through these rules, energy production in the CDT is related to the Scottish government's energy policy.

S3:   'So written into the plan—it was called the Delivery Plan—here were these ob-
       jectives, by 2020, do this, achieve that. So that's really been the basis on which
       we've gone forward. The funders have backed us, so a lot of our funding comes
       from the Climate Challenge Fund because of the Trust having these objectives.
       So, not only did the community want it, it also meant that we could all be here
       carrying out these programmes because it was in line with the government think-
       ing as well, which was to encourage communities to take this forward.
I: This was a Scottish Parliament programme?
S3:   Yes. Hence you've got things like Community Energy Scotland. It's the idea of
       communities taking charge of their own energy. Decentralized you know. Control
       being delegated. Empowered and those words' (CDT, Staff Member 3: 141–151).

Like the Land Reform Act mentioned in the earlier quotes, the Scottish government's energy policy here is described in its relation to cultural ideas, norms,
and concepts that have been shaping and are shaped by the Act. Prefiguring the
Act, decentralization, delegation of control and empowerment are culturally acceptable ends that are legitimated by and objectified in governmental practices.

More specifically, the quote relates renewable energy activities of the CDT
to community energy policies of the Scottish government. The government has
set carbon reduction and renewable energy targets for itself, which the Trust has
decided to mirror in its application for the CCF. The CCF itself has been set up

and is being funded by the Scottish government. While the carbon and energy targets are themselves rules that shape the government's as well as the CCF's activities,[16] the CCF—an organisation deriving from these rules—has also established a set of rules. These rules predetermine what organisations and activities can be funded by the CCF. The activities in the CDT are influenced both directly by the energy targets and by the set of rules defining the eligibility of community energy projects for CCF funding. That the CDT has become one of the flagship projects of the CCF shows how successfully the CDT has managed to live up to these rules.

On the political side, national politics acknowledged the importance of community energy in the 2001 Energy White Paper and politically institutionalized it by setting up FITs for electricity and the corresponding RHI for heat. The existence of the FITs for electricity has pervasively constituted the energy production practices of the CDT as they ensured the financial viability of the installation and usage of the solar panels. As shown, the production of heat through a biomass boiler has been pervasively involved with the RHI and other rules.

Rules like FITs or the RHI do not determine the production of community energy. Instead, these rules are always interwoven with the other elements of the context within which they are used. Rules like the RHI can, however, be very powerful aspects of these contexts. In the case of the CDT, the potential financial profits from the RHI were accompanied by the loans from the SDHL scheme, which provided an added financial incentive for installing a biomass boiler system.

B3:   'I was involved in the early discussions with the biomass. I collect the feeling that they were being run by the fact that there were deals available, and that we did not have—I felt that we were putting the cart before the horse' (CDT, Board Member 3: 438–440).

The RHI, a scheme through which the Scottish government aimed to motivate the production of renewable heat, is here explained to have been a decisive factor in the CDT's decision to install the biomass boiler. The interviewee criticizes the dominance of the incentive in the negotiations. In her opinion, the CDT should have been paying more attention to other aspects of the (then future) context of renewable heat production in the camp. According to her, the incentive has been given too much importance and has been made key in a situation where other elements, like renting out and making use of the huts, should have been decisive.

---

16   Or at least it should do so—so far the Scottish government has not managed to reach its own targets.

Of course, this interpretation of the situation is not shared by all board members. In its official statement, the decision to install the biomass boiler is defended with reference to the CDT's vision and aims. In its official announcement of the application, the Trust argues that '*[o]n balance the Trust decided that the opportunity to fulfil one of its key criteria for sustainability, minimizing the use of fossil fuels, was too good to miss and the application was submitted*' (CDT 2013: 14). Within this statement the RHI is described to provide an '*opportunity*', which enabled the CDT to realise its own '*criteria for sustainability*'. Instead of dominating the decision, the RHI is presented as a mere facilitator of the Trusts' own teleologies. The opposing interpretations of the situation are a good example of how the meaning of a rule depends on the meaning-giving processes of different actors within the same context.

The RHI has not only been important in the decision about the installation of the boiler, but continues to be the main reason the situation is ongoing despite its existing problems.

S4:   'So the Trust is in this position where, from a resource efficiency perspective, I suspect we'd probably be better off with closing down the boiler and using gas. Probably that would be the case. But because of the RHI—the government renewable heat initiative subsidy—it probably won't be what we'd do. So we're choosing to do something which is not as resource efficient because of the government's subsidy on the biomass stuff' (CDT, Staff Member 4: 485–490).

The incentive here is reported to play an important role in shaping the situation. In the quote, the incentive is described as the reason the system cannot be altered despite its inefficiency within the specific context. The subsidies provided by the incentive are described as the dominating aspect. These rules of the RHI—realised in the form of subsidies—have become the dominating element of the context instead of efficiency considerations. The dominance of the financial rules is not naturally given but derives from the explained dominance of the financial teleology in the CDT's momentary considerations and negotiations with regard to the usage of renewable energies.

At present, an insufficient number of people are using the huts or have decided to heat their huts. Consequently, not enough people have become consumers of heat yet. The inefficiency of the boiler results from this mismatch between the capacity of the boiler and the number of connected huts and consumers. The RHI shapes this situation, as it prevents an adaptation of the material system to the specific context.

S4:   'To be fair […] that's not what the RHI is designed for. It's designed for a system, which is running at the expected capacity and all the rest of it, so in which case it would make sense and be using less resources, I guess. But things like the fact that the boilers need to be new boilers. So we can't sell the big boiler and buy a

> smaller boiler because that's—the big boiler won't be worth anything. And that is
> because the—part of the regulation is installing new boilers. It can't be reused.
> And we can't say to somebody ((hesitates))
>
> I: Swap boilers.
>
> S4:   Yeah. Exactly. Somebody who's looking to expand—we couldn't swap the boilers,
>   which would probably be a sensible thing to do, efficiency-wise' (CDT, Staff
>   Member 4: 491–500).

The rules of the RHI force the CDT to not change the boiler and thereby prevent
the Trust from changing the material artefacts that—from a resource efficiency
point of view—do not fit into the context in which they are used. The problem
does not so much result from the RHI generally being a non-viable instrument
as from the fact that the incentive does not allow flexibility with regard to con-
textually used material artefacts. The problem is created by a mismatch between
the incentive and other aspects that constitute the context of renewable heat production in the CDT and the strictness of the incentive with regard to those elements of the context that are dictated by it. It appears that rules like the RHI are
related to and thus derive their—potentially problematic—meaning from their
interplay with other elements of the context within which they are applied. The
incident exemplifies how rules have prefigured the practices of energy production in the CDT, both by enabling these practices and by restricting and complicating these practice with regard to the context within which they are enacted.

Besides national and sub-national government rules, formal organisational
rules also significantly shape the practices in the CDT. For the CDT, rules associated with funding opportunities have especially come to play an important role
in the practices of the Trust. When the Trust decided to take loans to purchase
and renovate the camp and buy the woodchip boiler, it accepted the terms of
repayment. As already described, the necessity of having to repay loans and
mortgages has become one of the key elements in the CDT.

> B6:   'So all of that is out there; the potential is there. But we have to convert that to
>   reality and actually identify money-making projects. So that to me is our next task.
>   We have the mortgage, and that <u>has to be repaid</u>' (CDT, Board Member 6: 142–
>   144).

The emphasis in this quote is on the mortgage, which '*has to be repaid*'. Repay-
ing the debts is a necessity that does not leave much space for questions or negotiations. As transpires from the quote and as already described, the existence
of the rules dictating the necessity to repay the loans has not only severely influenced the teleologies and thereby the practices of the social world of the board
for which it has become a core activity, but also affects the activities of all other
social worlds in the CDT. The rules dedicated to the conditions of the loans and
mortgages not only demand that the money has to be repaid but also set a strict

timeline. It is not only the necessity of having to refund the loans, but also the temporal aspects that significantly shape the board members' practices.

B6:   'My own feeling is that we have a large project at hand which will take many years to come to fruition if it comes to fruition at all. The key issues and key elements within it are pretty much to look at income generation by 2015' (CDT, Board Member 6: 36–39).

For the interviewee, the *'key issues and key elements'* for the CDT until it has managed to repay the loans in 2015, are income generating activities. According to the quote, the deadline defines a temporal element in the context. As already described, the deadline associated with the terms of refunding has created a temporal hierarchy of the activities pursued by and in the Trust. It can now be added that both the necessity to create financial resources and the temporal hierarchy are heavily influenced by the existence of rules.

Another set of rules influencing the activities of the CDT includes the Trust's own rules for itself. Most of these rules formalize important teleologies of the Trust. Aims formalized within the rule set of the Trust can be said to be or to have been dominant elements within the teleoaffective structure of the CDT. Sustainability is an example of a teleology that had once been important enough to have been formalized in the rule set of the CDT. Based on the Trust's vision preceding the strategy revision process, nearly every activity within the CDT is asked to contribute to the sustainable development of the community. Defining the activities of the CDT as sustainable practices is primarily achieved by setting rules.

B3:   'I think it ought to be, but I don't know whether actually it will come about. It ought to be a focus. I mean, yes, it is. It's in the, it's usually written into the articles of this association. And it is written into the **directives** for each working group. It is there in every project. It underpins the—sustainability is certainly—is very much CDT's thing' (CDT, Board Member 3: 98–102).

When she says that sustainability *'ought to be a focus'* the interviewee hints that while rules play a role in nearly every practice, they are not constitutive of practices. The interviewee voices her doubts with regard to whether the Trust will actually become environmentally sustainable. Continuing with her argument, the interviewee's ambiguity with regard to the question unfolds when she describes sustainability as being *'there in every project'*. This argument is based on the fact that sustainability is mentioned in the directives and articles for each working group. The existence of rules is here used as an argument to interpret (and thus organise) certain activities. Rules here legitimate a certain interpretation of practices. The ambiguity exemplified by the quote mirrors the instability of rules with regard to practices. While the existence of the described directives is likely to somehow shape the activities pursued by different groups in the CDT, it is unlikely that they can ensure the sustainable character of each working group's activities. Rules thus do not dominate these practices. The example also serves to highlight how unstable the elements of the social site are, even when they have been formalized into rules. While rules might have been an element in the negotiations about the sustainability aim of the CDT, they have apparently not been dominating these negotiations.

Different sets of rules and regulations shape and influence the practices of and in the CDT at all levels. While both external and internal rules shape the context, they normally do not dominate the context or the practices. Even if rules like those relating to the payback of loans come to dominate the teleologies of one social world within the CDT, it has already been shown in Section 4.2.1 that the teleology inspired by them is challenged by other social worlds within the Trust. Rules might be made sense of in different ways by different social worlds within one social site. Rules are thus negotiated with the specific teleologies that motivate and organise a certain social practice within the CDT.

4.2.5 Human actors: Board, staff, and volunteers

Human actors are the most important kind of element within the social orders of the CDT. Environmental sustainability and its (changing) position within the social site of the CDT can be described in their relation to the human actors making up the different social arenas of the Trust. Especially in its early stage, the CDT had a pronounced environmental sustainability teleology. The fact that environmental sustainability became a core activity of the Trust back then is an outcome of the commitment of individual actors.

B7:   'the people who formed the first board of the Trust, who really stuck together for that first four or five years, the personalities involved in that had quite strong visions. So they helped to mould—also, they are very articulate young professional people whose passions were shared for sustainability and community empowerment' (CDT, Board Member 7: 161–165).

The first board—according to the quote—consisted of a group of people who not only had the same interests in sustainability and community empowerment, but also shared the same emotional involvement with these issues. They '*really stuck together*'. According to the quote, the social world of the first board was based on commonalities in the teleoaffective structure of its human actors and their high levels of commitment with regard to not only their interests but also the social world of the board. Furthermore, the quote indicates that these '*very articulate young professional people*' had a lot of theoretical and practical knowledge relevant to their engagement in the Trust. Being members of the board meant that these human actors were positioned centrally in the social order of the CDT. They were able to propagate their ideas and to motivate other actors to consider their interests and formalize these ideas and interests in the constitution of the CDT. The early social order of the CDT, especially its sustainability and community empowerment agenda, is related to the human actors who were decisive in the founding process of the organisation.

Nevertheless, the social practices of the CDT have to be constantly negotiated both within the social world of the board and staff and with other social worlds.

B7:   'So there's quite a lot of different things happening to shape that vision. But I just think that both in terms of the business plan for the trust and the vision for the camp, it was a very sound process. It was a <u>very</u> participatory, very open, very <u>engaging</u> process that created that. But I think that's changing now. And it's changing because priorities are changing in the village. Personalities are changing on the board' (CDT, Board Member 7: 170–175).

The quote exemplifies the relevance of human actors in two ways. First, the interviewee emphasises in the quote that environmental sustainability evolved out of a '<u>very</u> participatory, very open, very <u>engaging</u> process'. It is legitimated in terms of certain environmental reasons or rationalities but also with regard to its relations to relevant human actors in the CDT. Human actors are important elements not only because they create and realise the social order of the Trust. Instead, human actors also have meaning as sources of legitimacy. Furthermore, the quote explains that the position and meaning of environmental sustainability has changed *'because priorities are changing in the village. Personalities are changing on the board'*. The changed meaning and position of environmental sustainability are related to alterations in the teleologies and constellation of human actors. Consequently, while negotiations are based on the positions, relations, and meanings of human actors, any outcome of these constellations can only be temporary because interests and positions of actors are constantly changing.

Besides the board and staff, the working groups comprise important social worlds within the CDT. Local people participate in a working group when they are interested in its core activity. In terms of numbers of attendants and participants, gardening-related activities (allotments, orchard, horticulture groups) have been the most attractive groups. The individual reasons to participate in a group—the allotment group for instance—differ. Some people are interested because they do not own a garden despite living in a rural place. Some other members have a garden that—for different reasons—is not used or usable for growing vegetables or fruits. One commonly shared intention of most human actors for taking part in any of the activities of the CDT, however, is a social interest.

B1: 'I think there is a benefit in doing that in the end, otherwise you would die an old and lonely person. The benefit is that you meet people and you interact with them and you make friends and you share common goals' (CDT, Board Member 1: 31–33).

By engaging in the CDT, members become involved in a context in which people not only meet spatially but also socially. The working groups are social spaces in which people become related to one another because they '*share common goals*' and thus show commitment to the group's core activity. They are also positioned in a context that enables them to relate to one another socially—to '*make friends*'. Apparently, the social aspect involved in the activities is so prominent that it is assumed to be socially shared.

B3: 'One guy used to say: "She never stops working". And I thought that was a compliment. But it wasn't! It was a criticism ((laughing)). Some of them just stand and talk the whole time, particularly the men. They just stand and talk the whole time they're there, like it's just an excuse to get out of the house ((laughing)). They are all very friendly' (CDT, Board Member 3: 239–243).

The quote mentions two aspects in which social interaction is part of group activities. First, the interviewee describes the large amount of time spent on social interaction in the different working groups. Besides pursuing the group's core activity, the participants see their engagement as an opportunity to socially engage with one another. Second, according to the quote, this social aspect is deemed so important that non-participation in social interactions is met with irritation or even socially frowned upon. The social aspect does not only occur in the allotment group but characterizes all activities in the CDT. Irrespective of the group's specific core activity, social interaction is an important reason and motivation for people to participate in a certain working group or activity.

Like most activities pursued in the Trust, the production of renewable energy is a group activity. Energy production in the CDT is planned and run by a group of volunteers. While the members of this group '*share common goals*' with regard to the production of renewable energy for the community, their engagement is also a social activity in which they '*meet people and [...] interact with them and [...] make friends*'. The outcomes of renewable energy production in the CDT are consequently not only renewable energy and financial resources that can be put back into the community but also social contacts. While social interaction within the different groups generally, and in the Renewables Working Group specifically, is not always peaceful, energy production in the context of the CDT nevertheless comprises a social world that, besides commitment to the group's core activity, provides a social and time-space for interaction.

Being a working group of the CDT means that energy production is planned, organised, and run by volunteers. In terms of number of volunteers, the CDT has a very strong base of potential volunteer engagement. Around 600 of the members live in the community and are thus geographically positioned 'on site'. This proximity of human actors, however, does not ensure engagement in itself.

B6:   'Some will become involved because they are passionate about different things. Others will come in, do a little, and go away' (CDT, Board Member 6: 82–83).

The quote describes the problem of getting a sufficient number of people motivated to regularly spend a certain amount of their time on activities within the CDT. The volatility of both the numbers of people who are engaging at a certain time and the amount and stability of their engagement is especially unpredictable. Because of this volatility in the positions taken up by human actors in the context of the working groups, it is impossible for the Trust to pre-establish who is going to engage in which activity and thus what competencies will be accessible within a certain group. People do not join the CDT because they are sought out (as in hiring situations), but get involved if and when they are passionate about a specific activity, task, or project. The core activities of a working group thus not only need to be meaningful for the actors but have to remain meaningful in order to retain them in their position.

The volatility of human actors does pose severe problems for the CDT.

S2:   'Apart from the small amount of staff, everyone volunteers, and you only have so much time, and you're not <u>obliged</u>—you know, if you're in paid employment you're obliged to do things for a certain number of hours a week, whereas none of the volunteers are obliged to do anything. You don't have to' (CDT; Staff Member 2: 228–231).

The CDT is reliant on volunteers in order to organise, start, and continue its activities and projects. Only through voluntary engagement can the aims and visions of the Trust be realised. According to the interviewee, however, voluntary engagement causes two problems. First, volunteers have only so much time. Most volunteers need or want to spend large proportions of their time and energy in other contexts—like work, family, or other engagements. Consequently, they are not able to contribute as much time and energy to the Trust as staff members. Second, the quote points out the problem of the non-existence of rules. Unlike employees, volunteers are not obliged to spend a fixed amount of their time for the Trust. Implicit in this statement are two aspects of the non-existence of rules.

Because volunteers are not '*obliged to do anything*', they cannot be obliged to do anything at all or become engaged in a specific activity—which might be important for the CDT, but which does not interest them. The volatility of engagement is particularly problematic with regard to activities that demand long-term engagement and certain competencies, like running and maintaining the energy production equipment (Section 4.2.6).

As no rules can be applied, constant engagement needs to be secured by other means.

S1:   'Because people will go where their interests lie, and volunteers need rewarding. They do it for free, but they need some sort of incentive to get out of their armchair at night and turn up for a meeting, to sit at their computer all weekend, or to organise whatever it is that they're organising. Because they're motivated to do it, because they get something out of it, they enjoy it, or they enjoy the process' (CDT, Staff Member 1: 165–171).

As this staff member describes in the quote, motivation for engagement is related to private interest or motivation. The quote mentions 'individual' gain or reward as the main motivation for volunteer engagement. The statement that volunteers need to '*get something out of it, they enjoy it, or they enjoy the process*' indicates that a group's core activities need to be meaningful for a volunteer. According to the quote, groups are meaningful for volunteers not only because of the activities pursued in them. Instead, affective ends also need to be met. Being a volunteer activity as well, the production of renewable energy in the context of the CDT is based on being meaningful for a sufficient number of human actors to be constantly involved. Renewable energy production consequently is positioned not only within the context of the CDT, but also within volunteers' private lives, and is thus related to the working group members' private interests, as part of which it evolves. For these actors, the meaning of the working group is not just the realization of technical, environmental, or economic aspirations. Instead, the production of renewable energy in the CDT is also a 'social' social practice.

The production of renewable energy is intentionally related to the social site of the CDT and to other activities pursued in this site through the teleology of environmental sustainability. Despite the decision to remove it from the CDT's vision, the board and staff are nevertheless interested in promoting environmental sustainability within the different activities and working groups. Heterogeneous interests of human actors are integrated into the sustainability context.

S1:   'I think similarly, what you're saying about broadening out, the base of how you look at carbon consumption, "Actually, the fact that you grow your own vegetables, that's fantastic because that is saving x amount of carbon". If you can help tell other people that. So, I think it's about how you communicate. Even things like the Heritage Group. So I think, so the Heritage Group—that's one of the most active and sort of focused groups at the moment. But what do they do that's sort of carbon related? And actually we've had lots of conversations with people about how, during the war, people managed to reduce their consumption and people did live more lightly and they were more healthy and that kind of thing. I guess that's ultimately the Heritage and Futures aspect of the Trust is looking to the past and learning, even though what they did was pretty crap and landed us where we are, but also looking at the things that people are capable of doing and being inspired about that' (CDT, Staff Member 1: 472–482).

In this quote, sustainability is aligned with two different activities within the Trust—the allotments and the Heritage Group. In the first part of the quote, the interviewee describes her approach to make sustainability meaningful for individual people. Her general idea is that the crucial point is to make sustainability intelligible to individual human actors—in her words, '*I think it's about how you communicate*'. In her view it is important to get people positively engaged with sustainability by relating their individual activities and interests to the concept. Gardening activities, as they are practiced at the allotments, according to the quote are easily related to sustainability in an encouraging way. The same can be said for the other gardening-related groups, especially the community orchard and the horticultural group. These groups' activities are related to sustainability by not only intelligibility but also intentionality; gardening activities get related by thoughts and imaginings to sustainability. Gardening activities are indirectly related to renewable energy by intentionality in the sense that both activities are given meaning as sustainable practices. The different activities of gardening and renewable energy are furthermore directly related to one another through human actors. At the time of the fieldwork, most board and staff members were engaged in at least one of the gardening groups. All members of the Renewables Working Group are also engaged in other activities and working groups of the Trust. Within the context of the CDT, these human actors directly relate renewable energy production to other activities of the Trust.

The second part of the quote describes how the Heritage Group is also related to the concept of sustainability. The Heritage Group is elaborated upon in more detail, not only because it is one of the most active groups in the CDT, but also because it serves as a good example of how different activities are negotiated with regard to the environmentally sustainable teleology. Relating the activities of the Heritage Group to sustainability is first described as problematic in the quote. The interviewee then goes on to describe how the Heritage Group nevertheless does relate its core activities to the idea of sustainability. The group members relate their activities to the idea of heritage by describing how instances from the past could motivate and inspire activities in the present or the future. The aim to position the activities of the Heritage Group in the social site of the Trust is evident in an amendment to the Trust's principles, according to which the CDT aims to '*keep our eye on the future whilst learning from the past*' (homepage CDT). This statement exemplifies how the Trust negotiates the diverse social worlds and their core activities by relating the different activities to one idea, to which all groups can commit themselves.

At the same time, the Heritage Group strives to position itself within the CDT's general principle and its teleologies. On its webpage, the Heritage Group positions itself within the sustainability context of the CDT:

'Our plans for research around our collective stories from the past and the future [sic] and to be an economically sustainable community resource. Exhibitions will be diverse, covering the historical value of Cultybraggan Camp as a purpose-built WWII Prisoner of War; the rich social and physical heritage of Comrie village; the relationship between Comrie community and Cultybraggan Camp through time; the environmental work of Comrie Development Trust and how sustainable living today can learn from the past' (Webpage Heritage Group 2015).

As shown in the quote above, the past is argued to be a vital source of inspiration for sustainability requirements of today and the future. The idea is that activities of the Heritage Group today can uncover and utilize ideas and practices from the past that are of value for the future. In these statements, heritage is related to sustainable development. Heritage thereby is given meaning as a viable practice that is able to contribute to the realization of environmental sustainability. It is through the human actors' conscious efforts to negotiate and thereby stabilize the social site of the CDT that different activities are made meaningful with regard to the environmental sustainability practices of the Trust.

Human actors are the most important element in the social order of the CDT. First of all, the CDT's aim is to serve the community and thus the human actors who are the residents of Comrie. Second, the CDT and its activities would not exist without human actors who engage for and in the Trust on a daily basis. Most human actors who engage in the CDT are volunteers. This poses certain problems with regard to the predictability of the numbers, the dedication, and the competencies available for the Trust. The volatility of engagement is particularly problematic with regard to activities that demand long-term engagement and certain competencies, like running and maintaining the energy production equipment. As no rules ensure long-term engagement, volunteers only engage in the activities of the CDT when and if these activities are privately meaningful to them. Activities like the production of renewable energy in the CDT hence are private activities of human actors. Furthermore, activities in the CDT are always social activities; they provide a context for social interaction. Engagement in the Renewables Working Group thus can be argued to be a 'social' social practice. Because the actors who engage in the Renewables Working Group also take part in other activities of the Trust, renewable energy production in the CDT is directly related to these activities through human actors. Additionally, renewable energy production is intentionally related to other activities of the CDT's members through the concept of sustainability. Intentionality also relates different actors' activities to one another, for example if these are understood to serve the concept of sustainability.

4.2.6   Material artefacts: Cultybraggan Camp, energy technologies, and Hut One

This section shows how renewable energy production in the CDT is embedded in arrangements of material artefacts. These arrangements are not socially isolated but shape and are shaped by other elements of the social order and the social practices making up the social site.

Renewable energy production in the CDT is constituted by the existence of certain material artefacts. Without the necessary technological equipment, energy production would not be possible. While the production of electricity is based on solar panels, renewable heat is produced through a biomass boiler system. This system includes the biomass boiler itself, as well as the pipes through which the heat is transported into the connected huts. In order to make the system work efficiently, the biomass boiler needs to be connected to a sufficient number of huts via pipes.

S4:   'But it's actually quite expensive connecting up the huts. It is one of those things—digging the trenches and putting the—it's not cable, it's piping in there which is actually quite expensive' (CDT, Staff Member 4: 513–515).

According to the quote, the already described inefficiency of the biomass boiler system partly results from the fact that connecting huts to the system is too expensive for the Trust to afford on its own. Likewise, connecting the biomass boiler to places outside the camp would not be feasible.

B4:   'The cost is actually in the pipes. And the distance between places, so you could-you couldn't for example pump heat down here. From the boiler. Because it just wouldn't make financial sense' (CDT, Board Member 4: 46–48).

This quote explains the problem in terms of material artefacts and the costs associated with creating material connections between places. Connecting the boiler to the village would not be feasible because of the costs associated with creating the necessary 1.5-mile long pipe. The 'inefficiency' of the boiler is not a naturally given characteristic of the material equipment but results from its relation to other material artefacts and the costs associated with creating these relations. These costs are again problematic in the context of the CDT because of the already described difficult financial situation of the Trust. The estimation of the costs is likewise not naturally given but results from the specific context within which this estimation is made.

While connecting the boiler to the huts requires the material artefacts of pipes, the output of the system is also closely related to human actors and their practices. At the time of the fieldwork, very few people had decided to have their huts connected to the boiler system. While this was partly due to the high costs, it was also related to the kind of practices enacted in the huts. Users who practiced physical activities in their huts, especially small entrepreneurs like garage and workshop owners, users who used their hut as storage space, and users who used their huts only during the summer months—like the allotment and the horticultural group—did not require a heated space for their practices. Consequently, they were not interested in having their huts connected to the system. This situation also contributed to the 'inefficiency' of the biomass boiler system.

I: 'I've just been reading this feasibility study, and I think it's really about the number of huts you're providing for.
S4:   Yeah. Yeah. It's because we needed to be running at about 250,000 kilowatts of heat a year. We're at 100,000 or less. So we're at about 40% of where we need to be for it to be' (CDT, Staff Member 4: 510–512).

Part of the boiler's results from the misfit between the capacity of the boiler and the number of connected huts and consumers. The inefficiency thus appears not as a characteristic or outcome of the material artefact itself or its relation to other material artefact alone but is also causally related to the kind of social practices carried out in the camp.

The material artefacts of energy production in the CDT are related not only to the practices of potential consumers but also to those human actors who are going to run and maintain the systems.

S3:   'So the Trust, as you can see, with small staff, not full time, was not willing to take on any risky projects. You could say with biomass that it's been a bit of a shock that it turned out the way it has done, because it was sold to us as proven in many years of experience, very reliable boilers and everything else. So solar PV, more recent but has still been around for a while and price coming down, has lower maintenance—no maintenance. The idea of an anaerobic digestion, which needs a lot of maintenance, would put too much risk on the Trust, which is largely a volunteer operation, to be running the risk of something going wrong' (CDT, Staff Member 3: 113–122).

The quote describes the difficulties resulting from the interplay of technical equipment and the characteristics of the Trust, which is a volunteer organisation. Being a volunteer organisation, the Trust does not have the knowledge, skills, and capacities to deal with technologies that require much maintenance or are deemed risky. What constitutes a too high level of maintenance, is defined by the necessary amount of volunteer input. As it is not possible to constantly secure high levels of volunteer engagement, the material artefacts comprising a certain renewable energy production system need to be low-maintenance. Choosing a certain technology is thus based on the interplay between characteristics of the material artefacts and the characteristics of the human actors who would run and maintain the system.

The 'riskiness' of a technology, on the other hand, results from the type of capacities necessary in relation to the characteristics and features of the material artefacts. Risk in this quote becomes visible as an assessment of the situation, which again results from the interplay of a certain technology and the human actors and their capacities. This riskiness derives not only from the necessity to secure long-term commitment, but also from the kind of highly specific skills and knowledge needed for certain types of technical equipment. Being a volunteer organisation, the Trust cannot ensure the constant accessibility of the necessary competencies. Energy production in the CDT takes place in a context shaped by concerns about the competencies and capacities that are available. It is moreover based on volunteers as a certain type of human actor engagement. Being based on this type of human actor engagement influences the type of technologies deemed suitable for the context. Resulting from these considerations, new and innovative or more complex systems are dismissed for the context of the CDT. Instead, reliability and low and easy maintenance are estimated to be important features of technological equipment.

Energy production in the CDT furthermore takes place within the material artefacts of Cultybraggan Camp. The camp in fact is the most important agglomeration of material artefacts in the CDT. Since its early stages, most activities of the CDT have been focused on or have taken place in the camp. The solar panels and biomass boiler are installed in the camp and provide some of the huts with electricity and/or heat. The idea to produce and use renewable energy in and for the camp dates back to the beginnings of the Trust.

S4:   'I suppose there is the idea that—there's the whole idea of the camp being sustainable. It's the whole idea of having a district heating system there and have some solar power. That was another whole theme that was being looked at in those first two or three years of the CCF. It was also to look at those areas like the hydro and wind and—you know, to look at renewable sources that were potentially available and then focus on ones that could actually in practice be usable' (CDT, Staff Member 4: 448–452).

In the first two sentences of the quote, the production of renewable energy is related to the CDT's aim of developing a sustainable camp. A relationship is set up between sustainability, the district heating system, and solar power. Thereby, the technological equipment is related to and given meaning to the Trust's aim to develop the camp in a sustainable way.

While the second part of the quote identifies further potentially sustainable technologies, it does relate these to their '*practical usability*'. This practical usability results from the relation of the technological artefacts to other aspects of the context within which it is meant to be realised. With regard to what is described above, the practical usability of a certain technological installation is connected to those human actors who run and maintain the system. In addition, the practical usability of a certain technological equipment derives from its relations to financial aspects and the local environment. Both aspects are elaborated upon in more detail below (Section 4.2.7). For this section, it is important to note that the 'practicality' of a certain technology derives from its position in and thus its relations to various other elements of the social site into which it is embedded.

The decision to install the material artefacts of energy production in the camp can be explained in reference to the other elements of the social site. The decision to produce *heat* at the camp is related to the fact that the camp is not connected to the heat grid. It is inspired by the non-existence of other material artefacts that would provide the same service. Second, certain social practices that are projected to take place or that are already being enacted at the camp require the availability of heat. While the absence of a heating system or pipeline has causally contributed to the installation of the biomass boiler, it is also pervasively constituted by those social practices that take place or are projected to take place at the camp and that require warmth. These practices are mutually prefigured or even essentially constituted by the existence of a heating system— they could not take place at the camp otherwise. One example is a mushroom-growing enterprise in one of the huts. The mushrooms require constant maintenance of a certain temperature. Without the existence of the biomass boiler system, the owner of the business would not have decided to rent one of the huts in the camp. Furthermore, because of the already described costs associated with connecting huts to the system, the mushroom-growing hut is located close to the hut in which the biomass boiler system is installed. The practice of growing mushrooms here is essentially constituted by the existence of the material artefacts through which heat is produced and distributed.

The existence and specific material form of the renewable heat installations at Cultybraggan Camp, however, are a contested issue. The installations have to be negotiated with other material artefacts and their meaning within the social orders of different social worlds in the CDT. In the following quotes, the participants of the Renewables Working Group discuss the options for extending the range of solar panels at the camp. Because of the round shape of the huts, it is not possible to install panels on their roofs. This option is hindered by the mismatch between two types of material artefacts—huts and solar panels. Two alternative options are discussed. The first is to install the panels on a construction, heading the rifle range.

S2:   'They were saying that the kind of costs and the effort of maintaining it would be huge. The cost of building the structure would be huge, and probably make the whole thing quite unprofitable' (CDT, Staff Member 2, Meeting of the Renewables Working Group: 207–210).

Because of the necessary extra constructions, this option would be more expensive and require more maintenance work, as it would be a target for birds. This combination is infeasible due to extras costs resulting from the necessary combination of different material artefacts (roof construction and solar panels). The interaction of birds (non-human actors) and material artefacts—requiring more maintenance—is not practicable for the volunteer (human actors)-based CDT. The non-viability of the mounting option results from the relations of the material elements of the solar panels with other elements of the local and social context within which they are embedded.

The second option discussed by the Renewables Working Group is to install ground-mounted solar panels. While being the most feasible option with regard to the financial and maintenance considerations, this option problematic because of existing conflicts about the usage of the camp. These conflicts result from the necessity to negotiate different social practices that are being or will be enacted at the camp. Elements of the camp are positioned, related, and given meaning within these social practices in different ways—creating different social orders with regard to the camp. In particular, the mesh of social orders and practices created by the social world of the Heritage Group conflicts with various other social worlds and their relations to the camp.

S2:   'They seem to think ground-mounted is most feasible—I mean, it's going to take up a lot of space; I mean, really, how do you think that would go at the camp?
GM: On the camp itself?
S2:   I mean, to leave the heritage—and it's basically saying, "You're not going to use that for anything else. That's it"' (CDT, Staff Member 2 and Group Member, Meeting of the Renewables Energy Working Group: 248–255).

The quote indicates how material artefacts can become a contentious issue with regard to the different activities and teleologies pursued at the camp. Ground-mounted solar panels would require a lot of space. This kind of usage and the material artefacts taking up the space are not compatible with any practice other than energy production, as they would basically prevent them. Ground-mounted solar panels are not negotiable with any other social practice.

More specifically, erecting solar panels would conflict with the social practices and orders of the heritage group. In the quote, this problem is shown in the phrase '*I mean, to leave the heritage*'. Apparently, the problem is so obvious for all people involved that it does not need any further explanation. In fact, the conflict between the Heritage Group and some other social worlds like the Renewables Working Group or the different gardening activities was the most virulent conflict at the time of the fieldwork. The conflict derives from the problem that the Heritage Group advocates a different social order for Cultybraggan Camp, whereby the camp's material artefacts are given a meaning that is not compatible with many other ways of using them.

GL: 'So our group started off but I think we had slightly different angles to things. [...] And we are very keen to preserve the camp as it is, without too much change, because things have been changing an awful lot' (CDT, Group Leader: 87–89).

The social practices of the Heritage Group are organised in order to preserve the social order of the camp as a heritage site. Within the social order of the Heritage Group, the material artefacts comprising the camp should either remain unchanged or only change in a way that does not alter their historic character. For the Heritage Group, the meaning of the camp and the individual huts derives from their historical background. It is the core teleoaffectivity of the group to preserve this meaning. The historical social order of the camp potentially conflicts with the material necessities of renewable energy production. The interviewee does not oppose the legitimacy of sustainable development in general. Instead, the conflict results from the mismatch of the two different kinds of material artefacts—those that are 'historic' in her sense-making and those that cannot be positioned within the historical social order. The installation of solar panels, especially on the ground, would severely alter the appearance of the historical buildings in a way that is not compatible with historical sense-making of the buildings. The material elements necessary for the production of renewable energy cannot consequently be positioned in a historical arrangement of elements. The only possible way to meet the demands of the Heritage Group would be to use technical equipment that suits the historical standards.

GL: 'They should devise solar panels that look like corrugated iron, which is—I don't know—it would curve around' (CDT, Group Leader 302–303).

The Heritage Group—according to this statement—would not oppose the production of renewable energy if the material artefacts would leave the historical character of the huts undisturbed. The conflict is not based on the incompatibility of the social practices, but on the opposing social orders that are required, created, or used by these social practices. While using the same site, the two practices require a different arrangement of elements within this site.

Successful negotiation between the different social orders, however, is possible. An example of such a project is the refurbishment of Hut One into a Heritage Centre. The building has been refurbished in collaboration with an architect, using and showcasing four different types of insulation material.

S1:   'So Hut One is going to be completely stripped back, re-insulated with four different types of insulation. There's going to be a wee[17] window in the wall so that you can <u>see</u> the type of insulation there is. There will be an interpretation panel to say "This is sheep's wool. It has properties like this, and it costs blah blah blah" and tell a wee <u>story</u> about that. At the same time, the room will be filled as a visitor centre. Actually, what's really nice about that is that it's a really good example of what we wanted the <u>big</u> visitor centre to be. A Heritage and Future Centre. We want to look to the past and preserve the past, but look to the future as well and learn from the past and celebrate what we're doing positively now' (CDT, Staff Member 1: 709–717).

In the Hut One project, it became possible to position elements from a sustainable social order in the social order of the Heritage Group. The alignment of heritage and sustainability, however, was possible only because they could be integrated without obstructing one another. The insulation materials do not change the character of the hut, especially since they remain largely invisible. Having *'an interpretation panel to say "This is sheep's wool. It has properties like this, and it costs blah blah blah" and tell a wee <u>story</u> about that'* does not conflict with the historical exhibition and the information panels that constitute the buildings social order as a heritage centre. Information panels about insulation and even the biomass boiler do not fundamentally alter or obstruct the meaning of the historical installations. For the same reason, it is possible to heat the heritage centre with heat produced by the biomass boiler. Neither the boiler nor the pipes are visible and thus they do not obstruct the historical appearance of the hut. Hut One can serve the purposes of heritage exhibition and sustainable development at the same time, because elements from the different social orders (insulation material, pipes, the biomass boiler, and heat flows from the renewable energy social order versus general appearance and usage of the hut with regard to the historical social order) could be integrated with one another. Integration of the social orders at the same place also means that the respective social

---

17   Scottish dialect for small.

practices of renewable energy production and the Heritage Group could be successfully negotiated in this project.

As explained before, the Heritage Group does make conscious efforts to relate itself to the sustainability aim of the CDT. By developing the principle of learning from the past to enable a better future, the CDT has created a discursive element by which the two different social worlds can be made meaningful to each other. The Hut One project furthermore relates the social worlds of heritage and sustainability by means of material arrangements to one another. Connecting the Heritage Centre to the district heating system creates a direct material link and a constant material flow between the two different social worlds.

4.2.7 The local environment: wind, fish, and how they are made sense of

Renewable energy production, as a 'non-virtual' or 'embodied' technological activity, has to take place somewhere. As seen in the last section, material artefacts in particular are positioned in and related to other elements and their meanings within physical space. Energy production is further dependent on its environmental context because it needs to utilize aspects of this context, like sun or wind, to produce energy. The natural context needs to be favourable to the requirements of a certain technology.

Wind turbines are the technology most often employed by development trusts and comparable organisations in Scotland (see 1.3 Research on civil engagement in local energy production). In the case of Comrie, however, the valley is situated in a so-called wind-hole. This means that the quantity and quality of wind do not allow a feasible installation of wind turbines.

S4: 'Wind is an obvious one but the Trust land is not helpful for wind. So then you come back to the biomass and the solar for the camp. And then the hydro. I think it's being looked at, the little bit of water that runs through the hill ground, but it's not really suitable for it. So we're looking at some other possibilities for hydro, but they're all quite complicated around the flows' (CDT, Staff Member 4: 453–459).

It transpires from this quote that the production of renewable energy is constituted by elements of the natural environment that are relevant with regard to a certain energy production technology. Because they need to be directly available to the technical installations, natural 'things' like sun, water, or wind essentially constitute energy production from renewable sources. While the wind situation is not '*helpful*' for the installation of wind turbines, the flows of water in the area are unfeasible for potential hydro schemes. The technological artefacts for the production of renewable energy interact with their natural environment. The production of renewable energy takes place in a context shaped by the interaction of elements in the natural environment and technological artefacts.

Being a human practice, energy production in general and the interaction between nature and technology in particular take place as part of complex practice-order arrangements. It is not any 'objectively given' natural environment that makes a certain natural aspect like the amount of wind or flow '*suitable*' or '*helpful*'. Instead, the outcome of the interplay between a technology and the natural element is made meaningful with regard to other elements of the social order. The administrative, financial, and social aspects of the context are pervasively involved in constituting the 'practicality' of energy production. A good example of how the natural environment and its relation to a material artefact are embedded into a social context can be derived from the CDT's efforts to find a suitable place for a hydro scheme.

S2:   'We looked at the hydro at the hill ground, which turns out it was just too small, too little slope, there weren't any impassable barriers so any fish could get up, and it really just wasn't viable in terms of the costs' (CDT, Staff Member 2: 92–95).

The quote reveals that the non-practicality of the hill ground results from the relation of an aspect of the natural environment (the slope) to a certain non-human actor (fish). The low slope has created a natural environment for fish, which is considered as worth protecting.[18] The installation of a hydro scheme is not so much hindered by a naturally given physical misfit. Instead, the installation is made impossible by human sense-making of a specific situation (fish being considered worth protecting). It is with regard to this situation that the technological artefacts of renewable energy production are given meaning as endangering or destroying this element, perceived as worthy of protection.

An important element of the social order within which the natural environment is embedded are financial norms and ideas. Financial considerations are closely related to energy production and importantly influence a technology's meaning. Financial considerations define whether or not a natural environment is seen as feasible for a certain renewable technology.

S2:   'We heard of another site through one of the companies that could potentially be explored at the site of an old trout farm. And we visited that with Community Energy Scotland—to see if that had potential for getting funding. But it was found that, again, the cost of running it—you'd have to purchase the land, and the Scottish Land Funds don't allow you to claim FITs if they'd given you money for the land. And Community Energy Scotland could only give you grants and loans and

---

18   I do neither intend to criticize nature protection in general or this specific instance of nature protection. Instead, I want to make the point that it is not a natural condition in itself that leads to the non-viability of energy production technologies—it is much more an outcome of human conceptions of nature, i.e. defining what aspects of nature are to be protected.

> things if you were able to collect FITs. So basically any money that would allow you to purchase this, you wouldn't be able to—it would take so long to pay it back, and you wouldn't be able to actually make any money from the hydro scheme. You would just be able to use the electricity' (CDT, Staff Member 2: 100–109).

The quote shows that renewable energy production is not foremost estimated with regard to environmental elements. Within the existing system it is not enough to '*just be able to use the electricity*'. Instead it is necessary to '*make any money from the hydro scheme*'. The emphases on the words '*just*' and '*money*' show the decisive role of financial outcomes. Within the described complex situation, it is the financial outcome that defines whether the technology is viable and can be realised. Like environmental aspects, financial feasibility thereby becomes visible as an outcome of the complex relations of different elements. In this case, options for upfront investment grants do not match the requirements for FITs that define the financial income and thus the long-term financial viability of the technology. Financial issues are interwoven with natural environments; the output made possible by the natural environment is calculated in terms of the amount of money that can be earned by this output. This financial result of produced energy is furthermore calculated in relation to the necessary upfront investment and maintenance costs. As described in the quote, both upfront investment opportunities and financial output are an outcome of specialized policies. The Scottish policies that support community landownership and the FITs respectively clash with each other in this case. It is because of this conflict and not the natural environment or the potential of the hydro technology that the production of renewable energy through a hydro scheme is unfeasible at this place.

Cultural aspects like human perceptions of the natural environment are another aspect influencing the relation of local environment and renewable energy production technologies. (Sub-)cultural meanings of environmental aspects influence how energy production technologies and practices are perceived and assessed.

S1:   'Oh look. I was just going to invoke the 'W' word—wind farms, wind mills. [...] That's a dirty word around here. [...] One of the guys who is one of our volunteers came in and he's so into the whole Carbon Challenge thing. But SM2 mentioned wind mills, and he was like, 'phew' and he pulled a chair out and sat down and they were there for a good—I mean, all very good natured, but he's so anti-wind farms. Because he likes hill-walking and it spoils his view. So he's quite happy to have solar farms. It's not about land use, it's not about that. He's quite happy to have solar farms because you can't really see them, because they're flat. But he doesn't want to go up into the mountains and see these windmills' (CDT, Staff Member 1: 503–517).

Within the sense-making of the person discussed in this quote, the highly visible technical equipment—wind turbines—is destructive to what is important for him. This estimation of the effect of wind turbines results from his relationship to different aspects within the environment, like hills and viewpoints. In his social order of the local environment, wind turbines would '*spoil his view*'—a thing apparently highly valued by him. In his sense-making of 'unspoiled nature' he draws on existing forms of social sense-making. While the natural environment of the Scottish Highlands has been formed, changed, and influenced by human activities for centuries, the country is perceived to have 'unspoiled nature', which would be destroyed by the installation of wind-turbines. The cultural perception of the Highlands and their unspoiled nature—as voiced by the hill-walker for example—is an important aspect influencing the realization of renewable energy technologies in the CDT.

Besides the insufficient physical conditions for wind turbines in the Comrie valley, the anticipated conflict potential with regard to wind turbines and their effects on the local environment is a crucial reason that the CDT is not pursuing options to erect wind turbines on one of the surrounding hills or mountains. Wind turbines are not avoided for 'objectively' given natural environmental aspects, but because of human forms of sense-making with regard to this natural environment. For the social world of the 'anti-wind actor group' the meaning of the local environment derives from its perceived 'wildness', which has to remain 'unspoiled'. For the staff and board of the CDT, the meaning of the local environment is also constituted with regard to the Trust's economic and social teleologies. Apparently, however, the social world of the anti-wind group has been strong enough to dominate the negotiation process—so much so, in fact, that it was neither necessary nor possible to have a public debate about it. Wind turbines comprise an example of a situation in which diverging teleoaffectivities and understandings about energy production and its effects on the natural environment have to be negotiated among different social worlds. As the CDT consists of different actor groups with differing and even opposing interests and worries, renewable energy production here is an outcome of a highly conflictive negotiation process.

The outcomes of this negotiation process are unstable and constantly contested. Consequently, new arguments can destabilize a consensus at any time.

S3:   'It [a feasibility study about the different options for renewable energy production, AP] looked at other sources, for example wind. But the problem apparently is that this part of Comrie and the area roundabout is what they call "wind-hole." So there isn't sufficient wind. Now, in recent years, people have gone back to question that assumption. So it's not to say that sometime in the future wind wouldn't be looked at again. As you know, it's contentious—visually—but in recent years, people have come forward with smaller turbines' (CDT, Staff Member 3: 6–11).

The interviewee presents two arguments that evolved out of recent findings and developments. The first argument relates to the physical conditions and their interplay with wind turbines. While the wind conditions in Comrie and at the camp have not changed, members of the Trust have proposed new potential sites that could be tested. Second, recent technological developments have increased the technical efficiency of wind turbines. These developments have altered the relationship between the necessary amount of wind and the generated electricity. These two arguments have been introduced into the social arena in which renewable energy production via wind turbines is being negotiated, in order to offset the 'technical' or 'physical' arguments against wind. The second argument is directed towards the social world that argues that wind turbines would destroy the landscape. The increased efficiency of wind turbines also means that much smaller wind turbines have become possible. Being much less visible, small-scale wind turbines at least partly invalidate the argument that wind turbines would destroy the natural appearance of the local environment. The interviewee in the quote above uses these new developments to challenge the arguments of those social worlds that have argued so far against the installations of wind turbines for 'natural environment' or financial reasons. Especially with regard to the dominance of the anti-wind groups that have prevented an open discussion about wind turbines in Comrie, the quote shows that negotiation outcomes are unstable and constantly contested, for example by new arguments. These contestations might even change the position of social worlds and their arguments within a certain social arena.

Summarizing the argument, objectively given natural properties do not underlie the use of a certain technology. Instead, the perception of the natural environment as 'feasible' for energy production is intertwined with human perceptions of nature, with the interplay of the technology (as it has been designed) with this natural environment, and with other conditions that 'make' a certain technology feasible or non-feasible in the natural environment within which the CDT is situated. The natural environment does not precondition certain energy production technologies by enabling or preventing them. Instead, natural entities are related to energy production in their interplay with other aspects of the context, making them more feasible, more desirable, or less 'intrusive'. This also means that the production of renewable energy in the CDT is shaped and influenced by a number of aspects related to the natural, financial, administrative, and cultural contexts forming the social site within which renewable energy is being produced.

4.3 Interim conclusion: Sense-making of renewable energy production in the CDT

Within the CDT, renewable energy and energy production are mainly made sense of with regard to the aim to service the community. How and in what ways the community is served (best) is, however, a heavily contested issue among the different social worlds of the Trust. Renewable energy production is made sense of in relation to the community aim in two ways. First, renewable energy is part of the Trust's sustainability agenda. Reduction of the village's carbon footprint featured heavily in this agenda. Being part of the CDT's sustainability agenda, the production of renewable energy is not so much an end in itself but derives its meaning as a sustainable practice in combination with activities like carbon savings and energy efficiency measures. The sustainability agenda got so much challenged by other social worlds in the CDT that it had to be abandoned in a strategy revision process. While the social world of the board and staff still make sense of renewable energy production as a sustainable practice, they have altered the ways in which the related activities are made intelligible to other social worlds in the CDT. The activities related to this agenda are now made intelligible to the community members as social practices which enable money savings. Adapting its 'official way' of making these activities intelligible to the sense-making of the community members serves to momentarily stabilize the context of the CDT.

In a more direct way, renewable energy production in the CDT is meant to service the community by producing financial income. This income could then be put back into the community and be used to realise the members' interests. Again, energy production becomes visible as being not an end in itself but an instrument to realise the CDT's community aims. The meaning of energy production with regard to the CDT's financial teleology, however, is ambiguous. Because of the inefficiency of the installed biomass boiler, renewable energy production in the CDT has come to be made sense of as part of the Trust's debts. These debts are an obstacle to the CDT's ability to realise its community aim. The financial outcome of renewable energy production is shaped—among others—by existing rules (FITs, RHI), the interplay of the material artefacts of energy production with those human actors and their social practices which are meant to become 'consumers', and the financial costs of connecting these consumers to the heat grid.

Renewable energy production in the CDT is furthermore shaped by its relations to the human actors, material artefacts, and the local environment of the Trust. Being a community organisation, the CDT is dependent on volunteers. This situation shapes the practices of energy production in the Trust. The CDT can only install technical equipment, which can be maintained and run by a potentially changing group of people without professional knowledge or much time. Being enacted by volunteers as part of their private life also means that energy production in the CDT is a 'social' social practice, within which members from the community have the opportunity to meet and create social contacts with one another. Energy production thus is not only shaped by the human actors but also influences these human actors, i.e. their interactions. With regard to the material artefacts, energy production at the Cultybraggan Camp is limited by the fact that different ways of sense making of the camp exist. The installation energy production equipment has to be negotiated among the different social worlds of camp users. Last but not least, energy production in the CDT is shaped by the local environment of the Trust. The interplay between the local environment and energy production (devices) is shaped by existing rules (which for example regulate to be protected animals), norms (which define the 'viability' of energy production), and values of 'unspoiled nature'. It is not the local environment as a naturally given entity, which shapes the production of renewable energy. Instead different actors' ways of sense-making of the local environment have to be negotiated within the CDT. The outcomes of these negotiations are always only temporary and might be challenged by newly occurring arguments or (technical) developments.

4.4 Describing KEBAP

The culture and energy bunker (KEBAP) is a project planned and realised by a German grassroots initiative. The bunker is situated in Hamburg, Germany. The project is located in a so-called social preservation district in Altona-Nord. The aim of Hamburg's social preservation regulations is to counteract displacement processes caused by appreciation pressures. Areas subject to strong appreciation processes can become social preservation areas in order to maintain still-existing affordable housing and to protect the traditional social structure of the inhabitants. Being located in the centre of a growing city, the district Altona has experienced increasing gentrification and displacement in the last few years. Nevertheless, supported by the social preservation regulation, the area neighbouring the bunker is still mainly inhabited by lower and lower-middle income groups.

4.4.1   The bunker

Not much is known about the history of this particular bunker. So far, no organisation or initiative has attempted to collect historical material or information about the building. The bunker is more than 50 meters long and nearly 19 meters high. It is split into six floors, all of which are separated into more or less small rooms. It was built between 1941 and 1943 during the 'second wave' of the bunker building programme in Hamburg (www.geschichtsspuren.de, 29.10.2015). During the last years of the Second World War, it provided space for 1,650 beds and 165 additional seats. After the War, the bunker was put out of military service and in 1946/47 was used to provide a heated room for displaced or bombed-out people. Later, it was used as a storage space. In 1957, the first ideas to refurbish and modernize the building in order to use it as a shelter again were abandoned. Between 1970 and 1974, however, these plans were taken up again and realised (Hamburger Unterwelten, 28.10.2015; www.geschichtsspuren.de, 29.10.2015). The bunker became one of 2,000 shelter places in Germany still in service (www.bbk.bund.de, 29.10.2015).

Since its reconstruction in the 1970s, the bunker exists in its actual internal shape. It is now a 'twin bunker' with two identical self-sufficient parts, each of which provides space for 1,755 people (Hamburger Unterwelten, 28.10.2015; www.geschichtsspuren.de, 29.10.2015).

In May 2007, the Federal Ministry of the Interior decided to terminate the nation-wide shelter-room concept. Due to the technological development of war machinery, the bunkers were no longer able to provide comprehensive protection against modern warfare (www.bbk.bund.de, 29.10.2015). Since 2009, the reverse transaction process of the shelters and bunkers has been going on. The bunker of the KEBAP project was released from military service in 2014 and its legal ownership was transferred from the Federal Ministry of the Interior to the Institute for Federal Real Estate. This Institute was commissioned to organise and carry through the sale of the building.

4.4.2   Prologue to KEBAP

The person who developed the original idea for using a bunker to produce renewable heat later became one of the founding members of the KEBAP project. Having a university degree in communication design and being a trained joiner, the future board member had become interested in heat production and distribution for private reasons.

B2:   'I had been developing a plan for a district heating system in my parent's village.
      So I had the basic knowledge. And storage rooms for wood fuels are called "bun-
      kers". Because you need a fire proof room in order to store wood […]. Okay, and
      then 2009, in the beginning of 2009 I had the idea, while cycling through the city,
      I had this idea: bunkers for shelter, bunkers for wood fuel. Why not install an
      energy plant in a bunker, using wood chips' (KEBAP, Board Member 2: 61–72)?

The original idea for the bunker project was to use the building for a renewable
energy technology based on wood pellets which would require a certain amount
of space to store the necessary amount of wood-fuel. The idea to use wood as a
source for heat production, was inspired not only by the professional background
of the interviewee but also from the realization of how much waste-wood remains unused in Hamburg.

B2:   'And when I made inquiries, I learned that there is relatively much green and wood
      waste here in Hamburg, which at the moment is being transported away or being
      left to rot. […] Well, this does not need to happen. So, if using regional energy
      sources, then these should be as regional as possible' (KEBAP, Board Member 2:
      96–102).

Hamburg has approximately 850,000 trees in its parks, in green areas, and along
the streets. It owns about 3,400 square meters of forest in and around the city.
Caring for these trees yields thousands of tonnes of green wood and wood waste
every year. About 50,000 tonnes thereof are suitable for providing biomass for
energy generation (Landwirtschaftskammer Hamburg 2009: 20). Based on such
information and inspired by the realization that bunkers are *unused spaces in
the city*' (KEBAP, Board Member 2: 731), the interviewee developed a concept
for a woodchip-based biomass heat production system within the bunker.
About the same time, he also became active in a citizens' initiative for fighting
against plans of Vattenfall and Hamburg's government to build a new long-distance heat pipeline across Hamburg's city centre. Within the context of this initiative, he presented his concept to a group of other people.

4.4.3   Two conflicts about power: the Moorburgtrasse-Stoppen initiative and the re-municipalization of the heat grid

In 2009, Vattenfall, a Swedish state-owned energy company and one of the four
biggest energy providers in Germany, got approval to build a new coal power
plant in the outskirts of Hamburg. To transport heat into the city centre, Vattenfall planned to construct a new pipeline. The technological features of this pipeline (width of the pipes, temperature and pressure of the water pumped through)
would have given Vattenfall a quasi-monopoly over the heat distribution in large

parts of Hamburg. Furthermore, the pipeline was supposed to run through most of Hamburg's inner-city parks. During the building processes, most of the parks would have had to be (temporarily) torn down and old trees would have been cut down. While the destruction of the parks provoked an emotional response in people, what really outraged most was the quasi-monopole Vattenfall would gain by building the pipeline. Besides the existing public frustration about Vattenfall's new coal power plant in Hamburg, these pipeline plans aroused a lot of anger among Hamburg's citizens. A number of local citizens' initiatives were formed. Most of these started out as neighbourhood initiatives around the green areas and parks included in the pipeline plans. Together they created the umbrella initiative 'Moorburgtrasse-Stoppen'. Supported by environmental and other civil society groups—most prominent among them Friends of the Earth Germany (BUND)—these initiatives increasingly gained public attention and media coverage. Under the leadership of the BUND, different groups decided to take legal action against Hamburg's administrative procedure. Originally, the pipeline plans had been dealt with by the administration only in the form of a simple plan allowance ('Plangenehmigung'), which does not allow for any participation of citizens or environmental organisations. While at first the complaint was dismissed by the Administrative Court, this verdict was later overruled by the Higher Administrative Court in hindsight of the actual Federal Administrative Court's jurisdiction. In February 2010, the Higher Administrative Court declared that, because the damage to the environment would be much greater than originally projected, Vattenfall and Hamburg's administration had to implement a formal plan approval procedure (www.altona.info 5.3.2010). The requirement of this much more sophisticated procedure severely delayed Vattenfall's building plans. Also, it strengthened the position of citizen and environmental organisations in the conflict, which were now given the opportunity to officially take part in the process.

About a year later, in January 2011, Vattenfall submitted the necessary documents to the responsible municipal agency for urban planning and environment to officially start the plan approval procedure. From May to June 2011, the documents were publicly displayed. In this one-month period, about 4,250 formal objections were raised by different individuals and organisations (www.moorburgtrasse-stoppen.de 29.6.2011). These were to be discussed in a public hearing on November 2011. In the end of November, the SPD—the governing party in Hamburg—declared in a press conference that the pipeline plans would be put aside until the referendum about the re-municipalization of the heat grid in September 2013. If in this referendum the citizens would decide against the re-municipalization of the grids, the pipeline plans would be abandoned conclusively in favour of a new combined cycle power plant. If the referendum would decide in favour of the re-municipalization, the plan approval procedure would be resumed.

This plan is to be explained with the financial interests of Vattenfall. The re-municipalization of the grid meant the end of Vattenfall's quasi-monopoly over the long-distance heat grid in Hamburg. Before the referendum, not only was Vattenfall the biggest producer of heat in Hamburg, it also owned the heat grid. Unlike electricity grids, the heat sector has not been liberalized. This legal situation enabled Vattenfall to close the grid to any other potential heat producer. If the referendum would be successful, and the municipal grid operator would open the grid to other energy providers, the new power plant would not be financially viable for Vattenfall. In an effort to discourage citizens from voting for the referendum, Vattenfall threatened not to proceed with its plans to build the new combined cycle power plant if the referendum would be successful. The SPD supported Vattenfall's position by arguing that without waste heat from the new combined cycle power plant, Hamburg would be dependent on the pipeline from the older coal power plant for heat provision. This argument outraged many people as that coal power plant had been severely objected to by many citizens in the first place. They felt that they were now being threatened by the SPD to stop the referendum as otherwise they would involuntarily support a coal power plant they had objected to in the first place. Instead of succeeding in demoralizing the organisations involved in the re-municipalization campaign, the SPD managed to heighten the citizens' emotional outrage against their energy policy. Many people increasingly felt that the party was not acting on behalf of the citizens but was acting in league with Vattenfall. One outcome of this arrangement, however, was that while the pipeline plans were not terminated, they were delayed for the second time by the decision of the SPD to delay the plan approval process until the referendum (www.moorburgtrasse-Stoppen, 30.11.2011).

The referendum that caused this second delay had been formally requested by the 'Unser-Hamburg-Unser-Netz' (UHUN) network in July 2010. The referendum aimed to force Hamburg's government to buy back the electricity, gas, and long-distance heat grids. The UHUN network had been founded in early 2010 through the coalition of six partner organisations: the Moorburgtrasse-Stoppen initiative, BUND, attac, the protestant welfare organisation 'Diakonie', Robin Wood, and the Hamburg consumer advice centre. The network aimed to force Hamburg's government to re-municipalize the grids by regaining the concession rights from Vattenfall, which owned the heat and electricity grid company and E.ON, which owned the concessions for the gas grid. The UHUN initiative raised awareness for the political and juridical situation that granted Vattenfall a monopoly over the heat production and distribution. According to schedule, contracts for the grid concessions for electricity and long-distance heat would come to end on 31 December 2014. According to the contract, the gas concession would end in 2018 with a special right of cancellation effective till 31 December 2014. As Hamburg's government did not intend to regain concession rights of its own accord, the initiative aimed to force it to regain the concessions for these grids from the private energy companies by means of a referendum.

A public referendum in Germany consists of three parts. A first initiative ballot ('Volksinitiative') decides the start of the second step—the popular petition ('Volksbegehren'). A referendum ('Volksentscheid') can be held only if enough people subscribe in favour of it in the popular petition ballot. In each of these steps, a major percentage of the local population has to subscribe for the initiative. After having successfully collected enough subscriptions in the popular petition in only three weeks in June 2011, the referendum was scheduled to be held as part of the federal elections on 22 September 2012. In the summer of 2012, the Hamburg government, in reaction to the huge success of UHUN network activities, had already purchased 25.1 percent of the grid companies (www.unser-hamburg-unser-netz.de, 28.10.2015). Not satisfied with such a minority partnership, UHUN argued that this agreement did not meet the citizens' interests in the least. Instead of bringing the provision of energy—as an important aspect of basic modern human sustenance—back into public ownership, it would only provide Hamburg with a blocking minority. It was argued that a municipal operator subject to democratic control would be much more committed to social and environmental aims than a private company mainly driven by economic incentives. Furthermore, a minority ownership would not entail a share of Hamburg in the financial income of the grid operator. If, on the other hand, the grid operator would be a municipal company, the income generated through the grid would be kept within Hamburg.

The UHUN campaign and the Moorburgtrasse-Stoppen initiative did profit hugely from each other throughout the more or less parallel processes of working for the referendum and stopping the new pipeline. Public awareness and media attention raised by one of the campaigns usually also benefitted the other campaign as well. Apart from the overlaps in participating organisations and individuals, the two campaigns also had the same opponents—Vattenfall and Hamburg's government. In their campaigns, both initiatives were able to make use of the generally bad reputation Vattenfall already had environmentally interested people Hamburg.

M4:   'So, in Hamburg for years there have been conflicts between the energy movement and Vattenfall. Of course mainly because of the history of and the accidents at the nuclear power stations, and the missing communication. But also because of the new coal power plant and because of their mega-arrogant behaviour here in Hamburg' (KEBAP, Member 4: 162–166).

In Germany, Vattenfall is mainly known as a coal and nuclear power plant operator. Nuclear power in particular has an extremely bad reputation among most German inhabitants (Unger/Hurtado 2013: 8pp.). In the years preceding the two campaigns, the company's reputation had been further damaged by the way it had tried to cover up several accidents at two nuclear power plants, both located less than 90 km from Hamburg. In these two nuclear power plants, several accidents and incidents above Level 4 on the INES-scale happened in 2007 (Frater et al. 2008).[19] Vattenfall tried to cover up the accidents and released information about the causes and effects of the incidents only slowly and bit by bit. Environmental organisations, politicians, and the media heavily criticized Vattenfall's information policy (ibid: 221). In 2007 and between 2009 and 2011, both plants had to be taken out of operation because of incidents. While Vattenfall had only planned a temporary stop, in 2011 both power stations found themselves in the first cohort of nuclear power plants that were terminally put out of operation by the 2011 nuclear moratorium. The moratorium was decreed by the German federal government subsequent to the nuclear catastrophe in Fukushima.

Incidents like these did not contribute to establishing trustful relationships between the company and the inhabitants of Hamburg. Neither UHUN nor the Moorburgtrasse-Stoppen initiative needed to mention these incidents in their campaign centrally; as these are so much common knowledge in Hamburg that reference to the company as 'the coal and nuclear power plant operator Vattenfall' was sufficient. Instead, they emphasised that Vattenfall—being the biggest coal and nuclear power plant operator in Germany—was not likely to be interested in truly supporting the energy transition. Taking this view, the campaigns positioned themselves in opposition to the governing party, which, when purchasing 25.1% of Vattenfall's grid company, had also signed a cooperation contract with the company. The SPD argued that by cooperating with Vattenfall, Hamburg would be ideally positioned to realise the energy transition in the city.

Despite a heavily financed media campaign by Vattenfall and the Hamburg government, in which they promoted the actual state, the UHUN campaign succeeded in convincing a majority of the electorally registered inhabitants. With 444,352 votes (50.9 percent), the network won the referendum in September 2013. In February 2014, the SPD—which was governing in a coalition with the

---

19   In 2007, accidents happened in both power plants, which were heavy enough to fall in the legal scope of accidents reportable to the international Atomic Energy Agency. These accidents are put on the international list of incidents in nuclear power plant with a value more than Level 4 according to the INES-scale (https://nucleus.iaea.org/Pages/inres.aspx, 28.10.2015).

Green Party since September 2013—announced that Vattenfall's pipeline plans were finally 'politically dead' and would not be resumed, irrespective of the outcome of the formal plan approval process (www.moorburgtrasse-stoppen 21.2.2014).

4.4.4 Realising KEBAP

Most of the people who formed the first core group of the KEBAP project had met through the Moorburgtrasse-Stoppen umbrella initiative. For many members of this group, the Moorburgtrasse-Stoppen campaign had been the context in which they came to know about and started to engage against Vattenfall and Hamburg's energy policy.

Living close to one of the green areas that would have been affected by the pipeline, the second main promoter of the project (Board Member 1) had herself started one of the local initiatives that became part of the Moorburgtrasse-Stoppen network. She and the developer of the original concept (Board Member 2) met when he joined her group. Brought together in the local initiative started by the interviewee, she and BM2 found that they shared a common interest in preserving the park and in stopping the pipeline plans of Vattenfall and Hamburg's government.

In February 2010, the manifold activities of the different groups that acted under the Moorburgtrasse-Stoppen umbrella initiative were successful in forcing Vattenfall and Hamburg's government to start a formal plan approval procedure. For the people engaged in the initiatives, this success against the 'Goliath' Vattenfall was an enormously motivating experience:

> B1:   'And, for me that was an exhilarating experience. That it is possible. And, we did something. Because, these weren't people who were already active in any initiative, so, yes. It has really been work on the ground. And it really motivated me' (KEBAP, Board Member 1: 230–233).

The experience '*that it is possible*' for a citizen initiative, of which most members had no previous experience in political action, to succeed against a big energy company and Hamburg's government was an important motivation for the creation of the KEBAP project.

Having come to know and like each other within the local initiative, the inventor of the concept at some point had told the interviewee about his energy bunker concept. Apparently, she was immediately taken by the idea. Inspired by the success against the pipeline, they decided to organise an event in which different project ideas for local decentralized energy production—which would work as alternatives to the conventional energy production system—would be presented to the interested public. In this event, BM1 and BM2 together presented the concept for the biomass bunker. Although only a small group of people were present at the event, most of them were very much interested in the concept. They formed a group to discuss and work on the concept. This initial group wanted to take the action a step further. Instead of remaining in resistance, they aimed to create a constructive alternative to the existing, unsustainable heat system. In their words, they wanted to advance '*vom Wut- zum Mutbürger*': from angry citizens to brave citizens (informal talk with Board Member 3).

All of the people present at the event had some connections to the Moorburgtrasse-Stoppen initiative. They were all interested in the political issues surrounding the heat grid. Together, they developed the idea to abandon the original concept of constructing a micro-grid to provide the neighbouring buildings with heat, and instead to feed heat into the existing long-distance heat grid. The bunker itself functioned as an important inspiration, as it is located in one of the parks through which the new long distance pipeline would have run. Furthermore, the bunker is situated only about five meters away from the existing long-distance heat grid.

Wanting to be a stumbling block for Vattenfall, the members of the (at that time only) energy bunker project realised that if they became producers of renewable heat, they would be in a position to challenge Vattenfall's quasi-monopoly of the heat grid by way of legally forcing permission. This idea was an outcome of the fact that almost all of the members were not only active in the pipeline initiative but were also interested or even active in the UHUN network. The KEBAP group concluded that becoming a heat producer would put them in a legal position to challenge Vattenfall's quasi-monopoly on political as well as juridical grounds.

Together with Greenpeace Energy, the BUND, the consumer protection agency Hamburg, and other organisations, the group submitted an inquiry to the cartel office in 2010. The cartel office approved of their inquiry and requested Vattenfall to open its heat grid to other competitors. So far, the deriving possibility to legally force Vattenfall to open its grid exists only theoretically. The main obstacles to practical implementation are the technical requirements necessary for feeding into the high degree and pressure grid. Additionally, Vattenfall is in a position to prolong the formal process of creating a contract more or less infinitely and to set conditions that can hardly be met by any competitor. Due to the successful referendum and the fact that the heat grid will in future (very likely) be operated by a municipal company, this potentially severe obstacle to the realization of the project has been removed.

While the shared context of the Moorburgtrasse-Stoppen initiative provided a common ground for the group to think about and develop the bunker project, they also contributed to the concept from their different private and professional backgrounds. In the very beginning of the project, the special construction of the building as a twin bunker sparked ideas to combine the production of energy with other activities. The activities planned for the second part of the bunker derived from the private or professional interests of the original members. Two of the members knew about the severe shortage of rehearsal areas for musicians in Hamburg-Altona. They suggested using parts of the bunker as music rehearsal rooms. Another member, who is a project manager in his profession and an enthusiastic cook in his private life, had the idea to use some of the space to open a canteen in which regional food could be prepared at affordable prices by interested volunteers. For another member, a main idea from the beginning was to open up the building and the whole project to all interested people, especially in the neighbourhood. Being an artist herself, she wanted a multi-functional room in which local people could meet and local artists could display their work. Together, they developed the concept of an integrated culture and energy bunker project.

The group soon found supporters. In 2012, about 20 to 30 people were more or less regularly engaged in the project. The core group that mainly ran and managed the project during the time of the fieldwork consisted of six people. From the beginning, an important resource of the project was its network. During the time of the fieldwork in 2012, the institutional network of the project consisted of about 100 different organisations, groups, and initiatives from different sectors of society (universities, citizen projects, alternative groups, schools, research institutes). Throughout the process of the project, these relations have provided tangible and intangible help, support, and advice (for example soil for the garden, political advice) to the project. Most of the core group members brought their professional and private networks into the project. One member established contacts with different (urban) gardening projects and people. Another member, a teacher, has established contacts with schools and kindergartens. A large number of contacts derived from the members' prior engagement in the Moorburgtrasse-Stoppen and in the UHUN initiatives. Additionally, a number of members are active in the Right-to-the-City movement, which is rather large and influential in Hamburg's citizen initiative network.

Furthermore, many members are not shy about establishing contacts with people or institutions that they think would be helpful for the project.

M4:   'I joined KEBAP, because, the initiative for which I was working at the time, well a number of the people which founded KEBAP were active in it. And they more or less recruited me. And started telling me about it and writing emails. So, one or two of the members told me about it and wrote mails' (KEBAP, Member 4: 12–15).

The experience recounted by the interviewee shows how already existing contacts were utilized by KEBAP members. The next quote exemplifies how the members established initial contacts with organisations and institutions they thought could be helpful for the project.

TA: 'I have to think, how did we come to know KEBAP anyway? I think BM3 did at some point find us and then somehow asked if we would not like to do something together' (Technical Advisor: 124–126).

The interviewee is a lecturer at Hamburg's University of Applied Sciences. He agreed to contribute his expertise to the project and to ask some of his postgraduate students if they would be interested in doing their Master's thesis about potential types of energy production installations in bunker. Subsequently, two Master's theses were written about the bunker. The first one, in 2011, scrutinized the utilization concept that would be technically and economically realizable for the bunker, while enabling a sustainable energy provision for the district. The thesis paid special attention to the possibility of installing a heat-storage facility in the bunker. In 2013, a second Master's thesis analysed different technological variants of combined heat and power production for the bunker.

In 2011, the group commissioned its first feasibility study for the project. This feasibility study focused on the technical feasibility of the project. The Euro 2,000 needed for commissioning the study was paid by another social organisation—Germanwatch, a citizen organisation engaged in fighting for global justice and preservation of livelihood resources. In 2010 and 2011 the organisation supported a 100% regional campaign. During this campaign, they got to know about and came in contact with the KEBAP project and decided to sponsor the feasibility study.

M3:   'We did find KEBAP rather interesting. And then we collaborated with the project from time to time. And in this context we co-financed the feasibility study in order to create a basis for discussion' (KEBAP, Member 3: 26–29).

The other part of the financial capital for the feasibility study came from a private supporter. The feasibility study not only provided an 'objective' evaluation of the project but also brought two more members to the group—the contact person from Germanwatch and the specialist who conducted the feasibility study.

The feasibility study proved to be important shortly after it was finished, when it became known that another private investor was interested in buying the bunker. This became known when the group contacted the district administration to enquire about ways and potentials of gaining access to the building. The competing investor planned to refurbish the bunker into a tenement house. The investor was affluent enough to potentially buy the bunker from the district. His plan to refurbish the bunker into apartments was well received by the district government and administration, especially as flats are a perpetually scarce resource in Altona. Moreover, the project planned to not only construct studio apartments for artists but also a new side-building in which eight social housing flats would be erected. The situation started to turn in favour of the KEBAP project after a public hearing of Altona's district planning committee. In this hearing, both projects were presented to the members of the committee and the interested public in August 2012. An unusually large number of residents attended the hearing (comment of the chair, own notes from the event). Through their presence and the questions raised and comments made, it became clear that the majority of the present residents were in favour of the KEBAP project. Many people were worried that the concept of the private investor would further spur the already existing processes of gentrification[20] and displacement in Altona. KEBAP, on the contrary, was understood to open up new spaces for residents and to realise projects that favoured the people already living in the district instead of trying to attract new and wealthier people (own notes from the event). Subsequent to this event, KEBAP increased efforts to convince members of the different parties in the district (mainly by way of presentations in meetings of the different parties) in favour of their project. The Left Party, the Green Party, and finally even the SPD decided to support the KEBAP project. This decision put paid to the private investor's plans.

The experience, however, spurred an internal crisis among the members of the KEBAP group. The situation of having to compete against a professional investor created anxieties. Some members feared that in order to be taken serious by the political and administrative representatives, the project needed to have a cohesive concept for the project. This worry was particularly pronounced with regard to the energy concept. At that time, two variants were being discussed by the project's energy working group. The first variant maintained the original idea of installing a woodchip boiler. The alternative variant projected the installation of a gas CHP plant. Using a gas boiler would destroy the purely local circle but enable the project to become an important 'bridge technology' in the restructuring of Hamburg's energy system. Gas power plants are able to react within seconds to production and demand dips and peaks and thus offset fluctuation from other renewable energy technologies. Gas, however, is not only not local, but also not renewable. Woodchip-based methods are also questionable in ecological and social terms (especially land use) and have the disadvantage that they would require large numbers of heavy vehicles to pass through the neighbourhood in order to deliver the biomass. The neighbourhood, as the critics of

---

20   Gentrification at that point was a term constantly coming up in public discussions about general trends, and especially the exorbitant flat prices in Hamburg.

the woodchip variant argued, would also suffer more from emissions caused by woodchips than from those caused by gas. A second professional feasibility study did not solve the issue as it concluded that both technologies were generally feasible—both having different advantages and disadvantages. The conflict escalated because the members disagreed about whether the decision about the energy variant really had to be made at this point in time. Going against the idea that the project would seem unprofessional without a decisive energy variant, the opponents argued that this would oppose the fundamental idea of KEBAP as an open process. The crisis resulted in one member of the original core groups and two other participants leaving the project.

In 2012 and 2013, the project started organising events and taking part in street festivals, markets, and the like. While the bunker itself remained closed to the project, the members organised a range of activities and events around the building. Also, through contacts with Greenpeace, the group was able to purchase a trailer especially designed for the purpose of showcasing information on public events. The mobile was parked at the end of the cul-de-sac on which the bunker is located. In and around the mobile, activities like movie nights and information events were held. Also the mobile became the regular meeting place of the group during the summer month, since it was an open and approachable place in the neighbourhood. Additionally, many members continued to attend as speakers or panel members at events, conferences, or panel discussions on topics like energy transition, energy policy, community energy, renewable heat, remunicipalization, or the potential of cooperatives.

Members of the group increasingly came to be seen as interesting guests in talks or discussions on cooperative models, as the project had decided to set up an energy cooperative for the production and distribution of the energy produced in the bunker. They were (and still are) one of the first groups planning to set up a cooperative for the production of *heat* in an *urban setting,* and they also raised interest because of the innovative model planned for the cooperative. At the time of the fieldwork in 2012, the project planned to set up a cooperative in which most of the income generated would not be distributed back to the members (except an annual small adjustment of interest, which is legally required), but would be used to subsidize the cultural and social projects in the other half of the building. KEBAP plans to combine two different types of organisation. While a registered association, in existence since early 2012, will continue to be the organisational structure of the cultural and social part, the energy production will be operated by a cooperative. To ensure long-term coordination between the two organisations, the charity will be an obligatory board member of the cooperative, ensuring that the two parts continue to be one integrated project.

In order to set up the cooperative, a working group was established in the beginning of 2013. It started to create a business plan and a constitution for an energy-and-culture cooperative. It profited from the expertise of two group members, one of whom was trained in cooperative management, while the other had just participated in a large housing cooperative project. Additionally, the project had good contacts with some other large alternative projects in Hamburg, which had also organised themselves in the legal form of cooperatives. In February 2015, KEGA eG was founded and has been registered with the association of cooperatives in Germany on December 1st 2015.

In 2012, members of the group started an urban gardening project around the bunker. Though not included in the original concept, the gardening group has become the most active and 'visible' part of the whole project. Like all other activities in the project, the urban gardening project aims to be as sustainable as possible and to realise the project vision of a local, shared economy in which things are, as far as possible, not purchased but exchanged, self-made, or recycled. The raised garden beds are built out of old pallets, rain water is harvested in found tins, the soil was a present from another gardening project with which an exchange of seeds has also developed, a compost toilet has been built, and a self-made (very small) wind-turbine and biomass boiler have been constructed from different types of 'litter' and 'garbage'. In addition to the gardening project, a range of other activities have also taken place. Ideas for activities and projects come from and are organised, and run by an increasing and fluctuating number of volunteer members. The concept for the cultural part has more or less continually been changing since the first core group came up with the idea to include a cultural aspect in the project. To meet these changes, the project in 2015 has taken on a new concept, in which the new activities and key aspects of the project are integrated, while still emphasising the character of the project as an ever-changing and adjusting open process.

While the gardening project and the other activities of the group around the bunker were welcomed by local residents and organisation, they have also led to a situation in which the district government makes the group increasingly responsible for the cleanliness and orderliness of the area. Formerly a rather 'dark' and neglected part of the district, the area around the bunker is used by homeless people and drug addicts. While not having created any of the litter, the KEBAP group is continually asked by the district administration to clean the area, failing which the gardening project would no longer be formally tolerated and would be cleared away (mails 14.9.2015, 17.9.2015). Such misunderstandings between the project and different sections of the district administration are very frequent, despite the project's efforts to maintain constant communication with the authority.

In 2013, the KEBAP group battled for the creation of a continual meeting between the project and representatives of the authorities in order to discuss the actual situation in both institutions. The project had established regular contact with two members of the authority during the conflict with the competing investor. In 2013 these meetings were broadened and formalized, including site visits by members of the authority for urban planning and environment and other actors from district and city administrative institutions. The main issue in these meetings in 2012 and 2013 has been the acquisition of the bunker. KEBAP has dismissed the idea of purchasing the bunker itself. The project aims to re-create the building as an open space for residents and other interested people. An important aspect for the realization of this intention is to prevent privatization of the building through ownership rights. The project instead aims to convince Hamburg's government to buy the bunker and lease it to the project in the legal form of an emphyteusis contract ('Erbpachtvertrag').

After its release from military service, the bunker is now managed by the Institute for Federal Real Estate (BImA), which aims to sell the state's properties at the highest possible price. Hamburg's government, however, had no intention to purchase the bunker and is not willing to pay the amount of money charged by the BImA (about Euro 2 million). At the moment, negotiations are going on between the district government (which would take over the responsibility for the bunker from the city government), the two responsible city government ministries (ministry for urban planning and environment, and the ministry for finance and economy), the BImA (which is an agency at the federal level), and the KEBAP group. Harmonizing the divergent interests of these institutions is complicated.

DA: 'Well, as I said, this would be helpful for starting the project, if we would leave the whole thing in the ownership of the city, and then would pass it on to the project in the form of an emphyteusis contract. Well, we have to see if that is going to be <u>possible</u>. This is not clear right now. Because, as I said, there are different authorities involved. The purchasing authority of the city, which is the finance authority. The authority having the functional responsibility for our urban regeneration programmes, which is Hamburg's authority for urban development and environment. Then there are we as the district administration. Then there is KEBAP. So there are just very different interests, which have to be fused' (District Authority Officer: 160–170).

At the end of 2013, the district government started to push the KEBAP group to commission another, much more comprehensive feasibility study for the project. The district government made the continuation of their support contingent on the existence of such a study. Only if the feasibility study would estimate the project to be technically and financially viable, would the district continue supporting the project. The project succeeded in mobilizing financial resources from the district and federal funds. The first part of the feasibility study was published in September 2015. It proved that the energy concept—which projected a modular technological approach at that moment—was technically and financially viable. The core part is a biomass gasifier. Wooden biomass is to be not burned but heated in a so-called gasifier. The produced gas is then to be fed into two relatively small block-typed thermal power stations (BTTP). The advantage of this method is that hardly any fine dust is produced and it is carbon dioxide-neutral. The wood gasifier will be combined with one gas BTTP station and a heat-storage facility. While all these technologies have been long known, their combination and realization will make KEBAP a pilot project in Germany. The heat will be fed into the long-distance heat grid. Letters of intent have been contracted with prospective customer. Interested customers so far include different alternative cultural, social, and living projects spread throughout Hamburg.

4.5 Analysing KEBAP: 'vom Wut- zum Mutbürger'[21]

4.5.1 Teleoaffective structures: Resisting Vattenfall and creating a local economy project

The historic and social background of KEBAP are the Moorburgtrasse-Stoppen and the UHUN initiatives. As described above (4.4.3), Vattenfall had planned and acquired approval to construct a new long-distance heat pipeline through the Hamburg's inner district regions, thereby temporarily destroying a number of parks and green areas. This destruction of the local area for economic purposes of a private company was the motivational background of the KEBAP project.

B1:   'I found it, just, particularly scandalous that just this, in itself unbelievable project, the new coal power plant, was not to be connected to the existing long-distance heat grid, but instead that a new pipeline was to be constructed for it which would have guaranteed the monopoly. Just by the way, so by the size and diameter of the pipeline. And I found this so unbelievable, yes, I was convinced that one has to vehemently resist against it. And of course one has to look far beyond the local level that something like this does not happen generally. [...] And [...] because I got to know the BM2, I knew about his energy-bunker idea. And well, through this whole resistance the, so to say, potential came together, in my opinion. I

---

21   Board Member 3, informal talk.

thought it was time to not only be against it but to try something else. […] And, yes, as we were then able to in fact get seven people together, the opportunity occurred' (KEBAP, Board Member 1: 199–216).

The interviewee describes how, in her perception, the Moorburgtrasse-Stoppen initiative prefigures the existence of KEBAP. As for herself, engaging in the initiative provided her with knowledge, which increased her motivation to engage against not just this particular local project but Hamburg's energy policy and the conventional energy system in general. Prior to this sequence, she explains that, at the beginning of her engagement in the Moorburgtrasse-Stoppen initiative, '*I had not much knowledge about this whole energy issue*' (195–196). Her original motivation had been frustration about Vattenfall's power to intrude into her local area and thus her private life. Through her engagement in the initiative, her intentions expanded from preventing a local destruction to also include the teleology of looking '*far beyond the local level that something like this does not happen generally*'. The propositional knowledge she gathered through her resistance activities changed the focus of her engagement. Her increased political engagement was due to not only additional knowledge but also emotional motivation. Relating the aforementioned sequences to a later sequence from the interview, the importance of the affective element, entwined with her (changed) teleologies, become visible.

B1:   'Because one also has to say, we were active in the working group of my local park. This has also been—I might should mention this here in your context—an important step. Because this for example, very, very much encouraged me' (KEBAP, Board Member 1: 221–224).

Affective aspects in this quote are described as having motivated the activities leading to the creation of KEBAP. Having successfully set up a local initiative, the interviewee felt encouraged to take her engagement further. Unlike the previous sequence, which underlines the (subjective) importance of information and thus propositional knowledge for the interviewee's engagement, this sequence illustrates the relevance of affective motivations. In her interview, the interviewee believes that KEBAP was prefigured by the Moorburgtrasse-Stoppen initiative, whereby the latter has, in important ways, influenced the teleoaffective structure from which KEBAP was developed by its members.

Going back to the first quote, the interviewee therein explains that the initiative brought her in contact with BM 2, who had told her about his biomassbunker idea. As will be described below, the team-work of these two actors was essential for setting up KEBAP. The context of the initiative provided a setting within which actors which shared an interest in alternative energy policy and production could meet. In her next sentence, the interviewee extends the networking aspect of the initiative, stating that through it the potential had developed to start a project like KEBAP. While she does not explain the particular meaning of 'potential', the context of this statement indicates that the term is used as a catch-all phrase—to encompass her (changed) teleoaffective structure, the knowledge about both the energy system and the bunker project, and the coming together of a group of people who subsequently developed the biomass boiler idea into a project.

In a later sequence, the interviewee describes in more detail how the initiative led to the creation of the setting wherein KEBAP was set up as a group project.

B1: 'For me, this feeling was crucial, you know, to have the motivation to start such a project like KEBAP at all. And out of this, we [her and BM2, AP] together developed an event in the citizen hall. It was named "bundling energies". And through working together on this, one can say, BM2 and I very actively convened.

B2: November 2010.

B1: And on this event, [...] while having been a fiasco with regard to attendance figures, we had invited very good people and we got to know people there with whom we created a sustainable network. With all people who had been there. And in so far this, despite the little public interest, was the initial spark for KEBAP, because out of this it became obvious that one definitely has to start something to change things. And one has to create a model. And with this vigour we [...] continued working. And from there the seven other people originated with whom we brought KEBAP into being' (KEBAP, Board Member 1 and 2: 245–262).

In this quote, BM1 considers the Moorburgtrasse-Stoppen initiative as not only the background for her individual engagement but the source of '*the initial spark for*' KEBAP. According to this presentation, the Moorburgtrasse-Stoppen initiative is related to KEBAP by causality. It is causally related to KEBAP as it subsequently led '*people to perform actions and practices to take certain courses*' (Schatzki 2010: 139). Deriving from these explanations, it is argued that the original motivation for, and the first key actors of, the KEBAP project are closely related to the Moorburgtrasse-Stoppen initiative. The argument made is that resistance against the heat policy of Vattenfall and Hamburg's government is an important aspect of KEBAP. It provided a context out of which core aspects of the project's teleoaffective structure resulted and in which actors, who could connect to this teleoaffective structure, met. The subsequent parts of this section explain how this teleoaffective structure had to be constantly negotiated among the members of the project since then and how this has influenced the concept of KEBAP.

As described, BM1 and BM2 presented the biomass-bunker idea at the 'bundling energies' event. At this event, the idea of the biomass-bunker was received with much interest, and most of the participants started to engage in the project. Embedding the biomass-bunker idea in this context, however, spurred changes in the concept; most importantly the distribution concept. Originally, the concept had foreseen the construction of a new micro-grid.

B2: 'Back then I planned for about forty, fifty flats or houses. This would have been around four hundred flats. As I said this would have been with our own island-micro-heat grid, which one would have had to build trough the cellars and so on' (KEBAP, Board Member 2: 70–74).

After presenting his concept at the 'bundling energies' event and in the subsequent discussions, the concept was changed in order to better respond to the teleoaffective structures of the initiative, especially the initiative's aim to challenge Vattenfall's monopoly.

B2: 'So, why should one build a second micro-grid if there is a long-distance heat grid in existence? Through which one also could then make it a pilot project. Which would be politically very charged, to—like with the electricity grid—achieve the opening of the long-distance heat grid' (KEBAP, Board Member 2 148–152).

The concept was adapted in the evolving first core group in order to better fit the group's aims. Thereby two different teleologies were aligned—the creation of a local economy-based renewable and affordable heat circle and the interest of the Moorburgtrasse-Stoppen initiative to attack Vattenfall and Hamburg's existing heat system. The core idea was to not only establish a vital alternative to the existing heat provision but to also become a competitor for the heat grid and combat Vattenfall's monopoly on legal and economic grounds. KEBAP's concept was thus re-designed to fit the context in which this first core group had met and which provided the common ground for their engagement. This meant that the idea of the micro-grid had to be abandoned for the project to be able to fit the evolving teleoaffective structure of the first core group. According to the interviews and participant observation, the decision to abandon the micro-grid did not cause much conflict in the first core group, nor was it questioned at any point later in the process. The apparent relevance of the teleology to challenge Vattenfall's monopoly becomes even more pronounced when considering that feeding into the long-distance heat grid was for a long time deemed one of the biggest challenges for the realization of the project.

TA: 'And the crucial situation will now be Vattenfall's offer for feed-in charges. Does it work at all, or not? This will be the first question. They will very probably obstruct this and try to find a way to say that it doesn't work. And if it doesn't work it is going to destroy the business model' (Technical Advisor: 566–572).

In this interview, the technical advisor from the HAW defines the feed-in question as a *'crucial situation'*. Vattenfall, the owner of the heat grid and the only energy provider, did not have any incentive to allow other actors to feed their heat into the grid. In January 2012, the cartel office responded to an inquiry from Hamburg's consumer advice centre, the BUND, and KEBAP. In its official statement, the cartel office required Vattenfall to open the heat grid for external heat producers. In the inquiry, KEBAP officially appeared as a future energy producer and potential competitor for the first time. Being a competitor, KEBAP had the legal right to bring to court economic monopolies. Thus, in just the first year of the existence of the project, the members took the opportunity to 'use' the project as a tool in their continued resistance against Vattenfall's heat monopoly. Subsequent to the cartel office's response, the KEBAP members formally handed in an inquiry to Vattenfall in which they officially asked for the conditions and charges for passing through heat. The quote above explains that even if Vattenfall were to be forced to generally open its grid to competitors, it would be in a position to charge fees and ask for technical preconditions that cannot be met without destroying the economic model of the competitor. Most external interviewees agreed that this question was a crucial point for the project.

PE: 'So KEBAP, even if they carry through with this legally, Vattenfall will always
        have the upper hand. Thus you will never succeed without changes at or influences
        from the federal state level or at least the federal level, and this does not seem to
        be likely. Vattenfall is just able to put so many stones in KEBAP's way by saying:
        we need this form and that information, and—presto—somehow three years have
        passed and by then the project will be dead. And this is a strategy Vattenfall has
        already tried to use—delaying and not getting back and so on' (Political Expert:
        361–369).

While the inquiry at the cartel office presents an important event in which the teleological structure of the project resulted in distinct activities, the fact that—despite the obstacles described in the quote—the members of the KEBAP project never doubted the idea to feed their heat into Vattenfall's heat grid, is an indicator of this teleology's importance. Despite creating one of the biggest obstacles to the realization of the project, this intention has been such a central aspect of the project's teleoaffective structure that it remained unchallenged. The success of the UHUN initiative in 2013 changed the situation. While the intention of KEBAP now is not anymore to challenge Vattenfall in its position as grid operator, the aim remains to break the company's monopoly, as Vattenfall so far is the only heat producer feeding-in into the long-distance heat grid.

The hitherto given argument has been that the energy concept of the bunker has been subject to negotiations which resulted from the necessity to adapt the project to key elements of the group's teleoaffective structure. The concept was aligned with the diverging teleologies of different actors and the context within which these actors interacted. While hitherto the argument focused on the energy aspect, the teleoaffective structure of this first core group included not only energy-related interests and motivations but also ideas to combine local heat production with local cultural production in the bunker.

B2:   'So, and the initiative very much liked this concept. And then we <u>worked on it together</u> from there on. And <u>then</u> we, most of all, have come up with the whole cultural part and said, let's not just make an energy, but—well—a culture-energy bunker' (KEBAP, Board Member 2: 97–100).

In this quote, the interviewee relates to the situation in which the bunker concept was presented at the 'bundling energies' event. The participants were enthusiastic about the general idea to turn an old bunker into an energy production site. Working '*on it together*', the members of the first core group contributed their ideas and interests to the concept. During this process, the participants developed the idea to extend the original concept by including a cultural part. The integration of cultural production is not incidental or the outcome of participant's private interests, but related to the project members' interest in challenging the logic of the energy system.

B2:   'The original utopian dream thereby has been to combine a field which classically is thought of as deficient, namely culture, like culture always costs money but earns too little money […], with a field which is related to high sales figures, and which for exactly this reason is being taken over by companies and major corporations' (KEBAP, Board Member 2: 100–105).

According to the quote, energy production interests companies and major corporations because of its potential to earn high incomes. These companies produce energy for the sole purpose of producing financial income. The idea for the KEBAP project, developed in the 'bundling energies' event, intends to combine the production of energy with cultural production. According to the quote, this is deliberately done because cultural production costs money. While the energy part is still supposed to produce a financial overhang, through integrating the energy with the cultural part, this overhang can be used within the project and thus not be made available in the form of private capital. The combination of these two activities results in a sense making of energy production which does not intend to create private financial income. The intention behind this concept is to challenge the economic logic of those institutions which are dominating the energy production system.

     Cultural production, however, is not merely an outcome of deliberate considerations to oppose the logic of major corporations and political institutions. The decision to include culture into the project is also an outcome of individual affective motivations.

B1:   'And in this combination of the culture-energy bunker, for myself, I could only ever see a chance to do something which interests me. So beyond the pure production of energy' (KEBAP, Board Member 1: 212–215).

In this sequence, the interviewee explains her private motivation to commit herself to the project. Besides her political engagement, her motivation is also an outcome of her private interest and enthusiasm for culture production. The importance of this affective relation to the project becomes clearer when it is related to a later sequence from the interview.

B1:   'It is not the energy. Energy is one issue, and this is very clear in KEBAP. There are people who are attracted or interested by that. But there are also people who are not interested in that and which only—and actually I belong to them—who only can—or want to—engage with that by means of other strands. And I have to say that. If I would only occupy myself with the energy issue I would soon be fed up with it. Well, just because, not because I don't think it is important. But just because on its own it would be too abstract and technical, too difficult. And the motivation to busy oneself with it stems from something other. And I do believe that it is like that for some people in KEBAP' (KEBAP, Board Member 1: 397–406).

Taken together the two sequences indicate that for at least some of the members, the motivation to engage for the project does not stem only from political interests. Instead, long-term commitment is expected to require not only intellectual but also affective relations to the project. For the interviewee, continually committing to the project requires enthusiasm. While she accepts the importance of energy production, she believes it will not be able to provide enough motivation to maintain her interest in the project on a long-term basis. She considers the energy issue to be too abstract and difficult to keep her motivated. Her motivation to commit herself to the project instead derives in large parts from the '*other streak*'—culture production.

The two sequences do not indicate that the two parts draw interest or commitment away from each other. Instead, the argument in these quotes is that it is only the combination of these two distinct realms that motivates the interviewee's continued interest in the project as a whole. The 'fit' of the culture and the energy part is not negotiated in terms of abstract concepts deriving from the political intentions of the project. Instead, the combination results from the necessity of the project to have a day-to-day relevance for its members. For participants who are not passionate about the issues of energy production and distribution, the cultural part contributes an emotional attachment that could hardly be derived from the energy part.

M1:   'This is the great thing about the project that there are so many entry levels, and no matter from what direction one is coming, one can then also start to occupy oneself with other issues. That is why I think that even when you are into it already, you can maybe start being interested in the other things. One does always get the information anyway. All those different activities which are taking place there and information events and things like that. Yes, I am sure they create awareness, also, for the meaning of energy provision' (KEBAP, Member 1: 377–382).

According to the interviewee, interest for the different parts of the project does not create separate interest groups within the project. Instead, people who join the project because they are interested in the cultural aspect will constantly encounter information and activities from the energy part. Through this constant encounter, people might become not only aware of but also interested in the other aspect. This idea from the quote is supported by findings from participant observation. In all the meetings attended throughout the fieldwork, aspects from both the cultural and the energy part were discussed. Members who had joined the meeting because of energy aspects would also get information about and meet people engaged in the cultural part—and vice versa. While working groups are specialized in certain aspects of the project, they have to take the other elements of the project into consideration in their social practices.

According to the quote, the two parts can mobilize resources for each other. In the quote, the cultural part is described to create side-entries into the issues of energy production and distribution and thus to provide a way to approach people who are not initially interested in energy-related issues. The mutually supporting effect of the combination is also highlighted in another quote.

B4:   'So, first of all this combination is just what the people who created KEBAP wanted. And from where they continually draw motivation. So, if for example KEBAP would have developed like: nope, we don't want this energy thing, we only want a social space, then a third of the people or so would have backed out of the project, would not have continued. Or the other way around. Then people will just back off the project. It makes you stronger when different interests work in one direction' (KEBAP, Board Member 4: 314–320).

This quote describes the (affective) mobilization of people as an outcome of combining energy and cultural production. At the end of the quote, the interviewee also explains that by bringing together people with different interests, the project is able to increase its chances of being realised. The combination of the two parts enables the project to mobilize more people. These members, moreover, do not focus their engagement on one aspect exclusively, but instead work together to realise the project. By working together on the realization of the project, '*different interests work in one direction*'.

The members of the project understand energy and cultural production as two different ways of expressing or realising certain teleoaffectivities. While different teleoaffectivities are at first appearance tied to certain activities and sub-projects within KEBAP, the last three quotes show that the different activities are also related to one another. The two parts are presented not as naturally belonging together but as being brought together through and by the activities and teleoaffectivities of the project members. At the same time, the quote also indicates the necessity of negotiating the different interests of the paticipants. The project has to integrate and constantly negotiate the teleoaffectivities of its different members with one another and with the teleoaffective structure of the project.

M3:   'So this whole culture story. So this is indeed interesting and necessary. That's
        how I would see it. But I myself just have no idea about culture.
I: Necessary?
M3:   Absolutely. I think especially now that we are increasingly discussing this partic-
        ipation thing, it won't work without culture. Whatever one understands by culture.
        And in the end it can only work out, when those people who participate, have fun
        doing it. And this fun factor has to be cared for' (KEBAP, Member 3: 523–531).

The quote mentions the potential of the cultural part to attract people through an affectionate interest—having fun. The quote identifies affectivity as an important motivational factor. It describes an important function of the culture part within the project as a whole. At the same time, the creation of affectionate interest in this quote is emphasised with regard to the project's intention to be a participatory process. Attracting participation is understood as a valuable aspect of the culture part, because participation is highly estimated within KEBAP. Unlike the energy part, the culture part is able to attract participation from people who would normally not be interested in energy matters. In a context that greatly estimates participation, the culture part's ability to enhance the range of participants is highly valued.

On a more fundamental level, both types of activities—cultural and energy production—are related to the more abstract principles that form the ideological baseline of KEBAP.

B3:   'The new thing is this connection, which we are striving for deliberately, between
        the production of energy and culture. And, the closer you look the more connec-
        tions you see. There it is also about autarky, it's about self-determining entities, it
        is about, well, basically the essence of the grassroots movement. That you really
        do culture and energy with the people for the people. And, the one thing does that
        on the artistic, cultural level, the other is an energy production level. But, the ideas
        behind it, what it is about, this is well, I see many similarities there' (KEBAP,
        Board Member 3: 429–437).

The interviewee describes both energy and cultural production not as related by some kind of connection between the activities or in terms of their results but as expressions of basic principles and ideological aims. In this quote, the principles of autarky and self-determination are mentioned. The aim is to produce culture and energy themselves instead of consuming products and services provided by companies or public institutions. The interviewee goes on to describe these principles as being the essence of grassroots movement. These principles are also written down in the project's structuring concept. Self-initiative or autarky, as principles providing an important part of the teleoaffective structure, imply or create a strong opposition to the idea of being provided with energy by a company like Vattenfall. The idea to produce energy locally and in a decentralized manner acquires a particular meaning within this teleology of autarky and self-empowerment. Put the other way around, by producing energy in a local and decentralized way, KEBAP can realise its principles of autarky and self-determination.

Producing energy in a decentralized way is only one aspect through which the principle of autarky and self-determination is meant to be realised.

B1:   'We are attractive for certain people, because in our project you can <u>do</u> something. And through this doing, people are getting in touch with the energy issue. And this is a really interesting connection. But it does not remain there. And this is what—to be honest—<u>otherwise bores</u> me in energy conceptions. Because they so <u>quickly</u> become just superfluous. Then you've got a functioning system. As you have just been describing it, such a communal system. But this doesn't change the overall system. It does not really change anything about the way how people work together. And this is a <u>very important</u> aspect, this aspect of self-empowerment' (KEBAP, Board Member 1: 377–383).

In fact, this quote disagrees with the idea that self-determination or self-empowerment can be fully realised by setting up and running a decentralized energy production project. Creating a functioning system is described as quickly becoming unessential if this technological innovation is not combined with a different way of doing things. The interviewee thus argues for the importance of social innovations in the energy system.

While producing energy in a decentralized way is not seen to be enough to realise the idea of self-empowerment, it is also not merely an expression of this one principle. Instead, a range of teleologies are realised in KEBAP's energy concept. Ecological ideas and beliefs have contributed important principles to the project. Ecological concerns are implicit in specific aspects of the project, such as the idea of decentralization, and also underlie the project as a whole. It is not just ecological principles that underlie the energy aspect of the project. Instead, KEBAP intends to create a project within which '*the future core aspects for a societal transformation are not seen separately but are integrated in one project*' (structure concept: 5, translation AP). Different activities and aims like the '*decentralized production of renewable energy through renewable sources, nourishment with regionally produced whole-food nutrition, the re-vitalization of (perma-)cultural practices that recognize the availability of resources in order to combine and strengthen them and to enhance variability'* (ibid. 5pp.) are meant to be realised. Furthermore, the concept states that it is an aim of the project to transfer these principles into all areas of peoples' day-to-day lives, including mobility, social co-existence, consumption, and artistic practice (ibid. 6). Environmental objectives are not stand-alone issues; they are interwoven with one another and other interests. The production of renewable energy is meant to not only reduce carbon emissions and the exploitation of fossil fuels, but also create social profits—for example through the cooperative form, which enables a financial return of investment for the members and a participatory and democratic approach—and political benefits with regard to challenging Vattenfall and Hamburg's energy policy.

Neither the objectives nor their combinations have been stable in the project. Throughout the existence and development of the project, principles and their meaning have been introduced, changed, or abandoned.

B2:   'Things have changed throughout the progression of the project or have been
        added. Even in my own ideas and motivations. Because, in the beginning for ex-
        ample [...] we did not have this whole gardening idea. And we did not have this
        orientation onto local <u>economy</u> in this sense. This only <u>developed</u> throughout the
        progression of the project and it has been increasing in its meaning. And most
        importantly it has become an increasing motivation' (KEBAP, Board Member 2:
        309–317).

In this quote, the project is presented as a dynamic process, within which ideas and activities are constantly occurring and changing. The orientation towards local economy is one principle which was introduced at a later point in the project development and which has increasingly gained in strength. This process can be seen as a negotiation process. According to observation data, the process started in the realm of a video project through which the project could be explained. In this video, members explain in one sentence what constitutes the project for them. In the preliminary discussions on the video, one person mentioned that she would be interested in creating a roof-top garden on the bunker. This idea was immediately and enthusiastically taken up by a number of other members. Some months later, the idea came up that since the bunker was not yet available for the project, it could be a possibility to start an urban gardening project on the grass patch between the bunker and the street. Through their teleologies and bodily activities, the actors established a new social practice within the project. Furthermore, urban gardening and its underlying principle of local economy were made issues not only in internal meetings and discussions but also in external representations of the project.

B3:   'The aspect of <u>production</u> or maybe of self-generation, so this is an aspect which
        for example <u>did not very much</u> play a role in KEBAP. But which, now has been
        pushed very much to the centre through some people from the team' (Board Mem-
        ber 3: 410–412).

Though it was already in existence, the principle of local economy was gradually
made meaningful for the whole project and started to shape other social practices, like the planning of renewable energy production. Increasingly, the term
'local economy' and the idea of producing heat as a way to not only challenge
Vattenfall but also produce basic items of daily life in a self-generated way,
shaped the sense-making of energy production. More and more, local economy
got mentioned in internal discussions and external presentations, to make sense
of energy production as a way to become independent from large companies.

The garden exemplifies that the social practices pursued in the project, their
degree, and their reasons are related to the interests and motivations of the project's participants. As is obvious in the example of urban gardening and local
economy, interests and motivations are not stable.

B4:   'I think—at least that's my experience—that you can't define everything ahead in
        these projects. So, you might set some ideas and see how they develop. But, one
        always has to observe how the project, how the social composition, how people,
        develop. It all interacts. And without people it can't work. You need people who
        carry it into life, and you can't prescribe that to anybody. So I would be the same.
        Just imagine somebody would have an idea, "this now is an interesting project"
        and he wants to do it, and I could take this part or play that role, then this is first
        of all a projection. Where it is absolutely not clear if the person wants to do that
        at all. So, you can make an offer. And then you have to see how people want to
        participate' (KEBAP, Board Member 4: 58–70).

In this quote, the interviewee explains the relationship between the project, activities, and participants as an interaction process. According to him, this interaction in a project like KEBAP has to be balanced. In order to ensure this balance, the project has to be constantly adapted to its members and their interest.
Social practices within the project are created, maintained, and altered through
constant interactions between the project as an institution and the teleologies of
its members. Membership in the project is fluid. People join, contribute their
ideas and interests, and may stay or leave after a while. The person who originally introduced the urban gardening idea left the project after some months.
Her idea, however, remained and became an important part of the project. Because the teleoaffective structure of the project is composed from the teleologies
of its members, the interactions between the existing project and the changing
teleoaffectivities of its fluid membership mean that social practices within the
project are constantly changing. This makes KEBAP a dynamic process.

Members who were part of the project's first core group are very much aware of these changes. Some members—like BM2—felt that these changes have led the project so far away from its original idea that they decided to leave. Other members are happy with the way the project is developing.

B3:   'And this is related to the project's origins, that it has been developed out of the resistance against this pipeline. The first core group at least. Whereby we have proceeded very, very much from thereon, and that's great' (KEBAP, Board Member 3: 440–442).

While the intentions of the project members in the beginning were very much concentrated on creating an alternative to the existing energy system, their aims and intentions have developed since then. Social practices like urban gardening both initiated and signify these changes. The quote and the example of the integration of urban gardening show how new social practices are constantly introduced into the project. While on the one hand these practices shape the project, on the other hand they are always influenced by the ideas and intentions of the project itself. While urban gardening has strengthened the meaning of local economy for the whole project, the practices of urban gardening were also made sense of with regard to the aims of KEBAP—among others the aim to realise the interests of residents and members.

The interviewee above shows a positive attitude towards these processes. This presentation is in accordance with the project's self-presentation, wherein KEBAP is described as an open process with and in the quarter. These process, however, are not always conflict-free.

B3:   'And the other thing of course definitely is the group dynamic which sometimes develops. Where, people which maybe in other projects have been acting together and you have been working very close. And then suddenly you realise, particularly because you are engaged so much, emotionally as well, that this means irreconcilable differences and sometimes different mentalities as well. And this of course sometimes causes conflicts. And these not only strain the working process on the project but of course are also emotionally exhausting' (KEBAP, Board Member 3: 671–678).

The interviewee recites the conflict about the technological equipment which should be installed by the project for the production of energy. Underlying this question were conflicts about both teleologies and implicit general understandings (analysed in section 4.5.2). With regard to teleoaffective structures, this question resulted in a conflict between those actors who were interested in technologies that would realise the local economy principle and those who were more motivated by teleologies like producing affordable heat and contributing to future energy grids. While members tried to resolve the conflict using 'objective' numbers like estimated MW-hours or financial income, these numbers were rejected by other members, who argued that the financial income or MW produced would not be important factors in terms of the project's intentions (this will be described in more detail in Section 4.5.2). The conflict is an example of the interplay between the teleologies and technologies of energy production. It demonstrates that decisions with regard to a certain technology and its operation are not only the outcomes of technical or economic considerations but are shaped by the context in which they are utilized. The quote also implies that affectivities are an important part of this context. While emotional aspects might be especially visible in situations of conflict, it is shown in the beginning of this chapter that affectivities of individual persons—like anger about a Vattenfall's pipeline project or boosted self-confidence—were important for the creation of KEBAP. As these instances show, affectivities are important elements of the energy production practices in KEBAP.

Deriving from these considerations, it can be deduced that energy production in the KEBAP project is in large parts an outcome of its members' teleoaffectivities. Within the project itself, the different teleologies of the members are negotiated and integrated into the constantly changing teleoaffective structure of the project. The social practice of energy production both shapes and is shaped by the principles, interests, and affections. In KEBAP, the teleoaffective structure integrates, among others, the principles of self-empowerment and self-determination, ecological considerations, and local economy. At the same time, the social practices pursued in the project are heavily influenced by the historical and social background of the project and hence the teleology to challenge Vattenfall and the conventional energy system. Underlying this resistance are teleological and affective motivations, which have been transferred into the KEBAP project. Because of the constant interaction between the project as an institution and the fluidity of its members and their interests, the concept of the project has been changing frequently since its origins. One important change in this process has been the increasing relevance of the cultural part within the overall project.

4.5.2   General understandings: Demarcating KEBAP's social practices from those of the conventional energy system

The resistance against the conventional energy system and Vattenfall implies general understandings about the value or nature of practices in or of the conventional energy production system and general understandings about the practices enacted in KEBAP. The stand taken against Vattenfall and the energy system implies that at least certain aspects of these two entities are understood negatively. Among the most important aspects that are understood negatively by the members of KEBAP is the concentration of power.

B3:   'And it is about power. This is just, well for KEBAP, what always motivates me. Because I believe that those power shifts as we have them in the energy sector— well in other sectors as well [...]—that has caused a concentration of power on a few, who form an ominous ['*unheilvoll*'] alliance with political actors. This is obvious in the energy sector' (KEBAP, Board Member 3: 57–62).

The interviewee considers the existing energy system to be characterized by a concentration of power. As the interviewee engages against this concentration of power, it can be assumed that he values it negatively. This valuation is also signified in the term '*ominous*' ('unheilvolle' in German). Taking a stand against this ominous system is part of the respondents' motivation to engage in KEBAP. His motivation to engage in KEBAP might be called a 'resistance motive'. Implicit in his engagement for KEBAP is a general understanding of not only the opposed social practices which are deemed morally wrong, but also of the activities of KEBAP as being justified and (morally) right. The general understanding about the value and worth of KEBAP derives partly from the differentiation of the project from the energy system. Relating this quote to a sequence from another interview illustrates the importance of the resistance motive and the therein inherent differentiation.

M4:   'I was interested in the self-empowerment aspect. That citizens said: yes, independent of decisions made by politics and independent of the big companies, which have the power here, we do our own thing and thereby try to break these power structures. That was it, yes? So, I found it interesting that they wanted to break the monopoly of this mega-powerful company' (KEBAP, Member 4: 27–31).

Both quotes are from interviews with people who were members of the first core group. In their interviews, both explain that their motivation to engage in KEBAP resulted from their desire to engage against the existing power structures. Their decision to engage in KEBAP results from their opposition to what is being done otherwise. Both interviewees understand KEBAP as an instrument to challenge the existing system. General understandings about the existing system hence influence and shape both the engagement of people for the project and the way the social practice of energy production is being organised within KEBAP. This effect demonstrates the close relationship that might exist between general understandings and teleoaffectivities as well as between general understandings and activities. At least when practices of (other) actors are understood negatively, an important teleology for certain practices might be the desire to engage in a way that challenges or changes the former kind of practice. This kind or oppositional motivation is an organising element in many of the social practices enacted in the context of KEBAP. In fact, the resistance motive has heavily influenced the creation and establishment of KEBAP.

General understandings not only played an important role within the founding process of the project, they were also relevant in the conflict about the technologies and sources to be used in order to produce renewable energy. As mentioned in the case description, the original concept for the bunker included the installation of a biomass boiler technology to burn waste-wood from Hamburg's parks and green areas. During the members' occupation with the production of renewable energy, a CHP plant technology using gas as source came to be seen as promising for the project. In the second feasibility study conducted by the project, both technologies were evaluated by an external advisor. It turned out that both technologies were likely to be feasible, with the CHP plant probably enabling a higher income due to the smaller costs of gas. This result, however, did not resolve the conflict. Besides the teleoaffective reasons already explained, general understandings are a key to understanding the conflict. As explained above, the project is not dominated by one or two overriding teleologies; rather, the teleoaffective structure of the project is open for negotiation. Consequently, the conflict could not be resolved by considering a dominating aim of the project. Instead, different intentions were fighting for dominance.

General understandings provided important arguments with respect to the conflict.

TA: 'And in KEBAP—I call it a grassroots movement—things are of course discussed much more fundamentally. So, the concept of course has to be feasible in the end [...] but profit does absolutely not have any priority. Instead they discuss for example, [...] while finding the profit aspect of woodchips interesting, it is not of overriding importance for them. Instead the KEBAP members see it as a natural resource. And they discuss it against the backdrop or the problem of natural resources. So, is it really a re-growing resource? Or isn't this somehow like the woodchip-pellet production? Somehow completely non-ecological. And is this acceptable, and can you do it? And they compare this to burning a fossil fuel, like gas, in a CHP plant. And this discussion [...] is immensely important for them' (Technical Advisor: 148–170).

The quote describes how general understandings played a key part in the argumentation. According to the quote, the advantages and disadvantages of each technology were discussed mostly in terms of their ecological value. What was at stake in these discussions was the intentions of the different energy production practices as well as ideas about the general worth or value of these practices

themselves with regard to environmental objectives. Based on general understandings, both parties were able to argue for or against the technologies. The
key argument focused on the characteristic of 'decentrality'. As already described, creating a decentralized energy project was a key motivation at the beginning of the project. Woodchips produced from local sources were considered
to be able to realise this intention to the highest possible degree. While the process of burning gas would still happen in a decentralized institution, the source
itself would not possess this quality.

DA: 'The decentrality through which it is propagated will be a bit—so if one sees
       themselves as a natural gas project, a CHP—is of course not given any more. Be-
       cause, such a project you could do just as well in any other place. […] so with
       natural gas you can't talk about <u>decentrality</u>, because the natural gas does not
       originate from decentral places, and also might come from anywhere' (District
       Authority Officer: 390–398).

Two aspects of the material entity are described as reducing or even opposing
the decentrality motivation of the project in this quote (the technologies as material artefacts are analysed in section 4.5.6). Especially considering the history
of the project as a resistance project against the highly centralized energy system, it can be deduced that 'decentralized production' is generally understood
positively by the project members. The value of decentralized production can be
differentiated in a political/economical and an ecological argument.

The political and economic arguments against a centralized energy system
imply general understandings related to these social spheres and their underlying
logics. Centralization is associated with social inequality, political and/or economic exclusion, and thus questions of social power. The ecological argument
on the other hand is based on general understandings about the worth and value
of nature protection. Producing heat from non-renewable sources is generally
understood to be unsustainable or damaging to the natural resource reservoir.
Producing heat from coal, as Vattenfall had mainly been doing in Hamburg, not
only utilizes a non-renewable resource but also produces carbon dioxide emissions and consequently is understood to be destructive to nature, especially climate. To generally understand non-renewable energy as destructive to nature
implies theoretical forms of knowledge. The ecological argument is an example
of how general understandings include particular forms of representational
knowledge.

This example shows how general understandings, theoretical knowledge,
and teleoaffectivities are related to each other in the context of KEBAP. The
teleology of challenging the centralized energy system results from a negative
general understanding of centrality and its political, economic, and ecological
consequences. At the same time, general understandings about a certain way of

producing energy are influenced by the teleoaffectivities of the social site. If the aim is to challenge the system, a practice that enables this intention is likely to be generally understood as 'valuable' or 'good'.

The conflict about how and from what sources energy should be produced also entails that none of the available ways of producing or burning fuel perfectly match all the intentions or general understandings of the project. Burning biomass from wood waste, while being valued as fulfilling the intentions of decentrality and local economy, was criticized for having negative consequences with regard to emissions, traffic, and other issues. The third to last quote (Technical Advisor: 148–170) hints at one of the discussion points about burning waste-wood. In the argument mentioned by the Technical Advisor, the members of the project discussed the ecological consequences of burning wood. The argument centred on uncertainty about whether it would be possible to ensure that woodchips originate from local waste-wood. If they do not, the project would indirectly make use of agricultural land for the production of fuel instead of food. The project thus became entwined in the problem of globally decreasing agricultural land for the production of food (notes from meeting with the Technical Advisor). The argument about the wood sources thus entailed ecological and social justice considerations. As mentioned in the quote by the Technical Advisor, the fact that these kinds of considerations became meaningful has to be explained, taking account of the particular context. The Technical Advisor explains the relevance of environmental considerations by referring to KEBAP being a grassroots initiative and thus not primarily being interested in financial profit. Besides the ecological argument, the particular context of KEBAP also meant that considerations about the effects of the different technologies on the neighbourhood were also meaningful (section 4.5.7). Practices that have less negative impacts on the neighbourhood were understood as better, because they are less noisy, less polluting, or less destructive.

Within the KEBAP project, abstract senses about the ecological, social, local, and political worth or value of the projected energy production practices were meaningful. The different general understandings are not only an outcome of the project's teleoaffective structure but mutually influence it. Meaningful within the context of KEBAP are general understandings concerning the social, the economic, the political, the ecological, and the local sphere. These are specified in—among others—ideas about the worth and value of practices which contribute to issues like decentralization, renewable resources, non-CO_2-emitting production technologies, and local impacts.

4.5.3 Practical understandings: practically doing energy production

Within KEBAP, necessary practical understandings are related to the different practices pursued by the members of the project and the respective knowledge and skills that they can contribute.

B3: 'So I have learned a lot, and am still learning within KEBAP. Having learned sounds too completed. And I always find it fascinating and always try to use it immediately. And for me it entails three components. Whereby I could not decide which one is more important. First of course are just, technical questions. So, like those having to do with the issue of energy production' (KEBAP, Board Member 3: 711–716).

Planning and realising an energy production project requires a variety of skills and knowledge. While the practice of planning and projecting energy production entails theoretical knowledge about, among others, technological, juridical, political, and natural issues, it also necessitates certain forms of practical knowledge. The quote relates to the importance of acquiring theoretical types of knowledge. Learning about these issues, however, also entails practical knowledge or skills about how this type of knowledge can be accessed and acquired. In KEBAP, accessing and acquiring theoretical knowledge about the production of renewable energy has been done in two different ways. The necessary theoretical knowledge has been acquired not so much through autodidactic reading as through communication with other people.

B3: 'Whereby personally I am just very glad that we just have people in our team, who are originally concerned with this and who are able to help someone like me and explain things to me. So there I've got a constant learning process' (KEBAP, Board Member 3: 730–733).

The quote alludes to the specific way in which engagement and learning are understood within the project. Members may join working groups based on their interest and not their prior knowledge. Working groups thus not only include members who already have specialized knowledge in topics at hand. The project and its working groups provide a context for learning and the distribution of knowledge. Some of the most active members in the energy group, like BM1, BM3, and BM4, did not have any professional or other kind of specialization in the topic of energy production before they started KEBAP.

Planning and organising a project like KEBAP also requires interaction with different institutional actors from, among others, politics and public administration.

B3: 'Then, just, the differentiation between politics and <u>administration</u>, I wasn't aware of that so much before. So they can cooperate. But, as we have realised, they also can be very much pulling in different directions. And then you have to deal with both. And both of them want to be minded. And, that was new to me, and has been a learning process' (KEBAP, Board Member 3: 754–759).

In this quote, the board member recounts a form of knowledge that he gained not through theoretical learning but through practical experience. People might theoretically know about the differentiation between political and public administration institutions. The practical effects of this differentiation, especially the specific, practical ways of collaboration and opposition, however, have to be experienced. The interviewee also mentions that he became aware of the fact that the project necessitates cooperation with both political and public administration institutions. As the quote reveals, cooperating with these institutions requires practical knowledge not only about the particular ways in which these organisations 'work', but also about how to consider the interests and particularities of each institution and its members, as well as knowing about the collaborations and rivalries between different people and institutions. Accumulating practical knowledge about these institutions, their organisations, representatives, and—especially informal—ways of working is an important element for and within the social practices of the KEBAP members.

Practical knowledge acquired and required in the project is closely related to the activities pursued by the members. Besides energy-related issues, this includes activities like gardening, cooking, event planning, movie screening, and other cultural activities. As already described, cultural production is understood to improve the accessibility of the energy part. The cultural part, however, also contributes to practical understandings of the project.

B1:   'So, how can we combine that? What can people do there, or do there themselves somehow? And that's something I think is very important. To combine this abstract energy issue with something touchable, doable. And that is also something which for me is clearly showing to be true already. So I think it is very obvious that you can only interest people for this energy issue when you somehow do something, for example by gardening together' (KEBAP, Board Member 1: 364–371).

According to the interviewee, an important advantage of cultural activities is that they involve bodily activities. Energy production, on the contrary, is described as an abstract process that does not offer opportunity for practical engagement. Energy can be neither touched nor handled, nor are most (German urban) people able to produce the electricity or heat for their daily activities themselves. The modern energy production act is done by means of distant machines. Human beings only operate these machines or administer the production or distribution processes. This kind of energy production cannot be understood practically. People might learn how machines are operated, where the sources come from, and how the heat is being distributed, but they cannot practically or bodily participate in any of these processes. What can be encountered bodily is only the finished product—heat or electricity. This means that energy is not approachable for most people; it remains abstract and untouchable. Activities within the realm of cultural production on the other hand are mostly embodied or practical forms of knowledge. Gardening, cooking, yoga, making music, and experimenting are all bodily activities. People can see, touch, and smell plants, they can prepare and taste the food, and they can feel their body moving during bodily exercises or music making. Bodily activities are thus combined with bodily experiences. These are lacking in the energy production process.

The combination of energy and cultural activities is seen to provide a context in which participants can offset the abstractness of energy production by experiencing it in a context within which it is enacted simultaneously with more bodily types of activities.

B4:   'We have to be careful that the energy part does not become disconnected, because it now only allows access for specialists. So, the energy transition has to be maintained approachable. And that's why for me one quality of this space has to be that it opens up possibilities. For example, to understand how it functions and what options are available for everybody. Okay? Where can I save energy in a rational way? How can I be more efficient? What can I do in my own flat? So, not only in this bunker but when I leave and go home, to my own plug connector. So there has to be some knowledge transfer, a change of mentality. In other areas as well. So, like nutrition, eating, social interaction, culture' (KEBAP, Board Member 4: 511–524).

This interviewee's idea of the project is to create an open space where people can access and approach energy-related issues in different ways. By providing practical access for everybody, KEBAP aims to socialize or de-isolate energy production. Furthermore, by providing a space where people can access energy-related issues in a practical way, the interviewee hopes to enable a 'holistic' access to energy. For the interviewee energy production cannot be separated from energy consumption or efficiency. In the conventional energy system, people are socially and geographically distanced from the sites of energy production. Instead of differentiating energy production from the other two activity spheres by allocating it to the economic sphere and the two others to the private sphere, all three activities are meant to be dealt with in an integrated way within KEBAP. All three aspects are also explained to be related to people's daily activities. According to the interviewee, really understanding about energy necessitates first knowing about all aspects of energy, and second, not only learning about these theoretically but also practically comprehending energy consumption, efficiency, and production. The combination of the cultural and the energy part is seen as allowing people to learn about these issues both theoretically and practically, as energy production is being practically done within KEBAP. Furthermore, the combination is meant to enable people to understand energy as an integrated issue where they can directly experience the relation between their energy consuming activities and energy production processes. The interviewee suggests that acquiring practical knowledge of these issues can motivate people to also change their daily social practices outside the project. The combination of cultural and social activities with energy production activities in the KEBAP project is (among other things) meant to provide people with bodily experience about their social practices evolving amidst energy consumption and production.

4.5.4 Rules: Challenging, using, and creating rules in KEBAP

In KEBAP, two kinds of rules are important—those that influence the context from 'the outside', mainly rules that have been created in other social contexts, and those created within or through the social site of the project itself.

External rules can be differentiated according to the role they play within the context of the project. On the one hand, resistance to certain rules is an important element in many social practices enacted in the context of KEBAP. On the other hand, the project has to continually deal with existing rules that have to be obeyed or at least considered in the planning or enactment of certain practices.

As described, resistance against Vattenfall's pipeline—particularly its monopoly over Hamburg's heat grid—was an important motivation for creating KEBAP. This background implies resistance against a set of rules, as Vattenfall's monopoly was guaranteed through a number of formal rules. Concession rights gave Vattenfall the right to administer and run Hamburg's long-distance heat grid. The juridical fact that the heat grids are not liberalized in Germany enabled Vattenfall to prevent any other energy provider from feeding heat into the grid, which was quasi-'owned' by the company. The existence of these legal circumstances spurred the engagement of some members.

M2: 'What we have at the moment is a mixture of communism and planned economy. And the economic model of the energy industry is suspiciously close to what China is doing absolutely. A real market economy is somehow something different. […] And by now it works tolerably for the electricity grids, it works more or less for the telephone grids, it's slowly getting really good for the rail networks, even gas is somewhat okay, and for the long distance heat it just does not work at all. It does not work globally. And I did not want to stand for that' (KEBAP, Member 2: 41–49).

The interviewee is aware that the situation of the heat grid results from existing regulations. The interviewee '*did not want to stand for*' the situation resulting from the existing grid regulations. The quote also indicates the importance of not only formal rules but also normative convictions. For the interviewee, it is not just that the existing formal situation is contradictory to legal rules; he also legitimates his criticism through the normative convention of free markets. His belief in the existence and general acceptance of these normative conventions can be derived from the fact that he does not feel the need to explain the principle or rightfulness of the concept. For the interview partner, resistance against existing formal rules is an important element of the energy production and legal practices pursued in and by KEBAP. The social practices enacted in the context of the KEBAP project deliberately strive to challenge existing normative regulations of the heat grid.

KEBAP itself provides an important argument about the members' engagement for a changed regulative order.

M4:   'And this is a debate which mainly was a political and idealistic debate for most of the time. I mean, the liberalization of the long distance heat. But so far nobody has concretely claimed it. So, there was no project which claimed this for itself. Insofar there was no immediate pressure. And, what I thought was interesting—in the context of this initiative for which I was working—was the need to change this fundamentally, the ownership- and power-structures, and the enterprise policy in the long distance heat' (KEBAP, Member 4: 37–43).

B3:   'And then of course, mainly, now for the energy sector, the innovative thing is—especially for Hamburg in fact—that now here another heat producer is daring to approach the heat grid and takes the step to say, yes, we want to use the long-distance heat grid, just to provide our own customers. Something, obviously nobody else has been daring to do so far. And thereby destroy a monopoly' (KEBAP, Board Member 3: 509–514).

Part of the continuing non-liberalization of the long-distance heat grid, according to these sequences, results from the fact that so far no actor has practically claimed a change in the legal structures. In the absence of such a claim, the debate remains in the abstract sphere of purely ideological or political debates. As already explained, one of the motivations for the creation of KEBAP was to set up an actor which is in a legal position to make use of the rule of free markets and claim its right to feed-in to the long-distance heat grid. European and national rules concerning the organisation of free markets in Europe have opened this possibility. KEBAP has employed these rules to challenge Vattenfall's monopoly of the long-distance heat-grid. Put the other way around, the idea to create a project which—as a rival energy provider—could legally force Vattenfall to open its grid was made possible by the existence of rules. Here, KEBAP as a market player is made an instrument for challenging and changing the existing regulative organisation of the heat market. While many social practices in KE-BAP were inspired by the desire to resist the existing energy system, KEBAP itself is also an argument in activities through which the system is challenged. Resistance against existing rules is an element of the social practices both within and of the project.

Rules are not only constitutive of the project when they are resisted. Different external rules, which have to be obeyed, influence the practices within and of the project. Some of these rules are obstacles to the activities of and within KEBAP. Most important at the time of the fieldwork were rules concerning the ownership rights of the bunker.

DA: 'The federal state sells its estates through the BImA [...]. And this organisation is clearly focused on—if the purchasing party is a public institution as well—getting the highest possible price. In our case this could amount to the one or other million. And we have to see how this is educible' (District Authority Officer: 143–150).

According to the quote, the terms based on which the federal estate agency sells public property are critical for KEBAP. The rules pertaining to the ownership rights and terms of sales for estates belonging to the federal state require properties to be sold at the highest possible price. Neither KEBAP, nor Hamburg's authority are willing or able to pay the highest possible price, which is likely to be close to two million Euro. Without the bunker, however, the practices that are planned for the building cannot be realised. The members of the project have developed two plans to cope with the situation. During the time of the fieldwork, the strategy was to make use of another rule to force the BImA to lower the prices. According to one of the BImA's own rules, prices can be lowered if the potential buyer is a charity organisation and if the building is meant to be used for the public benefit. As both requirements are met by the KEBAP project, the members are discussing this option with the BImA. This strategy is accompanied by the project members' efforts to engage other public administration institutions and politicians with KEBAP. This strategy has so far been successful; representatives of the district authority and Hamburg's ministry for urban development and environment (BSU) have joined KEBAP in meetings with the BImA to discuss the possibilities. Furthermore, the information that the BImA has the option to sell properties cheaper if they are meant to be used for the public benefit was given to the KEBAP members in an informal meeting with two members of the district authority (own notes from the meeting). The project and its supporters in the public administration are thus trying to use a rule of the BImA to challenge another rule of the same organisation.

The concept of KEBAP criticises norms of private property. Instead of owning the building themselves, the members of the project want to convince

Hamburg's government to buy the bunker and lease it to the project through an emphyteusis contract. By leaving the bunker in public hands, the project members want to oppose the increasing privatization of sites in Hamburg that were once publicly owned. Instead of creating private property, the project wants to contribute to the establishment of common goods. The ownership question shows how rules are made sense of within certain contexts. In the KEBAP context, rules of ownership are associated with political statements. By using the rules associated with the idea of emphyteusis, the project members are able to make a political statement and resist the concept of private property. Because of this intention of the project, rules governing the question of ownership are made meaningful within the project.

A range of rules concerning the production and distribution of renewable energies is also relevant for the realization of the project. The 'Erneuerbare-Energien-Gesetz', EEG, (renewable energies act) constitutes the feet-in tariffs for the distribution of renewable energies and is essentially involved in the financial constitution of the project. At the same time, these rules are only relevant for and meaningful because of the social practices that are meant to be enacted in the project.

Besides these external rules, formal and informal rules have been created by and in the context of KEBAP. These rules are both expressions and elements of the social practices in the project. The most obvious of these rules are the formal ones that define the organisational structure of the project. The registered association and the energy-and-culture cooperative not only define the legal status but are also expressions of the project's general understandings and teleoaffectivities. Cooperatives are a legal form for organising common economic activities of different (natural and juristic) persons. Like other legal bodies, cooperatives set specific rules for how these economic activities are to be organised. The main principles are the principle of self-help in a democratic legal form, the identity principle, and the economic facilitation of its members (www.genossenschaften.de 2016). The principle of self-help in a democratic form describes the idea that members voluntarily decide to cooperatively pursue an economic activity that they could not have successfully carried out alone. Basic decisions are made in a general assembly where every member has one voice, irrespective of the number of shares owned. Because of this one-person-one-vote principle, cooperatives are protected from individual majority owners dominating their decisions and activities (ibid.). The identity principle pre-sets that the members are both owners and customers of their cooperative. The principle of economic facilitation of its members means that the facilitation of its members is the main aim, not the generation of the highest possible income return. This does not mean that cooperatives do not produce any profit. However, securing the property of its members is more important than the creation of a high return of investment. Consequently, cooperatives focus on 'safe' investments. In order to achieve the aim of facilitating its members, cooperatives thereby have to obey the rules of market and competition and have to act in an economical manner (ibid.).

These principles of cooperatives are translated with regard to the teleoaffectivities and general understandings of KEBAP.

B4: 'Cooperatives are a good way to include many people, so this has to do with the direct democratic element—that is good. It enables people with small amounts of capital to participate. That is good as well. It has the advantage that you can organise yourself—so to say—substantially. So in the spirit of the idea. Such, it is essentially ideal for KEBAP' (KEBAP, Board Member 4: 339–344).

B1: 'I think it is very obvious, because, cooperatives are the only legal form in which the idea of participation is realizable at all' (KEBAP, Board Member 1: 908–910).

In these quotes, cooperatives are described as the legal organisational form that best enables participation of as many people as possible. Both quotes explain those aspects of cooperatives that are relevant to the context of KEBAP and estimate these with regard to their suitability to the teleologies of the project. The legal form of cooperatives has been chosen because it best expresses and realises specific aims and intentions of the project. The cooperative form is made meaningful for the project because it enables KEBAP to translate its teleologies into a legal organisational form. The rules formalized in the cooperative principles are translated into the context of the project and interpreted with regard to this context. These rules thus are interpreted and generally understood with regard to the context within which they are made sense of.

Interpretation of rules means that (aspects of) these are altered in order to provide the best possible fit with the respective context.

B3: 'On a meta-level it is this connection of culture—self-determined culture—and self-determined energy production. I do not know any other example which puts these two side to side and also relates them. And what we also intend to do is to extravert this very much. That we are now putting both elements into the constitution of the cooperative and say: this is a culture-energy cooperative' (KEBAP, Board Member 3: 497–503).

The project's unique combination of culture and energy is an example of how existing formal rules can be adapted to a specific context. In the case of KEBAP the legal form of cooperatives is altered to accommodate the practices to be pursued in the project. The combination of energy and culture production in the project's cooperative constitution is meant to provide a legal form for the combined production of culture and energy. Much more specifically, certain elements in the cooperative framework formally institutionalize the idea that the energy production can financially facilitate and enable the cultural part of the project.

B2: 'Something like energy production, I mean, something where money is left over, which you can then transfer from one part to another, namely from the energy to the cultural part. After you've managed to pay off the system and to run it with all its costs in a way that it is cost financially self-sufficient. Then you can subsidize the cultural part with the profits. This has been the original idea of the concept' (KEBAP, Board Member 2: 107–111).

Financial cross-subsidization is one of the key relations between the energy and the cultural part. By formalizing the relation in the constitution, the energy and the cultural part are related to one another through rules. The idea of cross-sub-sidising the cultural part has been legally realised in the constitution of the en-ergy-and-culture cooperative. Mentioning this intention in the constitution and thus formalizing it as a rule not only realises the idea but also renders it independent of individual actors' aims and interests.

The members of the project are very much aware of the fact that the cooperative form does not by itself enable the realization of their ideas, but that a cooperative provides a legal form which has to be practically enacted to realise the underlying ideas.

B4: 'Cooperatives are a real form in which one can operate. They offer you just a way to organise. But in the end they are a legal form. And a legal form for now is something abstract, which is not yet filled with any contend. So you always decide what you want to put into this form. In my opinion you can use any form as long as you are aware that you are using a form. And that you do not fall for the forms' own momentum, the entelechy of a juridical form. Because it is tempting and it can be fatal, when you think that a form as such carries human, social emancipation in itself. It doesn't. Instead it is the humans who have to fill this form somehow and organise and work with it' (KEBAP, Board Member 4: 321–332).

The interview partner is critically aware of the fact that it is not rules but the interpretation and enactment of rules that realise ideas, principles, or intentions. Rules by themselves here are defined as enabling certain activities but not as equivalent to the activities themselves. In arguing that '*you always have to decide what you want to put into this form*', the interviewee explains that rules and their meaning need to be interpreted within a certain context. How these rules shape the social practices in the project is not naturally given, but results from sense-making of these rules in the specific context.

In the second part of the quote, the interviewee warns against being deceived by the momentum of a certain form. Rules do not themselves create a certain reality or kind of practice, nor do they constitute social emancipation. Instead, rules need to be actively made an element of the social practices of the actors in order to become 'real'. In a later sequence, the interview partner provides a practical example of this problem, where he talks about the legal form of a registered charity and its effects on the organisation of decision-making in the project.

B4:   'There happens a specialization, a decoupling of board and group. Here we witness exactly what I've been talking about earlier on. That suddenly the form starts to define the content. Only because the statutes of an association define the existence of a board, suddenly you have people who are ordering the operative ideas and which meet internally, whereby important communications primarily happen and are kept intern. And this is not re-connected to a group which suddenly is degraded into being members' (KEBAP, Board Member 4: 621–628).

The interviewee describes a conflict that occurred during the fieldwork. Members of the project complained that despite the project's original intention of providing direct democratic and participatory processes, certain key discussions and decisions were primarily being carried out among the board members. The existence of the board originally resulted from rules defining the workings of registered associations, which determine the existence of such an organisational entity. Unlike the working groups—the form in which all other activities in the project are structured—membership in and activities of the board are not defined with regard to certain activities. While the organisation of the working groups was established in a long participatory process, the existence and activities of the board were defined by legal rules alone. The members felt that the board and its activities were not based on the 'proper'—directly democratic—way of decision-making in the project. As the interviewee explains, the legal form had begun to define the content. This was especially critical as the form was not in accordance with the project's original intentions of participation and direct democracy. This problem shows that interpretation and realization of rules necessitate negotiations. What rules are given meaning and how, and thus how they come to influence social practices, have to be negotiated with regard to a certain context. The critical question thereby is whether it is the teleoaffectivities and general understandings that define the meaning of a certain rule or whether rules start to determine the social practices in a context. The interpretation of rules is likely to stir negotiations about different actors' teleoaffectivities. In the example of the conflict described above, the board members had tried to respond in a fast and effective way to challenges, which are constantly occurring—mainly communications and problems with political and public authority institutions. Their practice was thus organised from a desire for pragmatic decision-making. The members of the project on the other hand argued for participation and direct democracy. The conflict continued for the duration of the fieldwork and was only contemporarily stabilized when it was decided that the board had to communicate all events and ideas back to the assembly of members—the main decision-making body in the project. While this structure momentarily settled the conflict, it gave rise to continuous discussions about the independence and responsibilities of the board and other groups with regard to specific activities. What kind of decisions needs to be made by the members' assembly and what decisions can be made by the members of a certain group? These constantly evolving conflicts demonstrate the problem in the interpretations of a certain rule.

While rules are important for the social practices in the KEBAP project, they do not directly lead to certain activities. Instead, rules need to be interpreted and made sense of in a certain context. The context significantly influences what rules become relevant and how they are interpreted with regard to certain social practices.

4.5.5 Human actors: members and local residents

A range of different actors are related to one another, have meaning and identity, and are positioned within the context of KEBAP. In this section, the relations, positions, and meanings of project members, local residents, network partners, politicians, and actors from the district public authority are analysed.

The main reason for people to commit themselves to KEBAP is their interest in the project's ideas or in any of the activities enacted in the different working groups.

M1: 'I somehow found it interesting because there is an issue in it which is interesting for me. It fitted my interest, well, to finally find a way to do something in the neighbourhood—that was interesting for me. As well this idea of feeding heat into the long-distance grid.

I: What exactly? The technical or the legal aspect?

M1: Well, just generally to try it. The technical aspect as well but, just generally, to get through with it, against Vattenfall' (KEBAP, Member 1: 8–14).

Two 'types' of interest are mentioned in this quote. The interview partner combines his interest for the subject matter with his desire to engage locally. Both interests and their combinations are commonly occurring motivations among the project members. While not directly neighbouring the bunker, most of the members live within the boundaries of the sub-district. The spatial relationship of members and the bunker is mirrored in the project's self-description. In its publications, KEBAP is generally described as a neighbourhood project. KEBAP, however, is related to its members not only spatially. Being a volunteer project, the interest and engagement of its members essentially constitute the existence of the project. The members are intentionally related to the project, i.e. the project is meant to realise the members' ideas and interests. As shown by the quote above, the spatial relation is a meaningful aspect.

The relation to the locality, however, is ambiguous within the project. Though the project is locally related to a particular neighbourhood, none of the project members are residents in the neighbourhood, which is seen as a problem within the project.

B3:   'So, the most important group, actually, which is also most directly affected—we know them the least. And that's the local residents' (KEBAP, Board Member 3 at energy working group meeting: 175–177).

In this quote, the interviewee complains that the project does not have access to the local residents, who are understood as an important actor group for the project. Not having this access is meaningful within the context of the project, because KEBAP is defined as a participatory project that actively includes and represents the interests of the local residents. It is important to note that the quote is from one of the energy group meetings, within which the conflict about the two technical versions became most pronounced. During this conflict, the quote above is used as an argument in the context of this group. The legitimacy of the argument within the group is shown by not only the fact that it remains unquestioned, but that it remains an important and repeated argument throughout the whole discussion.

B1:   'So, I believe, we are at a point, where we can do that [present a fixed project plan comprising the decision for one energy version, AP] on this date. But then we are just an investor project. And those local people who are not there on that date, their opinion will be ignored. And we have to show that we are different. We have to make that effort. […] It is a process, a dialogue, a very long-term dialogue, in order to listen and find out what they [local residents] want' (KEBAP, Board Member 1 at energy working group meeting: 309–116).

In this later sequence from the group meeting, investment of effort and time to include the local residents and their interests appears as a key reason that KEBAP is not a normal investor-project. By distinguishing the project from '*just an investor project*', it is again (as with regard to the conventional energy system) defined by way of opposition. The inclusion of the neighbours' ideas and interests, and thus the creation of direct relations between the project and the local residents, are made an important argument within the context of the project. Within that specific context the (unknown) local residents are given meaning as sources of legitimacy of and for the project. As no 'local residents' are present in the project, their interests can, however, only be assumed. The ambiguity of the project's relations with local residents also derives from the fact that people who define themselves as not being local residents have defined the project to be a local project.

The ambiguity with regard to the local context does not only comprise the 'real existence' of the project's relations with the local residents. Rather, too much concentration on the local context is also understood as potentially dangerous.

B: 'Because, even when we are talking about local economy or local anchoring, it is
        not about separation. This is an effect which will immediately cause breakdown.
        In the moment where you put railings around the local, it is essentially dead al-
        ready. And, that's why it's so important, to—from the beginning—relate the local
        to something bigger' (KEBAP, Board Member 1: 556–562).

The quote describes the danger that local projects may become socially irrelevant if they keep themselves separate from '*bigger*' social processes. While being local is an important aspect of the KEBAP's meaning, it cannot solely be a local project. The interests and ideas pursued by the project have to be (made) meaningful both in the local context and with regard to larger social processes. According to the quote, it is necessary to embed local projects into social networks in which actors and concerns are related to one another and to larger social processes. In fact, relations to different social networks are important for and in KEBAP.

I: 'Would you say that networks are important?
B3:   Absolutely. Many of the people who are now active in KEBAP have come to us
        through networks. These are either already networks of initiatives, I mean organ-
        ised ones, as well as personal, private networks. And the origin of KEBAP was a
        network as well. So, we found us as a group, within this initiative back then' (KE-
        BAP, Board Member 3: 1037–1045).

With regard to its members, KEBAP can be described as originating from pre-existing networks. The Moorburgtrasse-Stoppen initiative not only shaped the project's teleoaffective structure but also provided the network through which the actors of the first core group met. Put the other way around, the project is related to the Moorburgtrasse-Stoppen initiative by not only intentionality (resisting Vattenfall and Hamburg's government's energy policy) but also by human actors. As mentioned in the quote, KEBAP is related to different social networks. Via the active co-engagement of members in different networks, the project is positioned within or related to different social orders. Examples of this are relations to organisations in the renewable energy sector, urban gardening projects, and the Right-to-the-City movement. KEBAP is related to these organisations and initiatives through actors who engage in both KEBAP and these movements. These people represent KEBAP within these initiatives and also transfer ideas and information from these initiatives to the project. KEBAP also becomes related to these initiatives whenever human actors from KEBAP take part in activities organised by other initiatives like discussion events, demonstrations, or petitions. Likewise, individual members of these initiatives are regularly mobilized to take part in KEBAP activities and events.

Through its networks, the project is also related to scientific social worlds. Apart from this dissertation thesis, students and scientists from different disciplines have participated in or have made the project part of their scientific work. Examples include a scientist and two students from technical environmental sciences, who contributed to KEBAP's development by analysing and evaluating technical versions of the project. Apart from this thesis, students from different social sciences have for example analysed the meaning of the project within the 'commons' debate. The engagement of these actors in and for the project transfers scientific knowledge, ideas, and ways of thinking into the project. In fact, many KEBAP members have started to describe the project as a scientifically interesting pilot project that can contribute to the creation of scientific knowledge. KEBAP, for example, has been described, and has come to describe itself, as a project in which transdisciplinary research is being conducted (Veciana/Neubauer 2013). Thereby, the project is not only being passively presented as a 'research object' in publications or conference presentations. Representatives of KEBAP have also actively taken part in scientific conferences or panel discussions. Through these interactions with scientific actors and institutions, scientific ideas, concepts, actors, and events have been made meaningful in the context of the project. At the same time, the project has acquired a meaning for itself as a scientifically interesting and relevant project. Thus, at the same time, it has started to be an actor in scientific social arenas and has made different scientific worlds part of certain social arenas within it.

An important way by which 'science' has been made meaningful for the project is its employment as a source of legitimacy. The fact that KEBAP was an object of research, and—even more—that scientific analysis supports the technical viability and social relevance of the project, has been used as a resource in—especially early—presentations of the project. This demonstrates the meaning of these networks as resources for the project.

B1:   'Indeed, it is so that in fact, the previous success of the project has indeed depended on this kind of strategy. Or, from, well, the involvedness in a larger network creates a power both internally and externally. So, concerning KEBAP and a certain general perceptibility and assertiveness to the outside, and of course the suggestion of power, this is really important. So, it's just a very important political argument. On the other hand, it is also important internally, because, to feel and understand yourself as being part of something bigger, are two important references. First, because it gives a certain strength. Second, also because it opens up certain perspectives' (KEBAP, Board Member 1: 543–552).

This quote describes different ways in which networks are turned into resources. First, being part of a bigger social movement or institution transfers some of the power of that movement or institution onto the project. It is no longer a small local project that has to mobilize awareness, legitimacy, and other resources itself; instead it is able to make use of the mobilizing potential of—for example—Greenpeace Energy or the urban gardening movement. Cooperating with organisations like Greenpeace Energy transfers some of the latter organisation's reputation onto the project. Also, members from Greenpeace Energy or other organisations have provided information and ideas to the project. Embedding KEBAP into the urban gardening movement provided the project with access to material artefacts like seeds, earth, and gardening tools, as well as with practical 'hands-on' support information, and advice from other institutions. Through these networks, KEBAP also becomes part of an even bigger network, which can be termed the 'alternative movement'. According to the quote, organisations from the alternative movement provide not only resources like information for KEBAP but also a way of social sense-making for the activities, ideas, and intentions of the project.

Different organisations from the alternative movement have been important for the project.

I: 'Are the alternative institutions, which have developed in the last years, important?

B4:   Yes, absolutely. There are now some more than there used to be. So there are just many competent partners like alternative project planning offices, which just know how to build up such an alternative project. And which also know about the difficulties and so on' (KEBAP, Board Member 4: 429–43).

The members of the alternative network include actors like artist collectives, squatted houses and project communities, and actors and institutions from the urban gardening, environmental, anti-nuclear, and Right-to-the-City movement. In the last years, the existence of the alternative movement has also created alternative versions of conventional institutions like alternative project planning and architecture offices, solicitors, and cooperative managers. These institutions work as resources for one another, for example through the exchange of goods, advice, and (moral) support. An important way in which the alternative network has become relevant for the energy production of the project is by providing future customers.

TA   'They say, "this is a cool project, and we just want to do this project. And we will see about the customers later. That is going to <u>sort itself</u>". And the <u>interesting</u> thing is: it <u>really</u> does. They actually find people, like this squatted house community, who say: "you are great. We want you as our energy provider". And I sit here and think: ((laughing)) look at it, it works, it can actually work like that' (Technical Advisor: 185–190).

The interview partner here describes how the alternative movement enabled KE-BAP to easily find future customers for the energy project. He recounts that he had originally thought that finding customers would be one of the biggest challenges for the project. He however, experienced that through its embeddedness into the alternative movement, the project mobilized customers without having to engage in customer acquisition. Information about KEBAP and its intentions had spread within the network and among actors, who supported the projects' interest to resist Vattenfall and other big energy providing companies. At the end of the fieldwork, the project had exchanged letters of intent with three institutions (a squatted house group and two alternative living projects). The existence of these relations indicates that the project has been accepted and is thus meaningful as a future energy provider in the alternative scene.

Apart from networks with neighbours, scientists, and social movements, KEBAP is also actively creating relations with actors in the public authority. The creation of social networks with the local public authority is an important resource for the project.

M4:   'I think it's good, that they for such a long time have engaged with the district authorities, and not on the federal level, because here they can gather power. When they have the support from there. The longer they can seek approval from there undisturbed, also from important politicians and the respective parties, the longer they can seek their support the stronger they will be, when at some point the Hamburg government decides to intervene. This will make it much more difficult for Hamburg's government level to intervene in a restrictive way, if the district government already said: no, we think the project is good. We support it. Then the big brother can't just kick his smaller brother's butt' (KEBAP, Member 4: 533–542).

Consent and support from public authorities is crucial for the realization of KE-BAP. The importance of this support could be seen in the conflict with the investor competing for the bunker. If the political actors would have decided to support the competitors' apartment project at that time, KEBAP would have had no chance of realising its project plans. Furthermore, the district authority is an important partner in KEBAP's negotiations with the BImA.

DA: 'Informal talks are surely important as well. From the perspective of the project developer, because he of course has to create a certain atmosphere and a certain understanding, or interest. And that is true for, as I would say, the administrative, but actually especially for the political side, because it is more likely to be open to being influenced in this way, maybe' (District Authority Officer: 182–195).

In order to ensure that KEBAP is developed in accordance with administrative rules and political aims, the project has to maintain relations with these two actor groups. In order to establish and maintain good relation with the public author-ities, KEBAP has to integrate aspects of the order of these social worlds. In the case of the public administration, KEBAP has to adhere to the formal rules per-taining to building plans, emission regulations, and special requirements result-ing from the location of the bunker in a social preservation area. In their inter-actions with the public administration, the project representatives have to ensure that the plans and projections of the KEBAP are developed in accordance with the existing formal rules and their interpretation by the district administrations.

In order to be accepted by political actors, KEBAP does not have to meet the formal standards of the administrative social world, but needs to be mean-ingful within the political social arena.

M4:   'I, for example, was there when the district council or the redevelopment advisory
       council were being bewitched. Which did work out really well. And thereby—
       that was pretty interesting—the aspect became relevant, this co-development of
       creating cultural spaces for local residents, to give them a room, and the energy
       aspect. I had the impression, this, as a package, was somehow very captivating.
       And, importantly, because of the social aspect we have many people in the project
       who are able to take people along. No matter if it's the development advisory
       council or the district council or a local resident. We have people who have high
       competencies in that and enjoy doing it. This was surely helpful for this very mun-
       dane energy part, in which these competencies hardly exist—also helped its ac-
       ceptance. And, on the federal state level, I know that there some of the KEBAP
       members continually remind politicians from this level about the existence of KE-
       BAP, and they are often interested' (KEBAP, Member 4: 68–83).

In order to be accepted and thus supported by relevant political actors, KEBAP has to become a meaningful actor within the district or federal state political social arena. Different aspects of project realization require support from polit-ical actors. One example is the acquisition of the bunker. If the political actors are not in favour of KEBAP, it is unlikely that they will agree to purchase the bunker and subsequently rent it to the project on a 99-year lease. Making KE-BAP meaningful for political actors implies a politically interesting project idea.

DA: 'Well, both [aspects] are interesting. Probably, from the viewpoint of the district,
       maybe, so to say, from political actors, I think the energy part. Because, then one
       would have, so to say, a showcase project. From the viewpoint of the Hamburg,
       maybe it's more the cultural part' (District Authority Officer: 413–416).

In this quote, KEBAP is described to be interesting for district and federal politicians because certain political interests can be realised through it. Creating a showcase project for renewable energy production is described to be a meaningful activity for politicians. It can be argued that KEBAP is not meaningful for these politicians as a resistance project against Vattenfall or Hamburg's government's energy policy. Instead, the project is meaningful for politicians as an innovative renewable energy project. The KEBAP members are well aware of the politicians' divergent interest in the project. Consequently, when presenting KEBAP to a political audience, the resistance aspect is only mentioned in terms of the historical background of the project (own notes from presentation at the Green Party).

A situation in which political support was crucial for KEBAP was when the project had to compete against another investor for the bunker. Prior to the public meeting in which the two projects presented their plans for the bunker to actors from the public authority and the local public, the conflict about selection of energy version escalated in KEBAP. As has already been described, one reason was a disagreement about the necessity to make a decision about which energy version the project should present. Some actors argued that the project would not be taken seriously by the political actors if it could not even present a project plan with a complete energy version. The opposing members believed that the project would contradict its self-definition of being an open and participatory process if the members of the project would decide about such a central issue as the energy version without including the ideas, interests, and voices of local residents. The escalation of the conflict (among others) appears to have been about the question of whether and to what extend KEBAP should adhere to (anticipated) political forms of sense-making or make these part of the project development. The integration of certain forms of sense-making into the context of the project appears as (partly) deliberate process that needs to be negotiated.

Different actors participate in and shape the context of KEBAP. The members of the project are the most important actor group. They are related to the project in many different ways; most crucially, the project would not exist without the voluntary engagement of its members. Other actor groups are relevant with regard to certain aspects of the project. Local residents, scientists, and members of the alternative movement contribute different resources. As exemplified by the local residents, actors do not even need to actively participate in order to be meaningful within the project. Other actor groups, like administrative and political human actors, can either support or hinder the realization of the project. Consequently, it is important to integrate these actors into the context of the project.

4.5.6 Material artefacts: the bunker and the technological versions

The material artefact(s) of the bunker shape(s) the creation and realization of the project in different ways. As explained above (Section 4.4.4), one of the later board members had the idea to alter a concept he had originally developed in his parents' village while cycling past a bunker. He had the idea that the material features of bunkers would make them suitable locations for the storage and combustion of woodchips. The material existence and visibility of the bunker causally led to the development of the project idea. Realising the potential of bunkers for the urban production of renewable energy was also based on his knowledge about the material infrastructure of Hamburg.

B2:   'The idea to use a bunker evolved from the idea to make use of a relatively little embattled space in the city, which is also as large as possible. And to use that in the most useful way, so that it would be of use for as many people as possible' (KEBAP, Board Member 2: 722–725).

The first positive material feature of the building, according to the quote, is that it is '*as large as possible*'. The concept of an energy and culture bunker requires a lot of room in order to install the technical equipment and realise the different social and cultural activities. The bunker is materially suitable for the projected activities of the project.

The quote also describes the bunkers as a '*relatively little embattled space in the city*'. This description relates the building to processes of gentrification, which have caused many social conflicts in Hamburg's central districts. The expression relates to the material features of the bunker, which have made it uninteresting for estate agents and private investors. Because of their thick walls and concrete ceilings, bunkers are hard to destroy or refurbish, making it unprofitable for investors or estate agents to purchase and refurbish them. Though they are situated in Hamburg's city centre, the material features of bunkers make them relatively uninteresting for conventional investors. As a consequence, bunkers are among the few sites in Hamburg's city centre that until recently had been excluded from gentrification activities. With the decreasing amount of space left for real estate investment and speculation, this situation has been changing in the last few years. Nevertheless, bunkers are still relatively little '*embattled space[s] in the city*'. The characteristic of bunkers as 'empty urban spaces' makes this kind of material artefact 'socially' suitable for the project.

The geographical position of the bunker was another important aspect of the material artefact.

B2:   'The Moorburgtrasse-Stoppen initiative made the suggestion to concretise the
        concept for this particular bunker. Which, as a detached building, a very big one
        as well—one of the biggest high-rising bunkers in this area—also is by its location
        predestined for this kind of use. Because it is situated relatively close to the
        planned long-distance pipeline and also is situated in one of the parks through
        which the original pipeline should have been laid. I mean they planned to dig up
        this park' (KEBAP, Board Member 2: 119–129).

The interviewee explains that the bunker has been chosen by the initiative because it fitted the intention of the first core group to oppose Vattenfall's pipeline
plans. The selection of this particular bunker was motivated by the specific context within which the project was being developed. More specifically, the material artefact of the bunker was chosen because it suited the teleoaffectivities of
the first core group. Within the teleoaffective context of the initiative, the position of the bunker in geographical space acquired a particular symbolic meaning.
This symbolic meaning becomes even more pronounced in the following sequence from the interview.

B2:   'Interestingly the bunker is also situated more or less exactly halfway—if you look
        at the map—between the—still under construction—coal power plant and the
        planned feed-in point. The pipeline would have been nearly eleven kilometres
        long, and the bunker is more or less halfway, very close to the planned pipeline.
        For us this was the crucial reason, to plan the whole project as a stumbling block
        for Vattenfall—which has apparently worked out alright' (KEBAP, Board Mem-
        ber 2: 132–138).

The position of the bunker is 'interesting' for the actors in the context of their
aim to challenge Vattenfall's pipeline plans. It also results from the spatial relation between the material artefacts of the bunker and the planned material artefacts of the pipeline. According to this sequence, it was this particular position
of the bunker that inspired the members of the core group to use the project as
'a stumbling block for Vattenfall'. Deciding to use this particular building has
only partly been inspired by technical or rational considerations related to the
production of energy. For most parts, it is the outcome of the specific intentions
and interests of the project members.

Nevertheless, technical considerations also played a role in the decision
process. As mentioned, the bunker's size and construction make it a feasible
space for the production of energy using woodchips—which require a large and
fireproof storage space. Furthermore, the bunker is situated only about five meters from the existing long-distance heat grid. This position in relation to the
energy grid 'materially substantiates' the project's plan to feed-in to the grid.
According to a project description on the KEBAP webpage

'KEBAP is optimally situated to feed-in into the existing long-distance grid. The position of the bunker is in principle even particularly well suited for the existing grid-situation, because it is at the start of a side-branch at the outer part of the grid. These particular parts on cold days have repeatedly suffered from supply shortages'.

According to this statement, the position of the bunker in relation to the energy grid provides a technical argument for the project. In this statement, KEBAP is made sense of with regard to heat supply security. Thereby the locality of the bunker is related to the heat grid, the existing energy producer, and the heat consumers.

To summarize this line of argument, the decision for this particular building has not been an outcome of considerations with regard to just one objective (energy production). Instead, the bunker appears as a site that successfully negotiates different aspects meaningful for and within the KEBAP project. The material features of the bunker specifically accommodate the diverse teleologies and requirements of the project.

This 'ability' of the bunker implies the negotiation of the cultural and the energy part. Like its position in relation to the planned pipeline and the existing heat grid, the particular construction of the twin bunker facilitated and inspired the idea of combining culture and energy within the bunker. The separation of the two parts and the space available in each part spurred ideas to use the second part of the bunker for something other than energy production. Throughout the existence of the project, different members have contributed ideas about how the building could be used. The first idea had been to refurbish half of the bunker to accommodate exercising rooms for musicians. The argument was that the thick walls of the bunker would prevent the transmission of any noise produced in the bunker to the outside. Because of its material features, the bunker was understood to facilitate the particular requirements of musicians. The potential meaning of the bunker as a rehearsal site for musicians resulted from the material facilities of the bunker in combination with the situation of musicians in the district. This particular potential of the bunker was 'seen' due to the particular interests of the group members. As already described (Section 4.5.1), other ideas like urban gardening were added to the concept. As with the rehearsal rooms, the material features of the bunker were relevant for all these ideas. Most ideas were mainly inspired by the general availability of space in the context of an alternative project. Certain activities, like urban gardening, however, resulted from specific features of the building, like the existence of a large (unused and sunny) rooftop or the unused grass patch at the southern side of the building.

The material features of the bunker have thus shaped activities and ideas within the KEBAP project in many ways. They are also related to the project in the sense that certain material/spatial aspects of the building were in specific ways made meaningful within the context of the project.

The negotiations of the first core group not only made the bunker meaningful, they also led to the substitution of the micro-grid idea with the aim to feed heat into the long-distance grid. Having altered the distribution strategy of the project spurred ideas about other potential sources for the production of energy. This development opened up a new social arena in the project. Within this social arena, different actors and social worlds negotiated the best technical version of the project. Underlying this material decision process was a decision about the project's aims and intentions. Originally, the concept was to install a biomass combustion technology into the bunker, which would supply households through a micro-grid. This concept was dominated by ideas like decentrality and renewable energy production—especially as the aim was to use woodchips from so far unused local sources. The decision to feed-in to the long-distance heat grid led to a discussion about whether a gas CHP plant might be better equipped to realise the ideas that were associated with this plan. Feeding heat into the long distance grid created ideas of designing the project such that it would become a much needed element in a changing grid infrastructure. The advantages and disadvantages and thereby also the relevance of the underlying ideas for the project were discussed in this new social arena, which soon became the most conflictive and critical one for the project.

The specific idea to use a gas CHP plant resulted from the integration of certain human actors into the project. One of the project members contacted a technical scientist from Hamburg's University of Applied Sciences (HAW) and enquired about potential collaborations. The scientist agreed to ask his postgraduate students if one of them would be interested to write a thesis about potential energy production technologies in the bunker. The first Master's thesis produced in this realm compared the biomass-boiler technology envisaged by the project to a gas CHP plant. The decision to use a CHP plant as comparative technology resulted from the institutes' particular research interests (informal talk with the postgraduate student). The first thesis about the project compared the potential of the biomass combustion technology and a gas CHP plant. While determining that both technologies would be technically realizable in the bunker, the thesis concluded that using a gas CHP plant would enable the project to play an important role in a future energy grid system based on renewable energies. Because of their technical characteristics, gas CHP plants could play an important role in a system of mostly highly volatile energy production sources.

In the KEBAP project, these technologies and their respective advantages or disadvantages were not so much discussed with regard to technical arguments as they were generally evaluated and discussed with regard to their suitability for the project's teleoaffectivities. This discussion spurred a conflict about the legitimacy and hierarchy of aims and intentions.

B3:   'So this is the point now, far from a purely technical solution. We are at the point again of what is KEBAP about? What do we stand for? What are our goals? That's what defines this question. Do we say, we as KEBAP, we mainly aim for climate protection? This is our main goal, our main intention. Then we have to align the technical installations according to that. When we say decentrality, energy cooperative, maybe also potential financial profits—no matter how we are going to use these—this is our focus. Then this affects the technical installations we want to have. I think this is what we have to discuss first, and then deduce from thereon' (KEBAP, Board Member 3 at KEBAP meeting: 16–25).

The quoted statement was made at the beginning of a meeting specifically dedicated to creating a discussion tool that would enable the group to select the technical system to be installed in the bunker. In this quote, the decision about any technical installation is framed as a decision about the project's main goals and intentions. According to the speaker, any decision about the 'right' technology necessitated a decision about the project's *'focus'*. The rationality of a certain technology is depicted as existing only with regard to a particular objective that is meant to be realised through that technology.

The 'opponent' in the discussion tried to argue that the decision with regard to the energy technology should be derived from an 'objective' appraisal of the technologies' respective characteristics in terms of certain criteria.

FM: 'What we should do is to start comparing. That we try to, as objectively as possible, grapple with criteria. And to compare. And then through this we will come up with one version. This is more convenient for the neighbours. And for the other version we will find that it is cheaper. Then we have many different criteria. And with regard to some version A is better, for others version B. […] then at least we know where we stand' (KEBAP, Former Member at KEBAP meeting: 147–152)!

In this statement, the former member of the first core group (who left the project because of this conflict) believes that the decision with regard to the energy technology involves objectively estimating and then comparing certain criteria. The neighbours' wellbeing or economic profits are, however, not naturally given criteria. Rather, their existence and relevance to the situation are closely related to the specific context of the project and—among others—the interests and values of its members. Any decision about the comparative valuation of criterion is also an outcome of underlying aims and understandings. The catalogue of criteria and the value of each criterion are related to the context within which these criteria are established and used.

The conflict shows that the decision with regard to a technical version is not the outcome of 'objective' or rational decision processes. Instead, the different technologies were estimated and valued with regard to their suitability for certain interests and aims of the project. The conflict furthermore shows that different ideas and concepts, like climate change, local economy, or decentrality, might underlie the decision for certain energy production technologies. In the context of KEBAP, the decision to use a certain technology appears as an outcome of a process in which different technological versions were negotiated in relation to a variety of different interests. Thereby, the heterogeneous teleologies of different actors had to be negotiated with one another.

M4: 'In such a process the pet interests of individual people can become much more pronounced. If one says, for me it's especially important that we produce heat without producing any CO_2, and another one says, no, for me it's more important that we protect the energy system, and support and protect that. And the next one says, I want it to be as cheap as possible so we can produce as cheap as possible. And all these interests encounter one another. And one person says, I want to support the new technology in order to create more diversity, and so on and so forth. And because all these people have an equal position, these different interests can be dealt with in another way within the discussion process. This can be extremely dangerous for such a group but also a very binding process if it's successful' (KEBAP, Member 4: 374–386).

The interviewee describes the conflict as a discussion process in which different people tried to push their ideas and interest. The quote highlights the range of arguments used in these discussions. By means of the different technology versions, people argued for the realization of such dissimilar interests as renewable or low/no-carbon energy production, affordable heat production, or technical innovations. The conflict shows the instability in consensus about the project. Decisions are the outcome of more or less conflictive negotiation processes. What is at stake in these negotiations is not only decisions about activities or installations and how these realise certain intentions of the project. What is also negotiated is the question of what the project is or should be about. The arguments used by the different members are not pre-existing 'naturally given' arguments. Instead, what is seen as valuable, rational, and technically of financially viable is only decided with regard to one's own interests. Whose and what interests are able to momentarily dominate the social site are negotiated within social arenas like the discussion about the different technologies.

After the exit of two members, the conflict about the energy technology versions calmed down. Instead of taking a decision on just the technological version, the actual concept used a mixed modular approach.

B4:   'Lately things have tended to a modular solution. So we now see; this is not a
        decision which has to be made now and forever. But, it's going to be composed
        of different modules, which should and could function separately and which also
        can be combined in changing ways, can be adapted. So, that's how one has to see
        it. Because this whole sector is changing rapidly. We have certain goals which we
        want to realise, but we don't know yet which technology in what combination in
        the middle or long term will be the best. So those technologies are developing so
        extremely fast, it would be foolhardy to say one would know already what's best.
        So now it has to be an open process, and an open discussion' (KEBAP, Board
        Member 4: 370–380).

The quote highlights the dynamic character of the discussion process. This dynamic character is related to developments both in knowledge and of the technical artefacts. Both are described to be changing so fast that a permanent solution would not be sensible. From this perspective, a modular approach appears
to be the best solution. This modular composition is more likely to be adaptable
to changing situations and requirements. The quote argues for adopting a modular approach as a way to maintain the project's ability to respond to changes
and developments in a highly dynamic field. By composing the system as a modular one, the project is more likely to remain adaptive to technical or social developments. Instead of a closed and self-contained system, creating one based
on the idea (and thus has the technical measures) of integrating different technologies, will improve the adaptability of the whole system. The composition of
the material artefacts thus depicts the teleology of wanting to create a project
that will work within the developing system of renewable energy heat provision.
Besides being able to respond to technical developments, the modular approach
is also able to realise the interests of the different social worlds that have come
to exist with regard to the question about the energy technologies.

The example of the different technological versions shows the extent to
which material artefacts are related to other aspects of the context within which
they are (meant to be) used. In KEBAP, the two energy technologies are meant
to function as realizations of the members' heterogeneous interests. Because of
the durability of these material artefacts and their central position in the project,
the decision about the technologies will (momentarily) stabilize certain ideas
and objectives in the project. Material artefact do not exist independent of, but
are very much socially embedded in, the context in which they are used. They
are mutually shaped by and shape the context within which they are embedded.

4.5.7   The local environment: producing renewable energy in an urban quarter

This section analyses how the local environment becomes or is made meaningful within KEBAP, especially with regard to energy production. The bunker is situated in a densely inhabited urban area. The position of the bunker—in close proximity to a residential area—affects the energy production part of the project.

M2:   'Next top: local effects. Effects on the neighbourhood and local environment. On the one hand <u>during</u> the construction process. What kind of burdens are created? Then we have the point tree cuttings. How do we handle that? Then noise pollution because of lorry traffic. Noise pollution caused by the operation of the system, not including lorry traffic or the construction process. Then we will have to think about the chimney, about further emissions polluting the air, other than CO_2. Can anybody think of any further burdens for the neighbourhood' (KEBAP, Member 2 on a special meeting to set up the criteria catalogue for the energy versions: 428–435)?

In this sequence from a meeting in which criteria for the evaluation and comparison of the two different technological systems were presented and discussed, the neighbourhood and local environment are presented as potentially affected entities. In the quote, KEBAP is related to human actors and natural entities (trees) that are spatially close, with regard to how these might be affected by either a gas CHP or a woodchip combustion system. Both technologies have direct negative effects like air pollution through the emissions, and indirect negative effects like noise from Lorries. These impacts were evaluated by the project members. The neighbourhood—the project's local environment—is made sense of as a limiting factor for the project, more precisely for the projection of the technological versions. This limiting character is not naturally given; it results from the context of the project within which the well-being of the neighbours is defined as a decisive factor.

The limiting character of the local environment also is related to the project by formal rules that, for example, regulate the amount of emission in residential areas.

DA: 'So certain aspects of KEBAP, certain aspects of the energy production within this environment, well surrounded by flats and parks, have created the idea that there are interests of the neighbourhood which have to be protected. And these then have to be elaborated in a planning permission process. Like emission protection regulations and aspects' (District Authority Officer: 129–135).

In this quote, the particular character of the neighbourhood is presented as creating certain requirements for energy production in the project. It could be said that the emission regulations have to ensure that certain relations between the project and the neighbourhood are *not* created. With regard to the planned energy technologies, air and noise pollution are potentially impactful and thus regulated factors. The local environment influences the social practices of energy production in the project because of the internal ways in which it is made meaningful by the project members themselves and through the existence of external formal rules. Because the energy production activities of KEBAP would causally lead to (negative) effects on the local environment, the local environment becomes essentially involved with the practices of the project in the sense that they shape and restrict these.

The relationship between KEBAP and its local environment does not include only negative impacts. Rather, KEBAP wants to actively change the meaning of the building for the local area and its inhabitants. Thereby the project wants to trigger changes that are understood to be positive for the local area.

B4:   'This is one of the few open spaces in this district, quarter. We have to be careful
        that we don't <u>burden</u> it, but, to the contrary, that we open up more possibilities.
        So, like a communal roof top garden. So, create <u>more</u> green rather than less' (KE-
        BAP, Board Member 4: 475–478).

In this quote, KEBAP is shown as wanting to change the bunker and its surroundings in ways that are deemed positive for the people living in the area. According to the quote, the project wants to change how the bunker is related to its local environment and thus the meaning it has for the local residents. This intention is also visible in a sequence from another interview.

B2:   'We want to proof that one can reintegrate spaces through a rededication, one can
        integrate them back into city life. Because, especially bunkers are often associated
        with death, horror, war, and terror. And we want to reverse that association to the
        opposite. From a black box to a colourful box' (KEBAP, Board Member 2: 881–
        890).

In this quote, the interview partner describes the intention of changing the symbolic meaning of the bunker and thus the way people relate to the building. While at the moment the bunker is an unapproachable, closed-off '*black box*', the intention of KEBAP is to make the building accessible to visitors, participants, and other interested people. Changing the symbolic meaning of the building is closely related to changing the material and practical ways in which people engage with the bunker and its surroundings. Before the project started its activities at the bunker, the place was a '*scary place [Angstort]*' for many locals (Member of the district fraction of the Green Party at a KEBAP presentation, own notes, AP). The building and its direct surroundings were a rather dark corner at the entrance of a park. This situation attracted homeless drifters as well as drug dealers and consumers (own notes from presentation of KEBAP to the district fraction of the Green Party). Through the project, this part of the street would come to be used by many different people, which would turn this lonely and dark place into a populated and much brighter area. The idea is that not only visitors but also passers-by and local residents would profit from this situation.

The intention of KEBAP is not only to change the meaning of the building through activities taking place inside the building. Instead, different visions exist with regard to how the construction of the bunker can be changed to open up the building materially. The urban gardening project taking place along the south wall of the bunker is understood as a first step to make the bunker an accessible place for local residents and other interested people. In order to build the raised beds for the gardening project, the members of KEBAP's gardening working group tidied up the area and in particular cleared it of hazardous remnants of drug-taking activities. In order to dispose of this special waste, the project members initiated cooperation with a local drug-counselling centre. Due to the urban gardening activities, the previously rather lonely and dark street and entrance to the park have become lively places.

The project members are ambiguous with regard to the possible effects of the realization of the project on the local environment.

B1:   'Well, I am interested in change, but change without displacing the people. Or, without prescribing them, what they have to do here. Well, just a sensitive change. But, of course, and that is absolutely obvious, we would be absolutely naïve if we would think we could start something like that here and it would not change anything. And I also do want to change things' (KEBAP, Board Member 1: 655–660).

This member of the project is interested in getting KEBAP actively involved in changing the neighbourhood. She hints that KEBAP will lead changes just through its existence. She describes change as something positive if it is sensitive to potentially disadvantaged groups. In many informal talks, this interviewee emphasised that she neither wants to displace the homeless people who sleep along the bunker nor wants to cause an eviction of the people living in trailers parked in the street. Instead, she and other members of the project aim to include these people into the developments. Nevertheless, the interests of the different groups sometimes conflict. While—as mentioned—most residents are in favour of the gardening project, some of the trailer inhabitants [Bauwagenbewohner] complain about the increased levels of activity and noise at their formerly quiet and dark living and sleeping place. The interests of the homeless people are even more incompatible with the interests and activities of the KEBAP project. In particular, the use of the shed—which has been constructed for the purposes of the gardening project—as a sleeping place is objected to by the project members. The district authority, on the other hand, has started holding the participants of the gardening project responsible for the cleanliness of the area. The authority has repeatedly threatened that it would stop tolerating the gardening project if the members do not keep the area clean. The project members are made responsible for litter and devastation that have not only not been caused by the members' activities but which are rather a problem for the gardening activities of the project itself (especially as the planted vegetables are meant to be consumed by the project members). The ambiguity is further increased as most KEBAP members do take the requirements of different actor groups seriously. While not wanting to displace the homeless sleepers and trailer inhabitants, the members also want to make the place a positive activity space for the other local residents and visitors. Additionally, the members want to use the place for themselves and their activities.

The described situations and effects exemplify the ambiguity of the project's relations to its local environment. Different actors relate KEBAP to its local environment in different ways. There is no single way in which the project is seen as interacting with its local environment. Instead, different actors interpret this relation and make it meaningful with regard to their own context. For the homeless sleepers and trailer owners, the darkness and solitude due to the bunker, park trees and hedges, and the absence of lightning posts made the area around the bunker a suitable sleeping and living place; the same elements are perceived negatively by most local residents, the public administration, and politicians. The social practices of the KEBAP members' evolve among complex relationships between things, material artefacts, and human actors forming the project's local environment.

4.6 Interim Conclusion: Making sense of renewable energy production in KEBAP

KEBAP has its origins in two citizen initiatives which aimed to challenge Vattenfall and the conventional energy production system. The analysis of the project has shown how much this background has influenced the original concept for the project. Both the teleologies as well as the affective motivations of the project founders were shaped by their engagement in two resistance initiatives. The teleologies of the project members have been found to be closely intertwined with their general understandings. The different general understandings are not only an outcome of the project's teleoaffective structure but mutually influence these. Hence, the project members generally understand Vattenfall's social practices as 'bad' or 'ominous'. Deriving from this Opposition to Vattenfall and the conventional energy system, citizen led renewable energy production is understood as a valuable social practice.

The creation of KEBAP represented the original core group's decision to take their engagement one step further by creating a valid alternative to the existing energy system. Setting up KEBAP was seen as a viable option to challenge Vattenfall in different ways. First, the members challenge Vattenfall on legal grounds as a competitor. For this purpose the project can draw on formal rules and conventions of free markets. More generally, the original concept for the production of energy had been designed with the intention to create a project which would be as renewable and as decentral as possible and which would also provide opportunities for active participation of citizens. While the last aim has remained relatively unchallenged, a severe conflict about how the two first aims could best be realised evolved. While some members supported the strictly decentral and renewable approach—which would best be served by a woodchip combustion system—the proponents of the gas CHP version argued for the value of creating a project that would be a relevant actor in a future smart renewable energy grid. What was at stake in this conflict was not only the best technical solution for the production of energy in KEBAP, but the projects' aims and intentions. The conflict exemplifies how material artefacts are made meaningful with regard to existing sense-making within a certain context. Advantages and disadvantages of certain technologies do not exist 'objectively' but derive from the context within which these are made sense of.

Neither the teleology of challenging Vattenfall, nor its combination with teleologies like self-empowerment or energy autarky has been stable in the project. Throughout the existence and development of the project, principles and their meaning have been introduced, changed, or abandoned. The orientation towards local economy is one principle which was introduced later and which increasingly gained in strength, until it became the dominant teleology. This change in the teleoaffective structure of the project is due to the composition of its membership. Being the most important actors within the project, the interests of the members significantly shape KEBAP. While the intentions of the project members in the beginning were very much concentrated on creating an alternative to the existing energy system, the aims and intentions of the project have developed onward since then. Activities like urban gardening both initiated and signify these changes.

From the start the project intended to produce not only energy, but also culture. This decision first of all derives from the aims and (private and/or professional) interests of the members. Cultural production, however, is in many ways related to the production of energy. First of all, both the production of energy and cultural production are understood as two ways of realising a local economy approach. Second, the integration of the cultural part also derives from the teleology of creating a project that would contradict the logics of conventional energy production on as many grounds as possible. While energy conventionally is meant to produce individually owned capital, income generated by the energy part in KEBAP is meant to be used for the economically deficit cultural part. Third, cultural production is seen as a way to offset the abstract and 'theoretical' character of energy production. The combination of cultural and social activities with energy production activities in the KEBAP project is meant to provide people with an opportunity to bodily experience self-made production.

Being a volunteer project, the project members are the most important type of human actors within and for the project. Deriving from the project's self-definition of being a neighbourhood project, local residents are another relevant group. The relationship of KEBAP to the local residents, however, exemplifies existing ambiguities. While on the one hand, local residents are an important point of reference and source of legitimation for KEBAP's activities, hardly any neighbour does actually take part in the project's activities. Political and administrative actors on the other hand are not meaningful as legitimating sources or points of reference within the project, while they are highly relevant in the project's daily activities.

The projects' way of sense-making also influences the way the material artefact of the bunker is perceived by the members. While, originally, the positioning of the bunker in close proximity to the planned heat pipeline was seen as one of the main advantages of the building, this position has come to be irrelevant, due to the changes in the project's teleoaffective structure. In the members' day-to-day activities, the grass verge between the bunker and the street, which provides space for urban gardening, has instead come to be a much more relevant aspect of the bunker.

4.7 Describing the IBA

With 35.3 km² in area Wilhelmsburg is geographically the largest district in Hamburg. Situated on the islands Veddel, Peute, and Wilhelmsburg the district is in close proximity to Hamburg's city centre. Because of its location, Wilhelmsburg was declared to be the most ideal industrial area for the ever-growing industrial harbour in the 17th century (Markert 2008: 191). Since then, the district has always been heavily influenced by the industrial harbour.

Because of its geographical proximity to the river, which is flowing through Hamburg, the image of the district has been strongly shaped by the storm flood of 1962. Of the 315 people who died in Hamburg, 222 deaths occurred in the areas of Veddel and Wilhelmsburg. Since the Second World War, the area had been mainly occupied by self-built wooden houses and subsistence farming. The vegetable gardens had weakened the stability of the dykes. When the storm flood hit, large parts of the dykes eroded immediately. Many of the wooden houses were not stable enough to resist the flood wave. People were buried under their collapsing houses or drowned inside them. The storm flood of 1962 has heavily influenced the Hamburg's and especially Wilhelmsburg's identity.

Major parts of the island are occupied by the harbour and harbour-related industries. A large part of the island has been free harbour area. As goods were not taxed during their storage or transfer period, fences and checkpoints can be found all over the district. Because of its proximity to the harbour, Wilhelmsburg and Veddel have always been occupied by dockworkers and other working-class residents. At the turn of the 20th century, industrialization, with its accompanying phenomena of rural-urban migration, resulted in increased levels of poverty and decreasing levels of education. Since then, the district has had the reputation of being a derelict district (ibid.) for the disposal of industrial waste.

4.7.1   The bunker

The bunker utilized by the IBA was built in 1943, mainly by PoWs. It is a so-called Flakbunker—it not only used as shelter for people but also hosted antiair-craft guns. For this purpose, four towers were erected at the four corners of the building. In 1947, the bunker was internally destroyed by the British allies. Six of the then eight floors collapsed due to the targeted internal detonation. Since then, the building has been in danger of collapsing and remained locked. As it was not utilizable for military services anymore, the bunker was handed over to the Hamburg's ministry of finance. When the building was reconstructed by the IBA, the destroyed six lower floors were turned into one huge room. The bunker covers a basal area of 57×57 meters. The main body of the building is 30 meters high. The towers are another 12 meters. The outer walls are up to three meters wide while the ceiling is nearly four meters thick.

4.7.2   Prologue to the IBA

In 1994, massive protests by the district's residents prevented the construction of a new waste-burning plant. The high levels of local public concern resulted from an earlier scandal, regarding a waste storage site from which highly toxic dioxin had leaked. The protests were organised by a newly established citizen initiative—the 'forum Elbinsel'. Due to their success in stopping the new waste plant as well as the growing dissatisfaction with political institutions in the district, the initiative quickly gained influence in the area. The forum was also involved in the establishment of the first citizen participation council, convened by the Hamburg's parliament (Markert 2008: 194). In 1996, the first borough citizen council meeting took place. The forum Elbinsel played both a supporting and a critical role in these meetings. The meetings changed the role of the forum. Its members no longer understood themselves to be in a purely protest initiative but also an active and creative actor within and for the district (ibid.). In the following years, the community council not only participated in architectural improvements but—more importantly—developed plans for an improved education system and training or job opportunities, as well as reactivation of the neighbourhood and integration. The council's slogan proclaimed '*a city borough helps itself*' (ibid: 195). In the first few years, the borough council was financially supported by the city government with three million Euros.

In the 1990s Hamburg's mayor came up with the idea to transform a publicly owned port area into a new living and working district to provide room for the growing city. Hamburg thereby joined a global trend of port cities, involving the transformation of formerly industrialized derelict dockland sites into exclusive waterfront areas (Desfor 2011). Because of increased demands for space— among others reasons because of the ever bigger container vessels and the amount of cargo—most industry harbour activities have been moving out of the city centres in the last few decades. Even where an industrial port is still in the city, the technological developments have enabled harbour activities to become quiet and 'clean', allowing for even higher class living, recreation, and business infrastructures to exist in close proximity. In fact, the reputation of ports has undergone a drastic change in the last few centuries. From being dirty, loud, and potentially dangerous areas, they have turned into much-sought-after interesting places in the city (ibid.).

How important this trend is thought to be for Hamburg is shown in a statement by a former mayor of the city.

> 'It is a onetime chance and a main locational advantage of Hamburg that its urban
> development does not happen along abstract development lines, but is guided by

the water. Sometimes I think that Hamburg is only just starting to realise the immense potentials for urban development deriving from its closeness to the water. Until now we have mainly valued the Elbe and port in terms of the regions' economic strength. But I think that Hamburg, with its architecture, its quality of live, its wellbeing-factor and its potential for growth has enormous chances to develop along the waterfront' (Programme Description Leap Across the Elbe: 12).

The quote shows the extent to which waterfront development is related to growth by political actors. Induced by the increasing roll-out of market ideologies onto the public sector, cities have become more and more reliant on recruiting private capital. The economy of societies of the Northern Hemisphere is increasingly dependent on third sector industries. Globalization enabled economies to outsource to the global south those mainly manual works that require little qualification, allowing them to save money because of smaller wages. Remaining in the Northern Hemisphere were those jobs in the service and refinement sector that require high levels of qualification. According to the 'human capital theory' the key to growth of companies, regions and nations *'lies not in reducing the costs of doing business, but in endowments of highly-educated and productive people'* (Florida 2003: 6pp.; likewise Bontje/Musterd 2009: 844). Many of the sought-after skills, like creativity, are ascribed to individual abilities. Companies cluster in cities in order to draw from

'concentrations of talented people who power innovation and economic growth. The ability to rapidly mobilize talent from such a concentration of people is a tremendous source of competitive advantage for companies in our time-driven economy of the creative age' (ibid: 5).

In their search for the best employees with these skills, are able to look globally. Globalization has also enabled these highly skilled individuals to go looking for and compare the best job opportunities worldwide. Consequently, a global competition for highly skilled employees has evolved. In this regard, companies need to create *'employer attractiveness'* (Berthon et al. 2005: 155) by offering benefits for potential or actual employees. Companies have discovered that many of these 'young urban professionals' do not decide in favour of a job offer only because of the job description, the reputation of the firm, or the wage. Instead, the location and thus the general living conditions have become an important aspect. Location has also come to be seen as the indicator of a company's prestige. Upscale and premium brands have moved their creative and representative headquarters to the hearts of *'creative cities'* (Florida 2003). Cities themselves profit from the presence of big enterprises, which are not only valuable tax payers but also offer highly paid job opportunities and thus generate affluent citizens. In order to be able to draw in big companies, *'what cities and regions should attract is not the creative or knowledge-intensive companies, but the people that work for these companies or those who might start such companies themselves'* (Bontje/Musterd 2009: 845). In accordance with the argument made above, in order to attract the 'creative class (Florida 2003), cities increasingly feel the need to increase their *'soft location factors'*

While different scholars have criticised this idea (among others Martin-Brelot et al. 2010), a 'vast number of leaders and spokesmen of local and regional (quasi) governments believe that a major thing to do is to become more attractive places to live for creative knowledge workers' (ibid: 843; likewise Peck 2005: 740pp.). In fact, the Leap Across the Elbe Programme exhibits exactly this intention.

> 'You just need a certain scale to be able to keep up in the international competition of metropolis. Second, and I don't want to cover that up, the fiscal aspect does play a role. A big city which provides many services and benefits is reliant on as many inhabitants as possible, in order to enable these services with the best possible quality' (Programme Description Leap Across the Elbe: 11).

Developing cities in a way that fulfils the requirements of creative and affluent (potential) employees seems like a win-win situation for both companies—which get a valuable argument to convince their favourites to take the job—and cities—which attract private capital to the city. As seen in the following section, these ideas and motives have shaped the processes and decisions leading up to the establishment of the IBA.

4.7.3   The 'Leap across the Elbe'

On the banks of the River Elbe, opposite Wilhelmsburg and Veddel, the plans for the construction of a completely new district had been developed in the 1990s. The then city mayor had come up with an idea to transform a publicly owned port area into a new living and working district to provide room for the growing city. The planned new district combined two important places. Geographically, it was located in the most inner parts of Hamburg, while also providing waterfront sites at the northern side of the Elbe. In accordance with the aforementioned global trends, the 'Harbour City' has mostly been planned for the requirements of affluent (future) citizens. After a master plan for the area had been developed, the publicly owned port site was divided into plots that were sold to different private investors. The new Harbour City extended the inner city area of Hamburg to the south. The Elbinsel, which so far had always been 'far away', came to be in close proximity to the city centre.

In the summer of 2000 Wilhelmsburg's bad reputation was highlighted once more. Different events drew nation-wide negative attention to the district. The killing of a six-year old schoolchild by two pit bull terriers, the public shooting of a man on the street, and the execution-style murder of a Polish ethnic German immigrant and her two daughters by a German were broadcasted nationwide. These tragic incidents depicted the district once again as a deprived city area with high percentages of unemployment, poverty, and (low skilled) migrants and low levels of education, income, and training or job opportunities (Result Report Future Conference: 4). These incidents made the Hamburg parliament decide that it was not enough to improve the district's infrastructure; it was necessary to do so in a publicly visible way.

Another reason is ascribed to the bad results of the SPD in the 1998 elections. Having traditionally been strong in the Wilhelmsburg, the party has been afraid of losing even more voters in the elections 2002 (Markert 2008: 197). In December 2000, the Hamburg Parliament agreed to hold a future conference for the district. Such a conference had never before been organised for one single district in the unified congregation of Hamburg (Result Report Future Conference: 3). Furthermore, the idea to organise the conference as a collaboration between representatives from different public authorities with volunteering residents from the district was exceptional. Between May 2001 and March 2002, the congress took place in the form of seven working groups: 1. Spatial masterplan; 2. Traffic; 3. Work and Economy; 4. Living and Accommodation; 5. Living together; 6. Schools and Education; and 7. Leisure and Culture. The working groups worked independently, developing ideas and strategies for possible developments in the district. All working groups consisted of volunteer residents and representatives of public administration institutions. Additionally, the working groups were given the opportunity to invite experts to particular meetings or issues. In a final presentation in March 2002, the future congress released its results in the 'White Book'. The seventh working group—Leisure and Culture— had come up with ideas for using different places in the district for recreational purposes. Among many others, the working group had developed ideas for the old Flakbunker in the middle of the district. Because of architectural, financial, and administrative reasons, the bunker could not be demolished. In particular, tearing down the massive walls (more than two meters in thickness) would demand an enormous financial investment if the bunker were to be demolished. Furthermore, the bunker was in the process of becoming a listed building at that time.

The massive, derelict, and dark façade of the bunker and the fact that it was completely unused were main reasons for the working group participants to request that something should be done with it urgently.

LO: 'It has always been, the Flakbunker, in its huge dimensions, a problematic build-
        ing. Because it always created a milieu and an atmosphere of: there is still a war.
        It really did not look nice and of course depressed the area around it, especially
        because it was so visible. Thus it has always been an issue: what do we do with
        it' (Member Local Organisation: 200–205)?

The White Book suggested various possible usages. These included the ideas to
utilize it as a climbing wall, an observation deck or even a café on the rooftop.
The White Book recognized that any restoration of the bunker would require
private investment and support from political and public administration bodies.
The White Book furthermore acknowledged that it was rather doubtful whether
the bunker's inside could ever be used (Result Report Future Conference). While
the future conference and the White Book are regularly considered to have been
important predecessors of the IBA, the book remained a non-binding collection
of ideas in the following years.

The future conference and its results, nevertheless, for the first time did
draw attention to Wilhelmsburg not as derelict and dangerous places, but as
promising sites for urban development (Markert 2008: 198). According to Markert, there are two important reasons for this. The first reason is the active involvement of citizens, who not only voiced criticism but also developed realizable ideas for the future development of the districts. The second reason is the
political changes during this time. While they were still the strongest party, the
socio-democrats had to hand over the government to a coalition of Christian-
democrats (CDU), the liberals (FDP), and a newly established conservative
party. The concept of the 'social city' developed by the social-democratic government was replaced by the newly elected Christian-democratic government.
In a printed matter, the new senate in 2002 proclaimed its new economically
motivated strategy (Hamburg Parliament Printed Matter 18/3023).

The growing-city strategy formed the background based on which the programme 'Leap across the Elbe was developed and passed by the senate in 2003
(ibid.). In the description of the programme, the Elbinsel was depicted not as an
industrial port area or derelict quarter, but as a promising site for urban development, for the first time in a political publication. The programme was set up
when Hamburg applied for hosting the Olympic Games 2012. While the application was not successful, the newly established political interest in Wilhelmsburg and Veddel continued. Within the strategy of the growing city, the relatively sparsely inhabited area, with its proximity to the water, was now seen as
a promising site for urban development (ibid.). In the programme, citizen participation is mentioned as an important feature that can help to ensure that new
developments take place in accordance with citizen's interests.

In a press release on July 22, 2003 the Hamburg parliament provided an update of the growing-city strategy. One of the ideas in this update was the establishment of a formal IBA, which would be '*a big challenge and a bracket to provide a geographical and temporal goal to the growing city strategy*' (Outline Hamburg Strategy: 74, translation AP). As the first step towards the realization of the 'Leap across the Elbe' and its concretization through a regeneration programme, the ministry for urban development and environment and the Chamber of Commerce organised an international design workshop ('Entwurfswerkstatt').

Neither the workshop nor the idea of hosting an IBA derives from a purely unbroken success story. Rather, the ongoing events are also a sign of problems. This is acknowledged in the design workshops' final report:

'Although several concepts for development plans have been worked out and a lot of small projects have been designed, discussed and implemented in the district in the past, apparent enhancement and a new designation has not yet taken place. This is in expectation of new ideas and visions resulting in the drastic social, economic and urban renewal of this part of the river basin and in a lasting strong north-south link' (Bruckbauer 2003: 19).

The event, which took place in the summer of 2003, brought interested citizens and students in touch with internationally renowned urban planners and architects. This unprecedented workshop aimed to actively include residents' concerns and ideas into urban planning projects (Markert 2008: 189). A second workshop took place a year later.

A number of actors who had already been involved in the citizen council and in the publication of the White Book took part in the future council. In particular, members from the Elbinsel forum were involved. Participants described the atmosphere in the workshops as enthusiastic and full of highly motivated teamwork between the different actors. In fact, ideas and strategies developed in the design workshops were taken up in the framework set up for the IBA. The 'memorandum for an urban regeneration programme' was published in 2005 by the ministry for urban development and environment (Programme Description Leap across the Elbe: 7).

The memorandum describes the motivation for choosing this particular spatial development instrument. The main motivation for having an IBA is its potential to bundle activities and long-term strategies related to urban development strategies. Furthermore, an IBA is seen to be able to provide a geographically and temporally bound, visible, and accessible form to the otherwise seemingly diffused and heterogeneous activities, deriving from the growing-city strategy and its urban planning programme 'Leap across the Elbe' (Memorandum IBA: 3)

'By implementing an international building exhibition, Hamburg commits itself to a presentable quality and competency in finding exemplary, internationally competitive solutions for urban development challenges and claims a leading role in the composition and organisation of big cities' (ibid: 7).

In its printed circular 18/3023, the Hamburg senate talks about the goals, conceptions, and projects related to the 'Leap across the Elbe' and requests the consent of the parliament to implement and finance the IBA. According to the printed circular, the necessary financial resources for each project cannot be secured through public sources only. Instead, private investors or public-private partnerships are described as the best way to ensure the financial sustainability of the IBA projects. Public funds would only pay for those few especially important projects that, because of their speciality or innovativeness, are not realizable otherwise. The printed circular estimates that a sum of around 100 million Euro will have to be provided by the city of Hamburg. This sum includes around 50 million Euro for 'lighthouse projects' that require intensive public investment, 20 million Euro for financial participation in a multitude of small-scale projects, about 9 million Euro for the preparation and presentation in 2007, 2010, and the final presentation in 2013, and about 21 million Euro for staff and material expenses of the organisation responsible for the realization. This money will mainly be taken from a special investment programme. The programme included a special section for financing the 'Leap across the Elbe' programme, which could be used for the IBA.

The printed circular proposes the organisational form of a limited liability company (GmbH). An independent organisation was preferred, because of the general inflexibility of public administration institutions, and because of the limited personnel resources and the administrative restrictions of the existing employees in these institutions. Second, all recent formal IBAs in Germany have used the organisational structure of a GmbH. While being independent of any political or administrative city body, the IBA would be held financially accountable to the parliament and advisory board.

After about two year's preparation, the IBA Hamburg GmbH was officially set up in September 2006 as an independent subsidiary organisation of the city, with a budget of 90.2 million Euro.

4.7.4   Realising the IBA Hamburg GmbH

In September 2006, the new employees of the IBA started working. At the very beginning of the IBA, in early 2007, the manager and employees went on a retreat together, where they were joined by some members of the IBA advisory board. The advisory board consists of international experts who examine the projects and support and guide the work of the IBA team. Apparently, at least one member of the advisory board had been very attentive to the results of the Fourth IPCC report, published in 2006. In accordance with the head of the IBA, he gave a talk about the IPCC report at the retreat.

S1: 'and there we also developed the third main topic, "climate change and urban de-velopment." This has originally not been on the agenda of the IBA-GmbH. But, both internally in the discussion, but also through the consultants which were there with us and who are still in the advisory board, there came the intervention that we have to broaden the range of topics to include climate change' (IBA, Staff Member 1: 25–30).

The team agreed that a modern urban planning process must pay attention to climate change. Thus 'climate change and urban development' was taken up as the third main topic of the IBA. Energy and sustainability had been a subtopic until then—not specifically related to climate change. For the participants of the retreat, however, it was obvious that climate change had to become a key concern generally for urban planning and specifically for an urban development process in a district which has historically suffered from storm floods.

Because climate change had not been a main theme of the IBA so far, there were neither any pre-settings to be realised nor many project ideas in the White Book or the memorandum to be tested and developed. Also, none of the team members had any particular expertise in the issue.

S1: 'Yes, nobody from the personnel was specifically designated for that topic so far. Thus, how can I say? It just happened. I said that I would like to do that. And then the responsibility was conveyed to me' (IBA, Staff Member 1: 32–34).

A small team was created to discuss how to integrate the topic into the existing projects or what new projects could be taken up. In cooperation with external experts, the team created and published a climate protection plan for the district. In this conceptual tool, they developed ideas to save carbon emissions through energy efficiency measures, energy savings, and the production of renewable energy in and for the district.

The production of renewable heat was deemed important because, until then, it was planned that the district would be provided with heat from a new coal plant.[22] Vattenfall and the Hamburg government set up a contract. Vatten-fall got approval to build the new coal plant in exchange for agreeing to extend

---

22 The same is also crucial for the existence of the KEBAP project.

its heat grid to Wilhelmsburg. Thus far, the district did not have any long-distance heat grid. Heat was generated by electricity, oil, or gas in every single housing block.

S1:   'This was an agreement which very heavily would have influenced our projects. And the area anyway. And, we could not accept it, because, well, to exhibit and show the future of building culture 2013 and then these buildings are heated with coal. That has nothing to do with future oriented and sustainable energy provision' (IBA, Staff Member 1: 582–587).

The team started looking for ways to produce renewable electricity and heat within the district, and to develop plans for a district heat grid. The activities and plans of the team were '*tolerated*' (Staff Member 1 IBA: 598) by the parliamentary advisory council. Simultaneously, the team contacted the energy provider and discussed its ideas and concerns with the company's management.

S1:   'And in the end, it happened that the contract was cancelled. Hamburg's commitment for Vattenfall was annulled. Well, this then went up to the international World Bank court. So, Vattenfall sued the city, not only because of this contract but other things as well. And in the end Hamburg and Vattenfall agreed on working together' (IBA, Staff Member 1: 603–608).

In their search for potential renewable energy production sites, the team took up an idea from the ministry of urban development and environment.

S1:   'Already in 2006 the ministry for urban development and environment had had the idea to turn it into a Solarbunker. To more or less initiate a second solar-powered village in Hamburg. In the middle of the 1990s such a […] village had been built in the east of Hamburg, which did not really work. But the ministry thought they could transfer the experiences they had made there onto the bunker' (IBA, Staff Member 1: 58–64).

In 2006, the ministry developed a concept to restore the area around the bunker and turn it into a solar village. According to the ministry's concept, the roof and southern walls of the bunker would have been used for solar panels.

The IBA-team asked some specialists for a re-evaluation of the ministry's concept. Soon it became apparent that the concept was not viable for two reasons. First, the amount of heat that could be generated through solar panels would not be enough to be feasible, partly because in order to deliver the heat it would be necessary to construct a new heat grid. Second, structural analysis showed that the bunker had no stability at all. Some years previously, actions had to be taken to prevent large chunks of concrete from falling off the roof and walls onto the street. Years of wind and rain had severely affected the structure of the material, and the missing internal constructions further increased the instability of the whole building. It became obvious that the bunker was a public danger in its actual condition.

Tearing down the building would have been extremely expensive. Furthermore, the building had by now become a listed building. Consequently, it was decided that the bunker needed to be restored.

S1:   'And this meant that the financing of the project would have to be secured via the city. This was successful because Hamburg is responsible for road and traffic safety. Thus the owner of the building, the fiscal authority, had to pay an amount of money which was estimated sufficient for a minimum effort restoration' (IBA, Staff Member 1: 187–193).

Restoring the bunker meant that the inner parts of the bunker—which had not been accessible so far because of the buildings' instability and the masses of debris—would become utilizable. At around 35 meters in height and 45 × 45 meters in area, the bunker's main interior room provides an immense space that could now be used for energy production technologies.

In 2008, the IBA organised an 'energy laboratory', where representatives from public administration, citizens, and experts were invited to discuss possible projects for the key topics of climate change and urban development. With some of the guests, the IBA established a further consultant board—the 'Advisory board Climate and Energy'. The board consisted of experts from different institutions and backgrounds who could support the IBA team with regard to its energy-related projects. On the first excursion to possible project sites, the IBA team presented their ideas for the bunker. Apparently, most members of the advisory board immediately advised the team that if they had so much space at hand and wanted to realise an innovative project with renewable energy, they must include a heat storage facility. Based on this idea, the IBA contracted technical experts to develop and calculate a concept for heat production and storage in the energy bunker.

S1:   'We added further experts which knew about storage facilities, which know about structural restoration, with the usage of buildings—so structural architects, architects, energy consultants, heat storage experts, and also control energy experts. With these we created a team, with the task to test different variants with small, large and extra-large storage facilities. And the highest possible amount of renewable energy' (IBA, Staff Member 1: 130–137).

After consultation with many professional engineering, architecture, and technical companies, a combination of different sources was developed. These included a biomass and gas CHP plant, solar and photovoltaic panels, and thermal waste from a nearby industrial site. These sources were to be fed into a 2,000 cubic meter heat store (which allowed the project to reduce the installed heat generation capacity from 11 to 6.5 MW).

The industrial waste heat became part of the project when, after the energy laboratory, a manager from a nearby oil refinery company contacted the team. The company had till then artificially cooled down its industrial thermal waste and released it into the air. The IBA team set up a contract with the company to buy its waste heat. In exchange, the company agreed to improve its air filter system to reduce the smell resulting from the industrial processes. In the White Book, the smell from the enterprise had been mentioned as a severe problem for the surrounding neighbourhoods.

Because the IBA had to be completed by 2013 and also because its competencies only included the organisation of the development processes, it was clear that an operating company had to be found to realise and operate the bunker project. In the beginning, the IBA presented the project to E.ON Hanse. The company, however, refused the project because of the incalculable costs of the restoration and refurbishment of the bunker.

In consultation with the city, it was agreed that the IBA would itself become the main building contractor in this special project. The money from the fiscal authority for the minimum restoration of the bunker was extended by IBA infrastructure funds designated to enable public infrastructure projects, and by further funds from the IBA. Having agreed on this allocation of costs, the IBA once again set out to find another energy provider who would be willing to pay for the installation of the technical equipment and the construction of the heat grid.

In 2009 a new municipal energy utility—Hamburg Energie had been established in Hamburg.

S1:   'And when we heard that there are preparations going on to establish a municipal
         energy provider, we very early sat together with them and asked them if they
         would dare to realise such a project like the bunker. […] It was an advantage that
         it was possible to talk as one city daughter company to another city daughter com-
         pany. And they did not get any funding from us but a bunch of liabilities and
         obligations. But, they were just happy, yes, in this early stage to get a handful of
         projects they could realise' (IBA, Staff Member 1: 517–531).

Hamburg Energie took on the project and became second building contractor. In this position, it was responsible for financing and constructing the energy concept developed by the IBA. A part of the money thereby came from the European infrastructure fund EFRE.

Very early in the restoration process, shortly after it became known that the IBA planned to restore the bunker, a heritage centre situated in the district got in touch with the bunker team.

HC: 'Yes, when we heard that this old bunker was meant to be refurbished as part of
        the IBA, we immediately got into contact with the IBA. And we said that we
        would like to cooperate. And would like to engage with the history of the bunker
        in the context of National Socialism and the Second World War here in this dis-
        trict. And this we did, and. Already in the beginning of the 90s we had made a
        book on the issue of "the district during National Socialism," and in there we al-
        ready had a small chapter on the topic of the bunker' (Heritage Centre: 14–22).

The heritage centre did a thorough research on the history of the bunker by visiting archives all over Germany and talking to surviving eye witnesses. Furthermore, the centre organised various school projects in which classes engaged in
certain aspects of the bunker's history. Once a year, the results of these projects
were presented as part of a neighbourhood festival organised by the heritage
centre and the IBA. Together with the heritage centre, the IBA team members
developed a plan to set up a memorial centre within the bunker. Funding, however, could only be secured to allow for a much smaller exhibition. The costs
also only allowed to create only few facilities for the public within the bunker.
A café was constructed on the upper floor, which allows access to a rooftop
terrace. Both the designing of the exhibition as well as the operation of the café
were commissioned through a tender process.

In October 2012, the bunker started to deliver heat into the grid. Most parts
of the bunker refurbishment were finished just in time for the final presentation
in March 2013. The exhibition and the café were opened in the same month. By
2015, the bunker is supposed to have delivered heat to 1,600 households. An
extension of the energy production would enable Hamburg Energie to supply up
to 3,000 households with renewable heat from the bunker in 2015.

With the end of the IBA, the contracts of all the employees terminated. The
manager of the IBA has now become managing director of an IBA follow-up
company. The two project coordinators of the bunker project have moved on to
other jobs.

4.8 Analysing the IBA: Creating 'a symbol for renewable energy production in the city'[23]

4.8.1 Teleoaffective structures: Negotiating the construction of a lighthouse project with reducing carbon emissions and the creation of a local asset

The first public document to describe the IBA is the 2005 memorandum. In this publication of the Hamburg government, the motivations and intentions for setting up a formal IBA are described.

> 'The Globalization of the economy, short-term location decisions and massive acceleration because of new information technologies are fast changing the ranking of cities and metropolis. Fiscal scarcity in municipalities and demographic changes also demand European cities to establish new fields for action for, and new governance forms in urban development. Who wants to succeed in the competition of cities, to constantly push forward renewal processes, and at the same time secure existing advantages of the city, has to accept the challenges of these processes and take them on with creative drive and sense of proportion' (Memorandum 2005: 6).

The quote starts by describing the challenges faced by cities in an increasingly globalized world. In describing these challenges, the Hamburg government talks about its intentions for the IBA. The IBA has been chosen as the instrument through which Hamburg's government *'wants to succeed in the competition of cities, to constantly push forward renewal processes'*. By starting a formal IBA, Hamburg's government defines the solution to these issues as being mainly infrastructural. More precisely, by setting up a formal IBA process, the authors of the memorandum have chosen to initiate *'a programmatic exhibition, the centres of which are urban planning innovations and new impulses for the building culture'* (ibid.). In the memorandum, Wilhelmsburg becomes a site of infrastructural developments that can create impulses for Hamburg within the global *'competition of cities'*.

One important strategy defined for the IBA in the memorandum is to create highly visible projects.

> 'Elsewhere already existing "dramaturgic" urban development strategies show that "lighthouse projects" are necessary. These make future topics accessible, give them a "prominent place" and develop a special, audience related, charisma. These are the projects that count. They form the picture and become symbols of Hamburg's urban development ambitions. They are able—when communicated in an engaged

---

23 IBA, Staff Member 1: 745–750.

and warm-hearted way—to attract media and a huge number of visitors. Consequently, these projects require the highest possible amount of care and concentration in ideas, realization and presentation' (ibid: 5).

According to the quote, the role of lighthouse projects is to function as '*symbols of Hamburg's urban development ambitions*' and to '*attract media and a huge number of visitors*'. Shortly after the IBA started, the bunker was defined as being one of these highly symbolic projects.

S1:   'From the beginning all the members knew, that just because of its **mightiness** in the district, in the quarter. These huge dimensions it just has. And, that this is just meant to become a lighthouse project. That it will raise much public attention, also from the media. And that of course it will have to qualitatively live up to this attention. But that it will be relatively easy to raise attention with the project. And it of course was meant to function as a symbol for renewable energies in Hamburg. This was part of the project idea from the beginning' (IBA, Staff Member 1: 745–750).

Deriving from the quote and the memorandum, attracting a lot of public and media attention and being defined as a lighthouse project means that the bunker project is seen as especially well-suited to function as a symbol '*of Hamburg's urban development ambitions*'. In other words, the bunker project was meant to create a certain kind of international prestige for Hamburg. Creating a globally interesting asset for Hamburg within the competition of cities thus appears as an important aspect within the teleoaffective structure that shaped the transformation of the derelict building into the energy bunker.

Besides the Hamburg government, the residents of Wilhelmsburg were crucial for and in the processes leading to the establishment of the IBA and subsequently to the energy bunker project. As described above (Section 4.7.4), the citizen initiatives leading to the future conference 2001/2 and the resulting White Book were the first steps of these processes. In these processes, the residents of Wilhelmsburg defined the district's social and infrastructural challenges and developed ideas for how the situation could be improved for the district and its inhabitants. In these processes and the IBA itself, the teleoaffectivities of the residents diverged from those of the city government.

LO: 'And in the final presentation of the IBA this was presented like: "Yes. Finally, a beautiful school has been built for the district." But it is only one school. And from the 50,000 inhabitants more than 20 percent are children under eighteen years. So, one school is not enough. And they did not say, well, it was presented as a super success of the IBA: And I thought: "but what about the three or four other schools"' (Member Local Organisation: 612–616)?

In this quote, a local resident complains that the IBA was mainly interested in the creation of lighthouse projects. In contrast, the interviewee would have been interested in general solutions for the problems and challenges that had been defined for the district. For the residents, Wilhelmsburg is their place of day-to-day living. Consequently, they were interested in—and in their initiatives and the White Book even asked for—developments that would improve the infrastructural and social situation of the district as a place of day-to-day living. For the social world of the residents, decent refurbishment of all schools in the district would have been more valuable than having one school redesigned to attract international public attention.

In the publications of the IBA, the opposing interests of the residents and the city government are presented as easily reconcilable.

> 'More than one hundred citizens worked in conjunction with the authorities on creating a vision of the outlook for the district, and in 2002 produced a White Paper […]. As a result, in 2004 the outlined its "Leap across the Elbe", and in 2005 drafted the memorandum' (Webpage IBA 2015).

The IBA in this quote is described as the outcome of a direct process in which the citizen initiatives have motivated the 'Leap across the Elbe' programme which resulted in the memorandum. The IBA is presented as resulting from a straight process within which Hamburg's governments' actors took up and realised the residents' aims and intentions. As will be seen throughout this section, the IBA has tried to align the interests of not only Wilhelmsburg's residents and Hamburg's government, but many other actors as well within the energy bunker project. In fact, throughout the IBA, negotiating the teleologies of many different social actors has been one of the programme's main tasks. This, for example, meant that within the teleoaffective structure of the IBA, the interest of creating lighthouse projects had to be negotiated with the teleology of creating locally meaningful projects.

The energy bunker project is one IBA project for which it has been relatively easy to negotiate the interests of local residents and the city government. Just making the derelict building a project of the IBA involved the realization of local residents' interests. In the White Book, the bunker is described as an ugly and oppressive building that can neither be demolished not be put into any use in its current state (2002: 168). Apparently, this characterization of the bunker was not new among the residents.

HC: 'Well, in the sixties and seventies, it was constantly being debated, "what do we do with this old block?" And, "can't we demolish it? Can we not make something new out of it? Can't we hide it a bit? It looks so ugly and oppressive"' (Heritage Centre: 23–27).

This long-term resident of the area describes the perception of the building as ugly and oppressive as a constant issue among the inhabitants of the Elbinsel for many years. In the White Book (2002), the members of the working group 'recreation and culture' ask for an *'attractive utilization; supra-local meaning; "nicer" appearance (i.e. painting, graffiti)'* of the bunker and suggests to use it for example as *'community/meeting centre, exhibition (historic), gastronomy, sports activities (i.e. climbing), ateliers'*. It can be derived from both the quote and the White Book that the residents did not come up with any specific idea about how the bunker could best be utilized. Instead, the interest transpiring from the quote and the White Book was to improve the appearance and the utilization of the bunker. By making the bunker one of its projects, the IBA already realised two interests, voiced by the residents.

Another way in which local interests were meant to be realised was by making the building accessible to the public.

S1:   'Actually, that was, one thing we wanted from the beginning. So, it **was clear**, this is a special building. This building has to be made accessible. This should be part of the project, and if possible not only during the IBA presentation year. So we are an exhibition as such. The topic "exhibition" and "display" is of course an issue in all our projects' (IBA, Staff Member 1: 739-744).

While underlining the relevance of an exhibition, the interviewee in this sequence explains that the intention of opening the bunker for the public—and thus also for local people—was one of the key intentions of the bunker project from the beginning. Combining the two teleologies, the quote describes how the bunker was made part of the IBA exhibition by making it accessible to the public. Opening the bunker for the public was not only intended for local residents, but is closely related to the teleology of attracting an international audience. The team members are well aware of the necessity of integrating the two teleologies—attracting international attention and fulfilling local interests.

S1:   'So this is the duality of an IBA. It not only has to create something for the local people, but it also is meant to have an exhibition character and to adhere to international standards' (IBA, Staff member 1: 375–377).

The quote describes the tension between local and international interests, which are both meant to be fulfilled through the IBA. The negotiating of these two interests has influenced the IBA's enactment of the social practices of urban development.

Other than the creation of an international lighthouse project and to meet the residents' needs, another teleology that influenced the realization of the energy bunker was to realise the IBA's key issue—'climate change and urban development'. This key issue did not derive from any of the issues and intentions raised through the citizen initiatives or in the participatory processes. The subsequent integration of climate change resulted from a discussion process among the team members of the IBA.

S1:   'In the beginning of 2007 the fourth IPCC report was published. Until then this topic did not play any role. In the memorandum from 2005 it does not play any role as well. And then, we said. So basically I would say, from the colleagues and the management, we can't ignore this issue. And we are an IBA, and an IBA always also has autonomous advisory responsibility' (IBA, Senior Staff Member: 26–35).

The subsequent integration of climate change into the agenda of the IBA is presented in this quote as a necessity, which derived from the release of the fourth IPCC report. The insights into the causes and consequences of climate change presented in the report convinced the team members of the IBA that a modern urban development process could not ignore the issue of climate change. Another team member further explains why the members of the IBA felt that they could not ignore the issue of climate change:

S1:   'And, for us as well it provided a reason to say "okay, we deal with construction and urban development. And this is an area where a large part of carbon emissions is being produced". Seventy percent of carbon emissions are in the end relatable to building, accommodation, and everything belonging to these' (IBA, Staff Member 1: 221–224).

With regard to the IPCC report, this team member argues that they simply could not ignore the report and its insights because construction and urban development are a part of those areas of human societies that cause the most carbon emissions by far. Because they were engaging in practices that cause such a high percentage of carbon emissions, the team members started to make sense of their activities with regard to the results of the IPCC report. The fact that climate change was one of the IBA's three key issues shows that the reports and the relations stated therein between climate change and urban development and constructions provided a highly meaningful way of sense-making for the team members. The way the IPCC report and its results are mentioned in these quotes suggests that a high legitimacy was ascribed to the report and its result. According to the quotes, neither the report nor its results caused any critical discussions within the team but instead were accepted immediately.

The results of the IPCC report and thus the issue of climate change were able to provide a meaningful concept even though climate change apparently had not been relevant earlier in the process. This was caused not only by the legitimacy ascribed to the report, but also because of cultural shifts that had taken place in the meantime.

AD: 'And then there was a new development, namely that one of these former <u>sub-topics</u>, which dealt with the question of "energy and sustainability", which in two-thousand-three, two-thousand-four we had more understood to be a standard-topic, became a completely new. Well, we did see it, but, well, more like a standard-topic; got a new <u>impulse</u>. And thereby this IPCC report, back then, which—I think—was released in 2006, if I remember right; which once again, raised <u>worldwide</u> attention, for this climate change and so on, yes; and also it was supported by this movie from Al Gore. Well, so this topic suddenly was talked about everywhere and <u>had</u> become a new accent' (Senior Manager Public Administration: 71–79).

In the first part of this sequence, the interviewee explains that issues which were dealt with under the rubric of sustainability as integral part of projects until 2004 now became stand-alone issues in urban planning as they were scrutinized under the rubric of climate change. Again, climate change is described as having provided a new way of making sense of the issues dealt with in the IBA. At the time of the citizen-led and participatory processes, climate change was not an issue, because at that time it did not exist as a widely dispersed way of social sense-making through which problems could be given meaning and solutions could be developed accordingly. In the second part of the sequence, the interviewee explains the appearance of the issue of climate change in the IBA as resulting from the worldwide attention received by this topic in the meantime. The integration of climate change into the IBA is not explained here in terms of new facts or knowledge to which the IBA responded but as an outcome of the public attention acquired by the topic as an internationally meaningful frame. By integrating climate change into its agenda, the IBA responded to an internationally existing form of social sense-making.

Making climate change a key issue meant that the IBA's social practices of urban development would integrate specific activities and ideas.

S1: 'The whole development of the key issue is corollary. So, in the end it was a very logic process, I would say. It is about <u>carbon reductions</u>. With that it is about renewable energies, energy efficiency, energy savings. This has corollary consequences. A) for new buildings; B) for the redevelopment project, and C) well, this question how can we produce renewable energies in the city? So, this was, in the end, not in the least an intellectual approach ((laughing)) which has to search for difficult justifications, but was relatively easy' (IBA, Staff Member 1: 253–262).

Having been made a core teleology, carbon reductions shaped the social practices of the IBA by providing specified concepts and ideas. Both the necessity for carbon reductions and the associated measures of renewable energies, energy efficiency and energy savings already existed as shared forms of knowledge which are associated with the concept of climate change. Consequently, according to the interviewee, the specifications of climate change into measures of renewable energy, energy efficiency, and savings appear as a '*logic process*', which did not require any '*intellectual approach*' on behalf of the IBA members. The integration of carbon reductions—more specifically renewable energy, energy efficiency, and savings—into the teleoaffective structure of the IBA resulted in the creation of the new core issue of 'climate change and urban development'. Producing renewable energy in and for the area emerged in these discussions as one of the key strategies to reduce carbon emissions.

After integrating carbon reductions and the three measures into the teleoaffective structure of the IBA, the team members started scrutinizing possibilities to realise these new teleologies.

S1:   'And basically we then discussed in two different directions. A) the question: how can we enhance our own projects with regard to this topic "urban development and climate change". […] And on the other hand, as well the question how can we, well, analyse, evaluate and develop the whole area of Wilhelmsburg with regard to this topic' (IBA, Staff Member 1: 38–45)?

Having negotiated the meaning of climate change, it became necessary for the IBA to find ways to realise this teleology. The quote shows that the relation between a certain element and the teleology of carbon reductions did not exist 'naturally'. Rather, the members of the IBA made certain elements meaningful with regard to carbon reductions by analysing, evaluating, and developing '*the whole area of Wilhelmsburg with regard to this topic*'.

The first way in which the key issue was realised, however, was the creation and publication of a 'climate plan' for Wilhelmsburg. Within this plan, the IBA developed a concept for transforming the district into a zero-emission area.

AD: 'We asked ourselves, isn't it possible to develop such an island in a way that in the long term perspective it might actually result in energy autarky? This is what basically is written down in the climate plan "renewable district." […] And the special thing about that is that we—together with the climate advisory board—calculated so to say a self-sufficiency concept for a regenerative self-sufficiency of the whole island' (Senior Manager Public Administration: 119–128).

After developing a concept to produce renewable energy in and for the district, the identification and realization of energy producing projects was required. Projects that were already part of the IBA were now re-evaluated and re-developed with regard to the new key issue. Additionally, the new key issue spurred ideas for new projects. At that time, the energy bunker was meant to contribute to the climate plan by becoming a solar bunker. Not yet aware that the bunker would have to be restored anyway, the members of the IBA included in their calculations only the option to produce renewable energy using solar panels on its outside walls. The bunker, however, was intended not only to produce renewable energy for Wilhelmsburg but also to create attention for the IBA.

S1:   'Simultaneously, we still asked ourselves "What new projects can we develop with regard to this topic?" In addition to those spots which were already part of the IBA area, well, potential project spots. And out of this we developed a search radar, thus we were scanning the area for what <u>could</u> be potential projects. And here these two specifically large and visible projects energy hill and energy bunker appeared' (IBA, Staff Member 1: 50–56).

The interviewee relates the energy bunker to the key issue 'climate change and urban development'. He describes how the bunker became a potential project only after the integration of climate change had created '*a search radar*' to find the project. Only after the development of the new key issue was the bunker noticed as a potential lighthouse project for the climate change and urban development topic, because it was understood to be '*specifically large and visible*'. The special value of the bunker for the key issue is not described as being particularly effective, but as being large and visible. This phrasing combines the two teleologies—carbon reductions and creating projects that would attract public and media attention for the city.

This interpretation of this interviewee's sense-making of the project is further emphasised and specified in a later sequence from the interview.

S1:   'From the beginning all the members <u>knew,</u> that just because of its **mightiness** in the district, in the quarter, these huge dimensions it just has. That this is just meant to become a lighthouse project. That it will raise much public attention, also from the media. And that of course it will have to qualitatively live up to this attention. But that it will be relatively easy to raise attention with the project. And it of course was meant to function as a symbol for renewable energies in the city. This was part of the project idea from the beginning' (IBA, Staff Member 1: 745–750).

As mentioned above, in this sequence the energy bunker is made sense of in the teleology of creating a lighthouse project. The bunker, however, is not only meant to be a lighthouse project, raising much public attention for the IBA and the city, but '*to function as a symbol for renewable energies in the city*'. Through this phrasing, the role of the bunker with regard to carbon savings is further specified in terms of the urban setting within which this practice is enacted. So far, hardly any renewable heat has been produced in urban quarters. By engaging in such an activity, the IBA was able to create an innovative project with regard to the question of renewable energy production. In the bunker project, the IBA combined the teleologies of producing renewable energies and the creation of innovative, internationally attractive projects.

At the same time, the necessity to produce renewable heat in the district was motivated by the fact that the district was meant to be supplied with heat from a new coal power plant.

S1:   'Vattenfall has not only been given the permission to build a new large coal power plant, but simultaneously has taken on the duty to extend its long-distance heat network [...] literally, into the IBA development area. And this contract would have influenced our IBA projects very much. And we could not accept that, because to represent the future of building culture 2013 and then our buildings are supplied with coal heat—that has nothing to do with a future oriented and sustainable energy supply' (IBA, Staff Member 1: 572–583).

In this quote, the necessity of producing renewable heat for the district in the district is associated not only with the intention of creating a future-oriented and sustainable heat provision, but also with the teleology of creating innovative and thus internationally interesting projects. The IBA '*could not accept*' the planned heat provision, because of its aim to '*represent the future of building culture 2013*'. The involved actors worried that:

P: 'If we now get long-distance heat from a coal power plant—the world is gonna laugh about us. If that is meant as innovative, carbon free energy supply for the district. Well, forget it' (Politician: 186–188).

Besides worrying about the IBA being unable to realise its self-imposed teleology of carbon reductions, this quote highlights the actors' fear that they would lose any credibility to '*represent the future of building culture 2013*'. In response to these worries and because of the static situation of the bunker, which required a restoration of the building (this is further elaborated in Section 4.8.6), the bunker became positioned much more central in the deliberations of the IBA members. The idea emerged to supply the neighbouring quarters with heat produced in the bunker. A company was contracted to evaluate different energy mixes in order to find the most suitable one for the site and the amount of heat that had to be produced. With these data, the IBA presented the energy bunker project for the first time to an expert audience.

S1:   'And here we presented the energy bunker for the first time. And with some of the invited experts we founded the advisory board "climate and energy". [...] With its members we made an excursion into the district and passed the bunker. And we told them our plans and that we are going to restore the building. And immediately all the experts agreed "if you have such a large space at hand and want to do a renewable energy project, you must integrate the topic of heat storage"' (IBA, Staff Member 1: 114–117, 120–125).

The reason heat storage was deemed such an important topic for the IBA was not only the available space—an otherwise scarce resource in urban settings, but also the fact that heat storage facilities had hardly been realised so far—especially in urban quarters. By implementing a heat-storage facility, the energy mix realised in the bunker would become a much more innovative concept. In fact, the IBA explains on its webpage that the '*project's most innovative feature is its large-scale buffer storage facility*'. This expression on the webpage is combined with data about the project: '*carbon saving of 95 per cent, or around 6600 tonnes of carbon per year*' and the explanation that the bunker '*creates local jobs and income*'. This description of the project again exemplifies the combination of the teleologies of serving local needs, raising international awareness, and achieving carbon reductions within the energy bunker project.

As the creation of innovative projects is an important aspect in the exhibition teleology of the IBA, the heat storage came to be seen as the '*big chance*' of the energy bunker:

S2: 'That we <u>must</u> have a district heating grid if we want to use the bunker as <u>heat</u> <u>storage</u> that we knew from the beginning. Because, the chance of the bunker principally was the storage issue. We realised already in the beginning that this was the big chance of the bunker' (IBA, Staff Member 2: 305–310).

In this quote, the heat-storage facility is described as the bunker's big chance of becoming an innovative lighthouse project. Integrating the heat storage would make the energy production in the bunker much more efficient. This meant that realising it would be necessary to connect a minimum number of households with heat. Because the district so far had no heat grid, the IBA needed to install a district heating system as they '*<u>must</u> have a district heating grid if we want to use the bunker as <u>heat storage</u>*'. The design of the whole energy system for the district in this quote is presented as being based on the IBA's decision to install a heat-storage facility in the energy bunker—a decision which has largely been based on the innovative character of such an installation, which is the '*big chance*' of the building to become a '*symbol for renewable energies in Hamburg*' (IBA, Staff Member 1: 745–750).

Shortly after the experts had advised the IBA to install a heat storage facility, a senior manager from a nearby industrial site contacted the IBA. He proposed that the energy bunker could use waste heat of the company he worked for. The members of the IBA were soon convinced by the proposal. When the IBA contracted a company to calculate the intended energy system options, the installation of a heat storage facility and the integration of as much waste heat as possible were main criteria.

S1: 'Well, there has been—always—the guideline not to construct the most economic, the most economic heat project, there, but, the idea was: it should integrate a large storage and it should use as much renewable heat or recycled heat, waste heat as possible. And then, it of course has to produce affordable heat. This was the hierarchical order set for the studies' (IBA, Staff Member 1: 137–142).

In this quote, the staff member of the IBA defines the energy bunker as a project not primarily intended for the production of financial income. Instead, according to the quote, the project mainly aimed to produce as much renewable energy as possible by using the heat storage and waste heat, and to produce affordable heat for the local residents. Arguably, one reason for the neglect of economic interests at this point was that no actor who would have brought an economic teleology into the negotiation process as yet part of the project. Technical artefacts and strategies, on the other hand, which were considered to realise the combined intentions resulting from the climate change key topic and the IBA's intentions to create architectural lighthouse projects, were at the top of the hierarchy. A third teleology, realised in the calculations for the best energy mix, was the development of local resources for the residents through the production of affordable heat.

The heat-storage facility, however, did dominate the project development.

TE: 'The question still is, if the storage facility in this massive bunker is relatively small, it does not look like much. Yes, and this is the IBA. This means one just wants to have something that can be shown. ((laughing)) If a relatively small storage facility sits within such a huge room, it just doesn't look very imposing. The question thus was: is it possible to find a reason for a bigger storage facility' (Technical Expert: 60–65)?

In this quote, a technical expert involved in the calculations for the energy mix recounts the role of the heat-storage facility within these calculations. According to him, the heat storage was not only the main element within any calculated system, but the system instead should be designed to allow for the heat storage being as large as possible. According to this interviewee, the reason the heat storage was meant to be as big as possible were the exhibition teleologies of the IBA. The heat-storage facility was not only meant to enable a much more efficient energy production, but also to be an exhibition object. Especially when related to the quote before, this experts' statement shows that no 'for-one-and-all' best solution exists, but that solutions are evaluated in relation to an actor's specific teleoaffectivities and intentions. Instead of creating the most economic system, the energy mix in the energy bunker was designed to realise the different specific teleologies of the IBA. This meant that while the integration of a large heat-storage facility was not necessary from a technological and economic point of view, the teleology of creating architectural and urban planning exhibition objects provided an important argument for a more visible storage system. The dimensioning of the heat-storage facility is an example of how the IBA members' negotiated their different teleologies not only within the project at large, but also in smaller and more detailed aspects of it.

Because it exceptionally well negotiated different teleologies of the IBA, the integration of the heat-storage facility is understood to as a successful part of the energy bunker project.

S1:   'What I find very important is that the heat storage worked out. That we were able to realise this topic there. Because, also, within the technical constellation this has been the aspect which is not really economically reasonable, but which has been pre-set by the IBA, where we said: we want this future technology to be realised there, even if this is not going to be feasible within the first, in the first year' (IBA, Staff Member 1: 1036–1041).

In this quote, the IBA's intention to realise the heat-storage facility is again presented as the overriding aspect of the whole project. This aspect was so important for the IBA that the resulting system did not even have to be financially viable immediately. While the IBA itself does not have to create income-generating projects, the IBA members otherwise emphasise the economic feasibility of all their projects.

S2:   'But, one has to say, admittedly we did subvention our energy-related project, but never on such a dramatic level that we would have secured the economic viability of the project. Instead everything we did subvention has always been a "Surplus" [French pronunciation] It was never said, "oh God, this is a decentral project, that's why it is economically non-viable". This has never been the idea, but it always had to be economically basically viable. Only when it was about a special feature like a Stirling engine, or something which, is very innovative, then we said: "okay we are supporting this". But we have never secured a basic economic viability. This would be disastrous, because, then everything would fall apart after the IBA has come to its end' (IBA, Staff Member 2: 481–492).

In this quote, the senior staff manager argues that the IBA would only subvention 'surplus' features that were deemed especially innovative. The quote thus again exemplifies the importance of creating innovative projects for the IBA. The teleology of attracting international public and media attention through the creation of innovative lighthouse project is here described to dominate economic considerations. While the quote already shows the IBA's ambiguous relation to economic interests, this ambiguity becomes even more pronounced when relating its statements to the earlier quote, according to which the heat-storage facility was deemed so important by the IBA that even its low economic viability was accepted.

The energy bunker has been shaped by the teleoaffective structure of the IBA in many ways. The whole project resulted from the integration of the teleology of carbon-saving into the programmes' teleoaffective structure. At the same time, the bunker was made a project of the IBA because it enabled the programme to realise its teleologies of raising international public and media attention through the creation of innovative lighthouse projects and the intention of realising local residents' interests. Also, the IBA's teleoaffective structure significantly shaped the way the project was planned and realised. To avoid a loss in credibility of its ambition to exhibit the future of urban development, the IBA needed to create projects that would produce enough renewable heat so that the district would not need to be supplied with heat from a coal power plant. Increasing and enhancing the originally planned energy installations on and in the energy bunker was one way in which the IBA was able to realise this interest. While the mix of energy producing installations in the bunker thus has generally been shaped by the IBA's intention to create a symbol for renewable energies in Hamburg, the integration of a heat-storage facility specifically exemplifies the IBA's intention of creating highly innovative projects. Both the integration of the system in general and its size in particular are significantly shaped by the teleologies of the IBA.

4.8.2   General understandings: is the creation of lighthouse projects a 'good practice'?

The energy bunker project of the IBA exemplifies how general understandings might vary decidedly among different social actors. The general understanding of the IBA's main teleology of creating innovative lighthouse projects also varies among different social actors.

P: 'Some people today use lighthouse projects to defame projects as withdrawn or something like that. So, I think, you sometimes have to test things, in new projects. And then you have to see, what of it can be broadened out? And this in the end is the function of an IBA. It does not have the aim that tomorrow everything is replicated, but to give special impulses. ((laughing)) To provide impetus not only for the city itself, but, also, in its effects, internationally. And this I think, is in its basic idea an important process and an important function' (Politician: 296–305).

For this politician, the creation of lighthouse projects is generally understood to be a valuable practice, as it creates '*some special impulses*' for Hamburg, as well as internationally. In order to create new impulses, according to her, it is necessary '*to test things, in new projects*'. It is this underlying aim of the interviewee that makes her generally understand the creation of lighthouse projects to be an '*important process and an important function*'. The interviewee, however, is aware that other people might be critical of the creation of lighthouse projects. These might be understood to be '*withdrawn*'.

The creation of '*withdrawn*' projects by the IBA is in fact actually criticized by local residents. An already presented quote shows why lighthouse projects might be understood differently by local residents.

LO: 'And in the final presentation of the IBA this was presented like: "Yes. Finally a beautiful school has been built for the district". But it is only one school. And from the 50,000 inhabitants more than 20 percent are children under eighteen years. So, one school is not enough. And they did not say—Well, it was presented as a super success of the IBA: And I thought: and what about the three or four other schools' (Member Local Organisation: 612–616)?

This local resident does not generally understand the creation of lighthouse projects as a valuable practice. Her evaluation of these projects derives from her specific position within the context. Being a resident, the development and improvement of the districts' infrastructure would have been her main teleology. The social practices in which the members of the IBA engaged did not realise her teleologies. Deriving from her teleology, her general understanding of the creation of lighthouse projects is critical. The politicians' teleoaffectivity, on the other hand, is not so much focused on the creation of basic solutions for the district as it is on the creation of impetuses for the city and at the international level. The position of an actor with regard to the context in which they evaluate a social practice is decisive for their general understanding of this practice.

Like the IBA, the general understanding of the energy bunker differs among different actors, depending on their respective position in the context of the IBA. The diverging teleologies of different actors required and provoked continuous negotiations and contestations. These were carried on throughout the entire process of refurbishing the bunker and realising the energy installations.

HE: 'But, at the bottom of the list, from my point of view, I would say the IBA has initiated interesting things, projects. Urban planning wise. And on the other hand, I would also say that concepts have been realised, but they have just not been designed by somebody who afterwards has to carry on operating the whole thing for 15, 20 years. So, I would say, project development yes, but in between it all, there has to be some realism about what will have to be discarded in the lifespan of the projects' (Hamburg Energie, Senior Staff Member: 16–23).

The quote from an interview with an employee of Hamburg Energie relates to the key conflict between the IBA and the municipal utility. The IBA was created as a temporary organisation and existed for seven years. Its main future perspective and temporal focus has been the presentation year, 2013. It is not officially responsible for the long-term duration and survival of its projects. The interviewee acknowledges that '*the IBA has initiated interesting things, projects. Urban planning wise*'. On the most general level, the interviewee thus acknowledges that the social practice of creating lighthouse projects for urban development processes is a 'valuable' practice. More specifically, however, the long-term efficiency is one of the main ends of any project for the municipal utility. From the specific point of view in the context of the energy bunker, the project's long-term efficiency constitutes a 'realistic' perspective for the interview-partner. His general understanding of the IBA's practices as not '*realistic*' arises from his specific interest in the project being viable in the long-term perspective. The sequence exemplifies that the same activity, if seen from the position of urban planning, might be understood and evaluated differently compared to the point of view of another social actor taking part—who evaluates the project in a different social context.

For Hamburg Energie, the question of the project's long-term efficiency is closely related to the economic viability of the project.

HE: 'I can say, because we operate the energy bunker today, the heat sold from it—we are reliant on selling this heat successfully for many, many years. So, while the project developer says: "Why? The thing is realised, the equipment is installed." […] The operator says: "Yes but it has all become much more expensive than it's been planned." It is and will continue to be a lighthouse-project. I think one of your questions has been: "Would you do it again?" My answer is no. You can only realise a lighthouse-project one time. It's very easy, yes? I still do stick to the project. You can do that kind of thing one time, but afterwards there is no reason to do it again' (Hamburg Energie, Senior Staff Member: 27–37).

The quote shows the ambiguity of how a project's success might be evaluated very differently from different perspectives. At the same time, the quote is an example of one and the same social actor having very ambiguous evaluations of the same project.

In this sequence, the interviewee compares the different perspectives of project developers and operators. According to him, the social practices leading to the project are understood to be successful if they are evaluated in the teleology of creating a lighthouse project. Evaluated as a lighthouse project for renewable energy production, the project has lived up to the aims of showcasing the mixture of innovative technologies and heat storage facilities as a technologically viable option for urban settings. More specifically, for the IBA, the social practices leading to the project are understood as successful because the project generated much more public and media attention than expected. According to the quote, the project is generally understood to be 'successful' or 'good' if the aims of a social actor are fulfilled by the project. For the municipal utility, the project is both a success and a failure. For the staff member of Hamburg

Energie, the project development is generally understood to be successful as long as it is evaluated through a symbolic teleology. However, the project and its 'success' are very differently evaluated from an economic long-term perspective by the same actor. The long-term viability of a project derives from the amount of money that had to be invested into the project to realise it. As the amount of necessary upfront investment has turned out to be much higher than expected, the projections for the projects amortization and subsequent income had to be abandoned. The project does not generate sufficient financial income to be 'successful' from an economic perspective.

It appears that the general understanding of social practices is not defined once and for all but is always related to their outcomes and the way these outcomes are made sense of by a specific social actor. If the outcomes of a social practice are not understood to be successful, the social practices leading to this outcome will be understood to not be realistic or 'good'. The general understanding of social practices might not only differ decidedly among different social actors; the same actor might have ambiguous general understandings of a project, its success, and thus the social practices leading to it.

4.8.3 Practical understandings and theoretical knowledge: dissolving the difference

Three arguments with regard to theoretical and practical knowledge will be made in this section: the importance of theoretical knowledge with regard to the creation of renewable energy production sites and the intertwinedness of theoretical and practical knowledge in the context of the IBA. These two arguments prepare the third argument: that the differentiation of theoretical and practical knowledge is dissolved in human actors' practices.

Firstly, the realization of the energy bunker exemplifies the importance of theoretical knowledge with regard to renewable energy production activities. The energy bunker project is an outcome of different types of theoretical knowledge. When deciding to include climate change as a third key topic, the IBA members deliberately made a certain type of theoretical understandings—scientific knowledge—part of the context. Both the general ideas as well as concrete aspects of the bunker project were designed in response to scientific knowledge about climate change, carbon reductions, and renewable energy production. The integration of the climate change issue into the agenda of the IBA had been an outcome of the publication of scientific knowledge about climate change, its causes and consequences.

S2:   'in the beginning of 2007 the fourth IPCC-report was published. Until then this
        topic had not been relevant for the IBA' (IBA, Staff Member 2: 26-27).

The quote highlights the direct relevance of scientific knowledge for the integration of the key topic 'climate change and urban development'. The topic was integrated into the agenda as a reaction to the publication of a scientific report. Knowledge about the relevance of carbon reductions and renewable energy production likewise derives from scientific understandings. Like the integration of the climate change topic, the design of the energy system in the bunker project is heavily dependent on the integration of scientific knowledge. Most of the practices enacted in the design, planning, and realization of the project require the existence of architectural or technical knowledge, for example. Additionally, the bunker project has been designed in response to scientific debates and state-of-the-art research about renewable energy production.

S1:   'Well, this is, I think, an area which is interesting for technicians. Because it is not
        standard yet. So, in Germany there are only a dozen heat storage projects, maybe.
        And, that's why it is most certainly useful to gain further experience' (IBA, Staff
        Member 1: 844–847).

The quote indicates that the bunker not only integrated knowledge about state-of-the-art research but was meant to directly respond to existing research gaps. The project has not only been inspired by theoretical knowledge but aims to further enhance theoretical knowledge about the production and storage of renewable heat in urban settings. The bunker thus is not only an outcome of theoretical knowledge, but also contributes to its creation and development.

        Scientific forms of knowledge were not only important in the establishment of the key topic and the planning and designing of the bunker. Theoretical understandings also provided important arguments for the realisation of the bunker in negotiations between the IBA and the Hamburg government.

S1:   'So, when developing our district-heating network, we of course have used the
        climate change argument and argued against long distance heat networks based
        on a coal power plant. Uncoupling heat from a coal power plant has a carbon
        output of—I don't know—250 to 270 watts per kilowatt hour. And this is of
        course much, much higher than when I use renewable energy. And that's why we
        didn't want this. This was our key argument' (IBA, Staff Member 1: 617–623).

In this quote the interviewee describes the IBA's strategy for negotiating the bunker and especially the construction of a district heating grid with actors from the government. The Hamburg government had planned to have the district connected via a long-distance heat grid to a newly created coal power plant. In order to stop this plan from being realised, the IBA strategically employed scientific knowledge. Within the context of the IBA, responding to research gaps and creating innovative solutions also has been a strategy to create a 'lighthouse project' and thus to raise international public and media attention for the project (see).

Within the negotiations with the city government, the IBA members not only used scientific knowledge about climate change. Instead, scientific knowledge about climate change has been intermingled with theoretical arguments about the IBA's aim to create local assets.

S2:   'All the energy we produce here ourselves, also produces jobs and income and in the end also taxes. Here, at this place, and not somewhere else. These are, well, relatively strong arguments. You can hardly be opposed to these' (IBA, Staff Member 2: 431-435).

The sequence exemplifies the intermingling of scientific knowledge about climate change with social and economic arguments. Like knowledge about climate change, these arguments, however, are also based upon theoretical knowledge about economic causal relations. Theoretical, and especially scientific, knowledge thus were utilized to realise the core aims of the IBA.

Secondly, within the social practices of the IBA team members, many different forms of theoretical knowledge are closely intertwined with practical knowledge. Designing and realising the energy bunker project did not only require theoretical knowledge about climate change and related issues, but also a wide variety of theoretical knowledge about urban development, architecture, and energy production technologies. Of course, the IBA team members did not possess all the necessary theoretical knowledge 'themselves'. In order to access the necessary theoretical knowledge, the IBA mobilized and contracted a range of experts. Among others, architects, technical engineers, structural engineers, experts in energy production technologies and the grid, technicians, builders, and construction workers were contracted and contributed their knowledge and skills to the realization of the projects. These experts provided the theoretical knowledge that shaped and influenced the general idea and the precise design of the bunker project.

Making use of these types of theoretical knowledge, however, also implied the existence of different forms of practical knowledge. The necessity for practical knowledge is demonstrated by the insufficient ability of the IBA to communicate with the local residents. Generally, the IBA tried to communicate its ideas and projects to the local residents. These efforts are, however, not deemed successful by the local residents.

LO: 'It's just a really big difference between the milieu of this IBA crew—these are all urban planners, architects and the like—and the normal people living here' (Member Local Organisation: 753–754).

In this quote, the interviewee, a member of a local organisation and a long-term resident of the district, describes the participatory efforts of the IBA as having suffered from a communication problem. According to the quote, there is '*a really big difference between the milieu of this IBA crew*' and the '*normal people living here*'. The interviewee relates this difference to the educational background of the people who were working for the IBA and the local residents. While this quote indicates the relevance of (missing) theoretical knowledge in the IBA's efforts to communicate with the local residents, relating this sequence to an earlier sequence from the interview highlights the relevance of missing practical knowledge on the side of the IBA team members.

LO: 'We were informed. The question is if it is possible to understand this information or not. Especially with regard to energy and technology it often is such a difficult language and information. And this caused misunderstandings. […] And if we requested to explain it again and they did it the same way just using different words. And then they said: well, we already explained that. Yes, but nobody understood it. So there often were formulation problems between the expert people and the local population. And you have to manage to translate that' (Member Local Organisation: 181–189).

This sequence describes situations within which the members of the IBA team and the local residents were literally not able to understand each other. Within this quote the problem is described as a lack of practical understanding of the IBA members of the local residents' way of expressing themselves. The quote highlights the IBA members' lack of experience in explaining their ideas and projects in a way that could be understood by the local population. Even when asked to explain an aspect or situation differently, the members were apparently not able to use an 'appropriate' language. While they had experience in talking to and understanding politicians and other city authorities as well as scientists, the team members of the IBA—according to the quote—did not have the practical knowledge to successfully communicate with the local residents.

The example highlights the relevance of practical knowledge when (trying to make) making use of theoretical knowledge. In fact, a multitude of practical knowledge has been necessary in the acquisition and employment any of theoretical knowledge. Theoretical knowledge about climate change, carbon reductions, and renewable energy production could not have been transferred into social practices without the existence of practical knowledge about—for example—how to speak, read, and write. More specifically for the context of the IBA practical knowledge about how to use a computer, make sense of an architectural drawing, how to engage in professional discussions, and how to 'make use' of social networks was necessary. The argument deriving from these considerations is that theoretical and practical knowledge are closely intertwined with one another. Furthermore, practical and theoretical knowledge are not only related to one another in that they are mutually necessary for one another. Instead, the energy bunker project exemplifies how fuzzy the differentiation between the two types of knowledge is.

The realization of the energy bunker did not only require the existence and combination of theoretical and practical knowledge. Instead, the project is also meant to contribute to the development of both theoretical and practical understandings about renewable energy production.

S1:   'And, apart from that, well, I think on a low-threshold level, it is a powerful renewable energy project for, for, well the interested population, so to say—I would say—as an example where you can see: this way, this is how it could work. This way we can use renewable energy in the city. So, beyond the hard technical and economic issues' (IBA, Staff Member 1: 862–867).

According to the quote, the IBA deliberately intends to contribute to the creation of theoretical and practical knowledge through the energy bunker project. People can visit the bunker and inform themselves theoretically about the production of renewable energies in general and the production of renewable heat in urban settings in particular. Apart from these theoretical forms of understanding, the bunker also creates knowledge how about '*we can use renewable energy in the city. Beyond the hard technical and economic issues*'. This statement indicates the idea of providing a practical example of renewable energies. This aspect is even more pronounced in a sequence from an interview with a senior staff member of the IBA.

S2:   'I believe that it is necessary to—if we really want to succeed with the energy transformation—to socialize ['Sozialisieren' in German original, AP] energy issues. So, in the sense of societalization, not in the sense of socialization, as it has been understood by the political left in the past. But more in the sense of "socializing" [term used in German original, AP]. We have to turn it into a social-cultural issue. And into an economic issue' (IBA, Staff Member 2: 638–644).

Turning the production of renewable energies '*into a social-cultural issue*' requires a '*socializing*' process. The energy bunker provides a way to socialize renewable energies by making energy production sites a part of people's day-to-day life. Instead of producing energy at a geographical and social distance from the consumers, the bunker provides a highly visible site of renewable energy production. It provides an opportunity to make people aware and have them experience energy production as an everyday practice. The energy bunker projects of the IBA intend to influence people's practical understanding of energy production and thereby maybe to influence how people relate to energy, energy production, and consumption.

The energy bunker does not intend to only be a practical example of renewable energy production for local people.

HE: At the moment, I believe, things are changing. With regard to that you can say that we are <u>innovative</u>. Because awareness and also the trend are more and more tending to develop awareness about, okay decentralized supply options will become more and more <u>necessary</u> for the energy transition. And for this 08/15 solutions are not going to be the nonplus ultra any more, but you have to have other ideas' (Hamburg Energie, Staff Member: 424–430).

In this quote the bunker is presented as a practical example of a decentralized energy production and distribution facility. Furthermore, the bunker is not an example of '*015 solutions*' or *'some heat concept on a green field*' (Politician: 78) but an example of how specific local infrastructures can be altered or re-created to enable the production of renewable energy in close proximity to its consumers. Besides producing theoretical knowledge with regard to the identified knowledge gaps with regards to heat storage facilities and urban energy production, the bunker project is also seen as a practical example that is able to create practical knowledge about the possibility and practicability of highly specific energy production solutions.

So far this section has argued the relevance of theoretical knowledge for the creation of renewable energy project. Also it has been argued that practical and theoretical knowledge are tightly linked within the energy bunker project. Taking these two arguments one step further it will be argued in the following that the differentiation of theoretical and practical knowledge becomes rather fuzzy within the social practices under scrutiny here.

An already presented quote from a team member of the IBA indicates that scientific knowledge about climate change, carbon reductions, and renewable energy production has already become so 'normal' for certain types of actors that it is perceived as practical knowledge by them.

S1:   'The whole development of the key issue is corollary. So, in the end it was a very logic process, I would say. It is about <u>carbon reductions</u>. With that it is about renewable energies, energy efficiency, energy savings. This has corollary consequences. A) for new buildings; B) for the redevelopment project, and C) well, this question how can we produce renewable energies in the city? So, this was, in the end, not in the least an intellectual approach ((laughing)) which has to search for difficult justifications, but was relatively easy' (IBA, Staff Member 1: 253–262).

It transpires from the quote that knowledge about the relations between climate change, carbon emissions, and renewable energy production had already been so much embodied by the team members of the IBA that the development of the key topic was '*relatively easy*'—implying that it did not require a lot of theoretical deliberations on their behalf. In fact, for the interviewee the development of the key topic and its derivate activities appeared to be '*not in the least an intellectual approach*' which did not have '*to search for difficult justifications*'. The way climate change as well as relations of climate change to other concepts and/or activities are explained in the sequence does better correspond to practical than to theoretical knowledge.

It appears that the analytical differentiation between theoretical and practical knowledge cannot be maintained. Instead, theoretical knowledge—for example about climate change—might become so much internalised by actors that they do not engage in conscious reflections anymore when acting upon it. Furthermore, even if social practices of renewable energy production themselves would largely depend upon practical types of knowledge, they would nevertheless require the existence of theoretical knowledge—as otherwise the technologies and sites of renewable energy production could not have been developed.

4.8.4   Rules: Making use of and creating rules in the IBA Hamburg

The IBA is a daughter company of the city of Hamburg. It was set up and capitalized by Hamburg's government. Basically, the IBA, as an institution, has been created through rules. Rules furthermore shape the activities of the IBA. These rules most importantly pre-set the spatial boundaries within which the IBA acted and also defined the time span of seven years. The limited amount of time available for planning and realising the project is partly responsible for the IBA not being able or willing to listen to the local residents' ideas and concerns.

LO: 'They did make an effort, in the beginning, to capture people's thoughts and ideas. But they are of course, also. Well, if they have come up with this concept urban regeneration programme, as—in brackets—a substitute for a normal urban development process, you needn't wonder if after latest half of their seven-year time was over, they felt the pressure like "Well, now this really has to shine and has to be international, and we can't cope with all the small and smallest issues of the local citizen initiatives"' (Member Local Organisation: 741–752).

The interviewee acknowledges that the IBA did try to communicate with local citizens '*in the beginning*' of the programme. According to her, however, the limited amount of time provides one explanation as to why participatory social practices decreased throughout the lifespan of the IBA. The example of the time constriction shows how rules might indirectly influence social practices by shaping the context within which these social practices are enacted.

The limited amount of time for which the IBA exited directly influenced the planning and realization of the energy bunker.

HE: 'In the end the time schedule, finishing it in time for the IBA, these were very specific requirements. And the consequence has been that the costs increased (Hamburg Energie, Senior Staff Member: 40–42).

In this quote, the member of Hamburg Energie explains that that the time restrictions increased the investment costs for the project. Increased investment costs mean that the project becomes much less financially viable than anticipated by Hamburg Energie. This is again one of the reasons the staff member in the interview also says: *one of your questions has been if I would do it again? My answer would be no'* (Hamburg Energie, Senior Staff Member: 33–34). Like with the participatory practices, the limited lifespan of the IBA has influenced social practices of planning and realising the energy bunker.

Rules from the federal state level shaped the realization of the energy bunker in different ways. Besides creating the IBA and dictating most of its basic conditions, federal state rules significantly contributed to the project's ability to be financially realised. After E.ON Hanse refrained from taking part in the project because it estimated that the financial risk would be too high, the IBA members needed to decrease the investment costs to be paid by a future energy operator.

S1:   'And that meant that A, the financing of the restoration of the building had to be ensured by the city. This was possible, just because the city was responsible for ensuring the safety of this liable to collapse building. So the owner—the fiscal authority—said "Okay, we will give you an amount x", which corresponds to what has been calculated as the minimal expenditures for ensuring the building's safety' (IBA, Staff Member 1: 187–193).

Rules in this case created a responsibility for the safety of the building. Because the fiscal authority of Hamburg was legally responsible for the building, it had to pay at least the minimum costs necessary for the restoration of the building. This situation significantly decreased the necessary upfront investment costs and thus made the project more likely to become financially viable. Rules in this case not only shaped the situation by paying for those social practices through which the building was restored, but also enabled Hamburg Energie to become involved in the project and thus to later start social practices of producing and distributing renewable energy.

Another set of rules that shapes the context in which the described social practices could evolve are national level rules like FITs and rules through which specific funding sources and incentives were created. These significantly contributed to the realization of the project.

S1:   'Well, we did use national funding sources, because—if you built a hat grid with a certain amount of renewable energy—you are eligible for KfW funding. And it is absolutely obvious that without funding you won't go places in the heat sector.

> Because the mechanism of an EEG is lacking. Which <u>in another way</u> supports investments into renewable energies' (IBA, Staff Member 1: 953–958).

According to the interviewee, the existence and availability of investment grants enabled the realization of the project in the sense that they contributed the necessary monetary capital to finance its realization. The investment grants ensured the financial feasibility of the project. Comparable to the money from the fiscal authority, the grants shaped the context within which the energy bunker could be successfully realised.

Besides being influenced by rules, the IBA itself created rules for the energy bunker project.

S1:   'We concluded a contract with Hamburg Energie, a quality agreement [...]. In this contract it was written down how the project is going to be financed. And the technical and time frame within which it has to be realised. So Hamburg Energie committed themselves to realise the technical concept that we developed in our report. And the IBA committed itself to restore the building so that the technical equipment could be installed' (IBA, Staff Member 1: 203–212).

Unlike the rules described before, the quality agreement directly influenced the social practices of Hamburg Energie and the IBA. According to the quote, the two actors agreed on a contract that obliged the IBA to restore the building. The social practices enacted to restore the bunker in a way that enabled the realization of the heat concept appear to have mutually been shaped by and to have shaped these rules and their influence. Hamburg Energie on the other hand had to realise the energy project according to the concept developed by the IBA. Based on the rules of the agreement, Hamburg Energie was not allowed to develop or realise the technical installations on its own accord or in a way that would suit their interests best. This influenced Hamburg Energie's practices of not only planning and realising the heat concept, but also running the bunker project.

HE: 'So, the conditions for the project did cause the project to be defined by requirements which in the end made the project much more expensive than we expected it to be' (Hamburg Energie, Staff Member: 43–45).

The requirements which became contractual in the partnership agreement, are described to have a lasting influence on the project. The conditions of the agreement forced Hamburg Energie to realise the project under requirements based on the teleologies of the IBA. These mentioned requirements and the teleologies established therein did not only influence Hamburg Energie's practices of planning and installing the technical equipment; they also significantly increased the upfront investment costs of the project. This made the project much less (not) financially viable. Consequently, Hamburg Energie now had to operate a project that is financially much less viable than desired by the organisation. Because of the financial experience with the energy bunker, the senior member of the municipal utility states that if asked *'if I would do it again? My answer would be no'* (Hamburg Energie, Senior Staff Member: 33–34).

Even without the special requirements of the IBA, the energy bunker project would not have been financially viable. In fact, only the funding sources that could be accessed through the IBA made it possible to realise the project at all. The unfeasibility of projects like the energy bunker, however, is not naturally given, but results from the existing rules governing the promotion of heat projects in Germany.

> PE: 'I agree with you; under the current framework conditions such projects are not replicable. Without massive financial assistance. But they show what is possible already now. What would be <u>basically</u> possible in Germany. In the end it's only about changing the legal framework conditions in a way that all the investment sources which the project got, are not necessary any more. So, to show it is not a technical problem, it is only a financial problem. And it will only be a financial problem as long as the framework conditions are as they are at the moment' (Political Expert: 466–474).

This quote describes the energy bunker project, and the social practices of renewable energy production enacted in the bunker, as being non-replicable in the current German system. While the bunker project itself showcases the technical potential of producing and distributing renewable heat in urban settings, the legal framework conditions in Germany are not favourable for the realization of any such projects. The project and the social practices leading to its realization are presented as not being replicable and thus 'normal'. Instead, the energy bunker appears in this quote as a project that could only be realised in the highly specific context of the IBA.

4.8.5 Human actors: staff, experts, politicians, and local residents negotiating their interests

Different actors shaped the social context of the energy bunker. The first type of human actors to be described are the staff members of the IBA. The IBA existed for only seven years. The working contracts of the staff members also lasted for only this period of time. The staff members of the IBA were thus elements in the context of the energy bunker only for the duration of the IBA. Within the context of the IBA, however, the staff members were decisive for the development of the energy bunker. The staff members were inherently part of the social site in the sense that they not only 'existed' through the social site but also shaped the context. The role of the staff members is especially important with regard to the key topic of climate change.

S1:   'And, well, because no one from the personnel of the IBA had been planned for this, it was, well how should I say? It just happened. I said I would like to do it. And then this task was assigned to me. And because of the belated or additional team creation or project development, there were especially few requirements from the city. Very few project suggestions which had to be taken on or assessed. And, accordingly, we had a relatively big scope of action' (IBA, Staff Member 1: 30–38).

The staff member describes how the subsequent integration of the topic 'climate change and urban development' altered the position and meaning of the staff members. Because hardly any conditions had been set by the city government for this topic, the staff members acquired a much more central position. This changed the staff members' identity from assessors of existing project concepts to the innovators of carbon-reducing projects. The staff members altered positions exemplify the instability of the positions and meanings of elements, as well as their interrelatedness to other elements of a context. By adapting their meanings and positions the staff members momentarily stabilized the new context.

Throughout the development and realization of the energy bunker project, the staff members continued to be important. Through their work, they shaped the context within which other human actors acted. The members of the IBA searched for and invited other human actors who could be of relevance to the project development.

S1:   'Professor K has supported us as a counsellor in organising the energy laboratory. So, with regard to the question how can we invite? How—most of all—can we invite foreign reference projects? Whom can we invite as experts? He has made suggestions, when we sat together, discussing this question. Also, he had a small consulting engagement with us. Professor K is an internationally working expert, who has done much work in Australia, China, and the US, and who was a member of the world council for renewable energies. Thus he could draw on a broad network' (IBA, Staff Member 1: 295–304).

The quote describes how experts were actively sought out and invited to take part in the IBA. The quote also illustrates how people were sought out to take part in the project because they were understood to be able to contribute resources to the project. The type and task of the experts can be related to the teleoaffective structure of the IBA. As described earlier, the intention for setting up an IBA was to stimulate international attention to Hamburg and Wilhelmsburg in particular. International reputation transpires as an important characteristic, as is emphasised throughout the quote. The IBA was mainly interested in experts and international reference projects. Both types of actors were understood to serve the IBA's teleology. They contributed not only their skills and knowledge but also their international reputation to the project. This interpretation of the role of experts is supported by a statement from the IBA webpage within which the members of the advisory board climate and energy are presented:

> 'The experts from the fields of science, politics, and practical application will accompany critically IBA's work, increasing the innovative strength of the projects and reflect and discuss the local efforts of IBA in the professional context internationally. The consulting committee can also support IBA in the international presentation of the projects' (www.iba-hamburg.de, 29.8.2017).

Like the interview, this quote from the IBA's webpage explains the importance of experts, because they not only contribute knowledge but also connect the IBA to international contexts. The context of the IBA, with its teleology of creating international public and media attention, influences the selection of experts invited to take part in the projects. As shown by the integration of the heat-storage facility, these experts have also influenced the IBA, and especially the energy bunker project, by contributing their ideas and interests.

Another important type of actors included politicians and members of the public authority.

S1:   'A second important aspect has of course also been the well, the IBA's network capacities into the public administration. Both on the council level as well as on the <u>city</u> level, including even the ability to contact the <u>federal minister</u> when things were getting tight' (IBA, Staff Member 1: 967–978).

Here, contacts with both the public administration and important political actors are described as an important aspect of the realization of the energy bunker project. The IBA was established and initiated by political actors at the federal level. Consequently, it was positioned within a political context from the beginning. Because of its background, the IBA has good relations to politicians and members of the public administration. The existence of these relations has contributed to the realization of the energy bunker project in significant ways.

P: 'Well, it is of course important, an IBA <u>needs</u> political backing. Whether projects fail or become a success, <u>this</u> depends on the backing. The energy bunker could have <u>failed</u>. If not—I might say—I had been particularly convinced that this is a project which has very much symbolic power' (Politician: 67–72).

The high-ranking political actor who makes this statement relates the success of the energy bunker project to his specific engagement in the project. Being a particularly complex and unusual project, the energy bunker needed political backing in different ways. The political actors were able to support the realization of the IBA projects by creating specific bodies within the public administration that condensed the activities of institutions, normally unconnected with one another. Additionally, the politician made sure that important actors from institutions relevant to the bunker project supported the project.

P: 'So that means, sometimes you have to bring different actors together for such an innovative project. Where you have to say: here everybody needs to contribute one's share, so we can try something new. And that's what I mean with political backing. That the political leadership <u>cannot do otherwise</u> but to position for the project' (Politician: 141–146).

The quote describes how political actors can and should contribute to the development of innovative projects like the energy bunker. Political actors are described as facilitators of the energy bunker project. The situation to which the quote relates has in fact been crucial for the project. In order to realise the project, the IBA needed to make sure that a sufficient number of customers could be supplied with heat produced in the bunker. Most of the houses in the vicinity of the bunker are owned by a municipal housing company. It was crucial for the project that the housing company would become a customer of the energy bunker. The housing company, however, was reluctant to become a customer. It was only willing to connect their houses to the bunker if the heat would not increase the tenants' heating bills. Apparently, the housing company was initially unwilling to become involved even in negotiations with the IBA. Only due to the engagement of the political actor did the housing company, Hamburg Energie, and the IBA finally begin to engage in discussions. The negotiations between the housing company and Hamburg Energie were complicated by the fact that Hamburg Energie needed to set prices that would make the project financially viable.

P: 'For the energy bunker project I had the chief of the municipal housing company, the chief of the municipal utility provider, the manager of the IBA and somebody else sitting at one table. And we said: "Okay! Now we have to huddle together and come up with an idea"' (Politician: 402–405).

Because all the involved actors were municipal institutions, the political actor was able to exert some influence on them. When he asked the managers of the different institutions to meet, they apparently felt obliged to do so. Within the social arena created by the political actor, the diverging teleologies of different social worlds were negotiated. The situation exemplifies the importance of political actors for the realization of the energy bunker project. Without the integration of important political actors into the context of the IBA, it is doubtful whether the project could have been realised.

The staff members of Hamburg Energie constitute the members of another social world within the social arena of the energy bunker. The importance of these actors lies in the time limitations of the IBA, which only existed long enough to plan and realise the project. Not being able to operate the energy system itself, the IBA needed to find a company that would be willing to run the energy production in the bunker.

S1:   'With this mix of technical systems we started looking for, A, the necessary resources for the restoration of the building. And, B, an energy provider, who would realise this project with us. And we first discussed this with another company, whether they could imagine doing it? But they chickened out because the risks associated with refurbishing the building were too high for them. For them it seemed too difficult to calculate the necessary investment. So then we decided that in this special case the IBA would itself be pay for the restoration. And B that another municipal daughter, Hamburg Energie, would realise the energy concept. As one of their first projects. They were only founded in 2009, I think' (IBA, Staff Member 1: 175–185).

The quote describes the problems of finding an energy provider that would be willing to realise the energy bunker project and become the project's operator. According to the quote, the IBA first encountered a situation in which it was not possible to negotiate the different teleologies of the energy provider and the IBA. For an enterprise aiming for and reliant on the generation of financial income, the project seemed too risky. The perception of the project's riskiness resulted from relations between the teleologies of the energy provider and the characteristics of the bunker restoration. For the energy provider, the necessary upfront investment was either incalculable or too high to secure a viable financial turnout of the whole project. Furthermore, the IBA had decided to create not the most economic but the technologically most innovative low-carbon project. The social worlds of the IBA and the energy providers were based on teleoaffectivities and intentions, which were contradictory in the context of the energy bunker.

The IBA developed two strategic approaches to deal with this situation. On the one hand, it decided to realise the refurbishment of the bunker—i.e. the part of the project that was deemed to be especially risky by the energy providers—on its own account. The IBA developed a compromise to meet the energy providers' requirements by decreasing the economic risk resulting from the necessary upfront investment. The second strategy the IBA engaged in was to look for a partner whose social practice of energy production would be better compatible with the social practices of the IBA. They decided to start negotiations with the newly created Hamburg Energie.

I: 'How do you evaluate success?
H2:   In terms of corporate objectives?
I: Yes, well is it more economic or things like: we have initiated something?

H2:   That is important. Absolutely. To realise ecological projects for our customers. To create services in that sector. This is an added value because everybody can sell electricity and gas. We can't compete with other companies price-wise. […] And to realise ecological services for our customers includes realising projects which are economically viable. We must not lose money. But we also do not need to get rich. I can say that. Others might have a different opinion about that, but we do not need to generate as much money as possible in the municipal context. These are the main objectives I would say' (Hamburg Energie, Staff Member 2: 279–291).

Being a municipal utility provider, Hamburg Energie does not need to generate a high degree of financial income. Instead, ecology- and customer-oriented ends are positioned centrally within the company's teleoaffective structure. While the hierarchical positioning of teleologies might be different, the teleoaffective structure of the IBA and Hamburg Energie are basically compatible. The identification and integration of an energy provider whose teleologies are compatible with those of the IBA was crucial for the realization of the bunker project.

Nevertheless, the realization of the energy project within the bunker required and provoked continuous negotiations and contestations, which were carried on throughout the whole process of refurbishing the bunker and realising the energy project. One important conflict evolved around the integration of the heat-storage facility.

S1:   'What I find very important is that the heat storage worked out. That we were able to realise this topic there. Because, also, within the technical constellation this has been the aspect which is not really economically reasonable, but which has been pre-set by the IBA, where we said: we want this future technology to be realised there, even if this is not going to be feasible within the first, in the first year' (IBA, Staff Member 1: 1036–1041).

The quote describes success as an outcome of successfully putting forward one's own interests in negotiations with other actors involved in the social arena established with regard to the heat-storage facility. For the IBA staff members integrating the heat-storage facility was a crucial aspect as it would immensely increase their chances of realising their intention to create an innovative lighthouse project. As the quote also admits, this aim conflicted with interests of the other actors involved in the project. The quote clearly indicates that the heat-storage facility—the project's best opportunity to become a highly innovative project—was not up for negotiation for the IBA. However, its financial unfeasibility made the heat storage a highly contested element of the project. The objectives of creating an innovation and generating income here appear as opposing aims. In order to realise the project, these conflicting aims had to be negotiated.

P: 'This density of very many <u>participants</u> and also mutual dependencies, possibly
        were the most difficult aspect of this project. Yes. And each actor had its individ-
        ual criteria. One has to obey its social criteria, the other one that he has to manage
        a viable company, like Hamburg Energie. The third, on the other hand—the
        buyer—that he organises affordable provision of energy for his tenants. Thus,
        these were the most difficult barriers, I think' (Politician: 412–419).

The quote summarizes some of the conflicts that resulted from different participants' social practices, especially their teleoaffective structures. These had to
be negotiated throughout the project's planning, designing, and realization processes. The IBA's teleologies of creating an innovative lighthouse project, fulfilling the local residents' interests, and realising the climate change key topic
had to be aligned with one another within the project. They furthermore had to
be negotiated with teleologies of various other actors involved in the context,
including the financial aims of the energy operator, the stipulations of the housing company, and the requirements of various experts, political actors, and residents.

    Residents are the last group of human actors to be analysed here. This type
of human actors was relevant not only with regard to the energy bunker but for
the IBA in general.

S1:   'The locality of the IBA, the Elbinsel, certainly, belongs to, with large parts of
        northern Germany, to the <u>endangered</u> areas—when we think of rising sea levels.
        What this **might mean** if a storm flood hits the island, is well known. Behind you
        is a picture, taken after the storm flood 1962, which flooded the Elbinsel. This is
        also a big issue for the old inhabitants. This has been inscribed in the memories
        of many people who have been alive at that time or where maybe the parents have
        been living and have **experienced it**' (IBA, Staff Member 1: 221–230).

In this quote, climate change is related to the IBA through the question of how
local residents might be impacted by the effects of climate change. The residents
are depicted as a vulnerable group, because they are living in an area endangered
by rising sea levels and potentially consequent storm floods. By reducing carbon
emissions and adapting the Elbinsel to rising sea levels, the IBA is presented to
be related to the residents by protecting them from the effects of climate change.
The residents are related to the practices of the IBA not only spatially—in the
sense that the IBA is working where the residents live—but also intentionally in
the sense that they provide a motivation and goal for the practices of the IBA.

    As already described, the IBA is also related to the local residents in the
sense that it is one of the IBA's teleologies to develop locally meaningful projects. While the ambiguity of this teleology has already been described, it provides another way in which the practices of the IBA are intentionally related to

the residents. Furthermore, the residents, or more specifically the residents' initiatives, have causally led to the existence of the IBA. It is rather doubtful whether the IBA would have been developed for the district without the long and intense engagement of the local citizens. The local residents are indirectly related to the energy bunker in different ways, in the sense that they are essential elements of the social order within which the IBA unfolds.

For the energy bunker project, residents are mainly meaningful as heat consumers. Without the existence of a sufficiently high number of consumers, the project could not have been developed as a district heating system. Because the neighbourhood of the bunker consists of apartment blocks and houses, local residents are the main type of consumer for the project. Furthermore, the bunker was made part of the IBA because the transformation of the 'black block in the park' had been desired by the inhabitants of the area for a long time. The bunker is not only spatially but also causally related to the local residents as initiators of the transformation process and as consumers of the heat. A third way in which actors are meaningful for the energy bunker is as visitors to the finished project.

S1:   'Apart from the <u>masses</u> which are now visiting the café. But this of course is another important aspect. Well, it's functioning much better than we expected. […] So, also <u>thence,</u> that this is an **attraction** for the residents—and not only the local residents but many are visiting the island <u>because</u> of this project. That's working out brilliantly' (IBA, Staff Member 1: 890–892, 898–901).

Local residents and people from further afar are made meaningful in this quote as visitors to the project. By visiting the energy bunker project, the residents acknowledge it as an exhibition object. Since it is a showcase project, the number of people visiting the bunker is an important marker for the IBA's understanding of success. While people could not engage in practices of visiting and viewing the energy bunker project without the IBA, the bunker project would not be a successful exhibition project of the IBA without these visitors.

In all three described ways in which residents are made meaningful for the energy bunker project, they are, however, only meaningful as passive actors. While the IBA defines itself as a governance project, no active participation of citizens has been realised in the bunker project. Residents have not been engaged in the planning, designing, or realization processes of the energy aspect. Also, citizens do not have a chance to become actively involved with the realised energy bunker. For the energy bunker project, citizens are only meaningful as consumers of the produced heat and as beneficiaries of the project and the IBA in general.

The lack of integration of residents as active participants in the activities of the IBA is perceived ambivalent. While many residents are satisfied that the social and infrastructural problems of the district are finally being tackled, they are critical of how the IBA is acting.

HC: 'So all these problems, this is a district, people here are relatively far off from these exclusive events. And, well, so just these stereotypes and how these people act, made the residents distancing themselves. And, well, I can't give any particular example at the moment, but for many people here the feeling has started to creep upon them: they act so cocky, and always these glossy brochures and big media brouhaha, and acting as if people—before they arrived—had still been sitting on trees. Right? So this idea of "We now colonize this here. We now tell you how to do things right"' (Heritage Centre: 759–770).

The interview partner explains how the activities, especially the communication activities, of the IBA alienated the IBA from the residents. The feeling described in the quote is one of not being taken seriously. The residents felt that their identity in the context of the IBA was mainly one of objects that had to be developed instead of being subjects in the development process of their own neighbourhood. The quote supports the argument that—at least in projects like the energy bunker—the residents are mainly meaningful as passive consumers or beneficiaries. According to this perception, the residents did not feel themselves to be meaningful as active participants in the context of the IBA. While in the former quote the interviewee reports the feelings of other people as she has perceived them, she later gives an example of the experience of actors or activities from the district not being taken seriously.

HC: 'And the guide from the IBA, who sat with us in the bus and explained all the projects. We all, realised, well, we sat there with some people who have been in the district for a long time. Among other a politician from the Green Party, which I have known for more than 30 years, who has also been active in our organisation. And we realised that, yes, he [the guide] always withholds exactly those background stories, eh? So, many of the projects are realised within the IBA, have been continued by the IBA, but have originally been developed before. And they don't explain that in the stories they tell on these guided tours. And this impinges negatively, eh?! This is exactly the picture people have in their wary mind: Yes, they think they have invented it all anew. While we have been always already working on and we have made efforts to improve the district' (Heritage Centre: 796–808).

The interviewee relates to experiences she made on a guided tour to the projects of the IBA. During this tour, the guide described all the projects they passed as developments of the IBA. For the interviewee, this manner of presenting the projects excludes the engagement of the citizens and residents, who often developed the first ideas for certain projects and sometimes even concretized them, for example in the White Book. By not mentioning the historical background of these projects, the IBA rejects the meaning of the residents as active initiators and developers of ideas and projects. Instead, on this guided tour, residents were mostly presented in the role of passive beneficiaries of the IBA's activities.

The example serves to demonstrate the IBA's ambivalence towards the role of local residents. On the one hand, the active role of the residents is highlighted as an important part of the history of the IBA. Moreover, the IBA has initiated and realised many projects in which residents were actively involved. Most importantly, however, the IBA realised many ideas of the residents.

LO: 'And the IBA has—you have to admit that—developed some ideas which have evolved out of the local citizen initiative. And has developed them onward. And, well, has realised them, I mean the ideas and wishful dreams, which in the beginning had really only been idle dreams' (Member Local Organisation: 726–730).

The interviewee describes how the IBA was able to realise those '*ideas and wishful dreams*' of the residents that could not have been realised by the residents themselves. The IBA thus did not create and realise projects that were unrelated to the ideas and interests of the local inhabitants. Instead, many of the IBA's projects were carried over from the citizen initiative. As already described, the activities of the IBA also served to alter the image of the district. Many residents are grateful for the existence and the activities of the IBA.

On the other hand, a number of projects of the IBA were developed without the residents' being given an active role. Some residents complained that even when officially involved in the activities of the IBA, they were only allowed to agree with the activities.

HC: 'That there wasn't any real citizen participatory board. Because those sitting in there, which have been elected to sit in there, can only voice criticism about plans or goals, but, well, the decisions have already been made then. They do it anyway. You can discuss things a bit "Well, how do you like it"? "Hm, ((laughing)) We don't like it at all". "Okay. We heard that. Continue" ((laughing)). That may sound rather sloppy but I often heard that' (Heritage Centre: 778–795).

The quotes indicate that while residents were positioned within the IBA, they were not able to define their own position or relations within this context. Residents mainly acted as sources of legitimacy for the IBA, not as active participants or decision-makers. The residents' positions and relations, as well as their meaning, appear to have been mostly defined by the IBA, not by the residents themselves. This situation has led to an ambivalent attitude in many residents towards the IBA. While they are satisfied with the existence and activities of the IBA in general, they often criticize their position and meaning in the social order of the IBA.

Similarly, in the context of the energy bunker, residents never appear as bearers of knowledge or expertise or in other roles in which they could actively contribute to the development or realization of the energy project. Residents are (only) important as points of reference—as consumers and visitors—and as legitimating sources for the project. They do not appear as active participants in and initiators of any bunker-related activities. The energy part of the project has been developed for them but not with them. The role of initiating and realising the project is instead taken on by the IBA, Hamburg Energie, and the invited experts.

4.8.6   Material artefacts: the bunker and the technological equipment

The material features of the bunker have shaped the activities of the IBA and Hamburg Energie in different ways. Before the bunker was inspected and found to be in acute danger of collapsing, the idea had been to install solar panels on the building's outer walls and roof. The need for restoring the building significantly altered the plans for the building.

S1:   'This focus on the restoration, changed the way how we looked at the project. [...] And at that moment it became obvious that we must restore the concrete pillars and then we will have an unbelievably big room, which for example could be used for energy plants' (IBA, Staff Member 1: 95–96, 104–106).

The interviewee explains how the restoration of the building created a new space in the district that had not existed before. Restoring the building made available the space inside the building. Because of the teleologies of the IBA to produce renewable energy and reduce carbon emissions, the bunker had already been meant to be turned into a site of energy production before. The new space inside the building thus did not completely change the meaning of the building for the IBA. However, the availability of extra space made it possible to significantly extend the amount of renewable energy which could be produced by the project. This meant that the project became positioned much more centrally within the key topic 'climate change and urban development'.

The material aspects of energy production and especially heat storage necessitate the availability of relatively large spaces. Large unused spaces are a scarce resource in many urban settings. The (re)created space within the bunker provided such a large unused space. The size of the space spurred ideas about the integeration of a heat storage facility.

S1:   'If you have so much space and want to do a project with renewable energies, then you simply must integrate the heat storage issue. And you can integrate that because you have so much space and so much room which can be utilized for that' (IBA, Staff Member 1: 123–126).

The heat system developed and realised by the IBA is described in this quote as being intertwined with the bunker and especially the size of the space provided by the building. The bunker is essentially involved with the energy system, as without the building the technological installations, the heat storage facility especially could not have been realised.

By positioning the energy production technologies in and on the bunker, the material characteristics of the bunker have become materially and spatially related to the production of renewable energy. The project members have to deal with the special material features of the bunker. These were designed to separate the inside—and thereby protect it—from the outside world. This specific characteristic on the one hand makes bunkers especially suitable to house energy production technologies.

TE: 'With a CHP, again it is like that, the big walls are good. Because, they of course provide natural sound insulation. Okay? In this regard the measurements are important. And for the CHP it is brilliant if you somehow already have something wherein you can put it and it may make as much noise as it likes. Because these things are really loud! CHP's are really very loud. […] as a standard, big machines don't have that [acoustic insulation]. For them I have to think of something else for sound insulation. And if I already have a structure which provides the sound insulation, that's brilliant' (Technical Expert: 804–814).

The big walls of the bunker shelter the CHP plants from the outside world. This is an advantage for the project because CHP plants of the size installed into the bunker cause a severe level of noise. The wide walls surrounding the plants make any further constructions to subdue the noise unnecessary. The bunker's walls here function as a structure that provides sufficient sound insulation. The existence of sound insulation is important because the project is situated close to a densely inhabited quarter. The constant noise created by the working of a CHP in this specific environment would be perceived as a 'sound pollution' of the area. Consequently, the special requirements of the locality create a context within which the wide walls are understood as an asset, since they are able to protect the outside world from the noise produced inside the bunker.

The walls are also perceived as a positive feature in the specific locality because they are able to protect the inside from the external world. The closed environment of the bunkers provides an extremely high level of protection for the expensive materials and technologies from vandalism or robbery. Questions of safety are a more or less constant concern in the project. In fact, these issues were raised throughout the building process in nearly every building owners meeting (notes from participant observation). These specific concerns seemed to derive from the local context in which the project is situated. Being a so-called 'problematic quarter', the district is seen as prone to problems like vandalism and robbery. Due to the material features of the bunker, however, these concerns only needed to be discussed with regard to how entrances and openings could be guarded.

The following quote reveals some further contextual contexts that 'make' the bunker a suitable place for energy technologies in an urban setting.

TE: 'You need to have the possibility to store the heat somewhere. For that you need a space. I have to place the parts somewhere. And space, in such a densely build city is always a scarce resource. At the same time there is non-utilized space available. That's classic! These bunkers are a classic example of non-utilized space. Yes? Or, for only insufficiently utilized space, or for difficult to utilize space. Because, of course, if you want to use it with humans, you would want to allow contact with the outside world and not to be enclosed in a pitch-black room. And technological equipment doesn't care about that. That it is sitting in and enclosed in pitch-black. That means, I have an enclosed room, which presumably is well-equipped for technological equipment, but which so far has not been used widely in that way' (Technical Expert: 785–795).

In the first part of the quote, the interviewee explains a 'classical' problem for the installation of energy production technologies in urban settings—the non-availability of space. The production of energy requires material artefacts, which have to be placed somewhere. In most urban settings, space is a scarce resource. According to the interviewee, bunkers are, however, an example of empty spaces or '*insufficiently used spaces*' in urban settings. The interviewee explains this under-utilization with the specifics of a bunker. The dark and enclosed rooms of a bunker are not hospitable for most human activities. The performance of activities carried out by human entities would require cutting openings into the bunker walls—something that is extremely expensive. Energy production machines on the other hand do not require air or light to carry out their activities. The requirements of technological artefacts and the specific characteristics of bunkers in urban settings are characterized by a 'positive interplay' between the bunker and the technological equipment. The specific material characteristics of bunkers and their special relations to human activities and to the surrounding urban settlements have made them suitable for the production of renewable energy.

On the other hand, the same material aspects that are favourable for the placement of CHP plants also create certain obstacles.

HE: 'You have three times the costs; when you say, I want to realise that in the energy bunker, the costs are three times as high as somewhere else. [...] This is due to the infrastructure of a bunker, of course again, this can't be generalized, but that are very specific requirements' (Hamburg Energie, Senior Staff Member: 168–178).

Refurbishing the bunker meant having to cope with up to two-meter thick walls of concrete. Each new window, door, or other opening of the bunker to the outside world involved immense costs. This dramatically increased the costs for every cable or pipe for which a hole had to be drilled. The materiality of the bunkers as a context for the necessary technologies of energy production increases the financial resources needed and the amount of work associated with the installation of the technologies. The material context of the projects is thus closely related to the amount of resources that have to be mobilized in order to realise the projects. The 'feasibility' of the project is not a natural outcome of the material artefacts; it relates to the amount of resources that have to be mobilized within and for the specific context. While they function as a resource in relation to some characteristics of energy production equipment and its interplay with the local environment, they lead to additional costs in other aspects. Deriving from the building's original purpose of separating the inside from the outside world, it does not lend itself easily to any new openings. This aspect is relevant, for example, when feeding the industrial waste heat, the heat from the solar panels, or the energy from the photovoltaic panels—all produced outside the bunker—into the heat-storage facility. Also, the heat stored in the bunker has to be distributed to its users, which necessitates further openings to connect the living spaces of the consumers with the heat-storage facility inside the bunker via pipes.

The materiality of the building prefigures the realization of energy production technologies, in the sense that it makes certain activities or material arrangements either easier to realise or more complicated and expensive. The prefigurative relation between the different entities is ambiguous and complex, in the sense that the same arrangements can be made easier or more complicated depending on what aspect of this relation is considered. The walls or any other features of the bunker themselves do not in themselves positively or negatively prefigure the production of renewable heat. The characterization of the material artefacts is only established in relation to the context within which they are located and being used.

What also transpires from the quote is that the bunker is not the most feasible site for energy production with regard to financial deliberations. Because of the materiality of the bunker, *'the costs are three times as high as somewhere else'*. While the project still has to be financially viable, the quote from the staff member of the operating company underlines that energy production in the bunker has not been realised for financial reasons alone. Rather, the bunker is not only a prestigious project for the IBA and the Hamburg government, but is also a symbolically important project for the newly established Hamburg Energie.

The restoration of the bunker not only enabled the practices of the IBA and Hamburg Energie, but also altered the meaning of the building for the local population.

I: 'Has your perception of the bunker changed?
LO: Yes. If I have people visiting me here and we talk about the bunker. People are much more positive about it. And also because I can say "There is a café and we can enjoy the view from up there", this motivates positive feedback. While in the past it was like: "I live near the bunker that's a black block, just standing there. A ruin". That was the only story to be told about it. [...] Now another story has been added' (Member Local Organisation: 138–147).

The interview quote from a local resident shows how the integration of the bunker into the context of the IBA has changed how not only residents but people in general relate to the building. The restoration and opening of the building has causally changed the way the building is embedded into social practices of residents and visitors. Prior to its restoration, the building had been a closed space—inaccessible and even potentially dangerous for the local residents. The practices of the IBA and Hamburg Energie have not only removed the dangerous elements of the bunker, but also created new spaces like the café, the exhibition, and the energy system, which can be accessed, visited, and used by everyone.

4.8.7   The local environment: developing Wilhelmsburg

The social practices of the IBA took place in the locality of Wilhelmsburg. The local context influenced the activities of the IBA and other actors in many ways. At a general level, the local context has shaped the activities of the IBA in many ways. It was the derelict infrastructural and social situation in the district that started the processes leading to the IBA. After the IBA was established, the specific local situation in the district was one of the reasons for the integration of the key topic 'climate change and urban development'.

S3:   'Well, I mean, it [the IPCC report, AP] featured highly in the media at that time. And, I mean, we were as an IBA, as a young, so to say, newly established IBA on the lookout, basically, for those topics, which were of relevance for this place. Because, every IBA lives through its place' (IBA, Staff Member 3: 83–86).

In the quote, a staff member describes how and why the members of the IBA realised the importance of the IPCC report for their activities. The quote indicates that climate change was actively related to the locality of the IBA as a topic that was defined as being '*of relevance for this place*'. Climate change was not (only) understood as a meaningful scientific concept for a global phenomenon but was made meaningful for the specific local context within which the activities of the IBA were meant to take place. Making sense of the place in terms of its relations to climate change transformed the way the members of the IBA perceived of entities in this locality.

S1:   'the locality of the IBA, the Elbinsel, certainly, belongs to, with large parts of northern Germany, to the <u>endangered</u> areas—when we think of rising sea levels. What this **might mean** if a storm flood hits the island, is well known. Behind you is a picture, taken after the storm flood 1962, which flooded the Elbinsel. This is also a big issue for the old inhabitants. This has been inscribed in the memories of many people who have been alive at that time or where maybe the parents have been living and have **experienced it**' (IBA, Staff Member 1: 221–230).

In this quote, the local context of the Elbinsel is related to the effects of climate change—especially sea level rise and storm floods. This relation is further emphasised by a description of how throughout its history the locality has been hit by storm floods—which are one of the main future effects of climate change. By making sense of the place with regard to climate change, the locality becomes meaningful as potentially being vulnerable to its effects. In fact, the island has been hit by numerous storm floods throughout its history, with the storm flood in 1962 causing more than 300 deaths. The quote, whose argumentation resembles the official presentation on the IBA's webpage and publications, presents sea level rise as a locally meaningful effect of climate change which has been negotiated and made relevant for the context of the IBA. By relating the activities of the IBA to the protection of the local residents, the interviewee explains the necessity for the IBA's activities with the specifics of the local environment. The quote explains the integration of the key topic 'climate change and urban development' with the vulnerability of the local environment.

More specifically, the local context has also shaped the activities of the IBA with regard to the energy bunker. In particular, the positioning of the bunker in a densely settled urban area has influenced the refurbishment of the building. Primarily, the urgency of developing the bunker was increased by the fact that the bunker had itself become a hazard in the neighbourhood area.

HC: 'I think that started around two thousand, two thousand and one that parts of the concrete started to come off the building. And this of course caused a huge uproar, because on the site below the bunker had been a playground. And then the public administration put a huge net around the whole bunker to ensure these concrete parts stay on the bunker and would not fly around the area' (Heritage Centre: 31–36).

In this quote residents are depicted as imperilled by the worn-out material artefact of the bunker. The dangerous situation results from the proximity of the derelict building to the residential area. These are positioned so close to one another in geographical space that humans could potentially be harmed by parts falling off from the bunker. The danger of people being harmed because of the current material state of the building provided an important argument for the integration of the bunker into the programme of the IBA. Furthermore, the state of the building and its position in an inhabited area was the main reason the fiscal authority had to provide the financial resources to restore the building.

Like the building, the technological equipment installed therein has been influenced by the local environment. The technologies are not only situated in the bunker but (thereby) are also realised in the local environment within which the bunker is situated. Basically, the local environment of a city dwelling provides a context in which the production and distribution of heat 'makes sense' to the extent that it provides consumers.

Positioning energy production within a densely settled urban district—however, also creates certain requirements in relation to the safety, health, and well-being of the population.

S1:   'It was obvious that we of course could not burn wood waste in the centre of a residential neighbourhood. The state agency does have to give legal emission permission, of course, for this project, because it produces a certain amount of thermal output. The state agency has insisted that we markedly underwent the legal emission limits. I think, about 30 percent, something in that range. By way of additional filters. Modern filter facilities. So, from this perspective, the project is also exemplary' (IBA, Staff Member 1: 936–943).

Being positioned in a residential neighbourhood limits the range of technologies that can be used for the production of renewable energy. Air and sound pollution are two ways in which energy production is negatively related to human health and well-being. Human well-being in this case determines the type of energy production equipment in urban areas, in the sense that these activities must not cause any reactions that are perceived as negative for human well-being.[24] Residents—as a main part of the local environment—are inherently part of the social order as they prefigure and constitute the realization of certain energy production technologies in the bunker.

The interplay between energy technologies and residents is, however, not only a negative one.

S1:   'But, yes, I think it is definitely good to use solar heat or solar electricity. It is just the most urban compatible, just because no sound, no transport has to be created.

---

24   I do not intend to question the truth of negative relations between emissions and human well-being or question the rightness of any measures taken to prevent any negative reactions. Instead, I mean to say that the characterization of these emissions and pollution are an outcome of today's knowledge and social power structures.

And because of that for us it has never been a question if that [the solar and photovoltaic systems] should be realised. And, of course, the visual effect was also
desirable. It is not the main intention, but, that it can immediately recognized from
the outside that energy is being produced there, is a desirable side effect' (IBA,
Staff Member 1:821–828).

In the first part of this quote, solar panels are described as the type of energy
production technology which is most compatible with urban environments. Solar panels do not emit anything that is perceived to negatively impact the physical and mental well-being of humans. Quite to the contrary, solar and photovoltaic panels are described as creating a positive relation between the project and
its local environment in the sense that they make the bunker intelligible as now
being a site of renewable energy production. The intelligibility of solar and photovoltaic panels enables them to signify and explain the new meaning of the
building to its local environment.

TE: 'We now engage in the energy transition; thus we need more of these [solar panels,
        AP]. They are always positively connoted [...] Consequently, the energy bunker
        has all these, externally visible, solar panels. Heat output wise you would not need
        them. Yes. Well, they are okay. They do do something. But, just relatively little.
        But with them you can immediately see what kind of building this is. Otherwise I
        would have to write on the outside: energy bunker. But these are strongly visible,
        and because there are so many of them, it does look like something' (Technical
        Expert: 747–755).

According to this interviewee, solar and photovoltaic panels visibly position the
building in the context of renewable energy production. This relation first derives from the intelligibility of solar and photovoltaic panels as parts of the German energy transition. Renewable energy production units like solar panels are
a materialization of the cultural concept of an energy transition. They integrate
the IBA bunker within the context of the contested but powerful cultural concept
of the 'German Energiewende'. Furthermore, the quote explicitly relates the decision of installing solar and photovoltaic panels to their positive connotation in
society. Solar energy production technologies are used not only because they are
intelligible as renewable energy technologies, but because they have a positive
meaning. According to the quote, they are not necessary from a heat output and
thus a technical or economic point of view. Instead, the decision to install them
on the bunker was largely based on the intention to create a positive image. Solar
and photovoltaic panels have deliberately been made part of this arrangement
because of their positive relation to the human entities, such as neighbours, visitors, or passers-by, who are more or less constantly part of this arrangement.
The local environment has not only influenced the social site; the social site
itself has been deliberately and consciously designed to interact with its local

environment. The example of the solar panels also illustrates that energy production in the IBA bunker has been designed not to create the highest possible financial income, the most heat, political resistance, or public benefit. Instead the solar panels constitute a decision with regard to energy production that has been heavily influenced by the aim to create a certain positive image for the building and thereby the IBA, Hamburg's government, and Hamburg Energie. The production of renewable energy in the IBA energy bunker is related in different ways to the local context in which it is positioned. The context has both enabling and constraining effects on the realization of energy production. It is inherently part of the social site of the IBA because it prefigures and constitutes the realization of the bunker project.

4.9 Interim conclusion: Renewable energy production as a lighthouse project

The key teleology of the IBA has been the creation of projects that raise international public and media attention for Hamburg's urban development ambitions. Being more complex and heterogeneous than that, however, the energy bunker results from negotiating this teleology with the self-imposed aim to reduce the district's carbon emissions, thereby taking part in international efforts to mitigate climate change. Intertwining these two teleologies created a context within which the bunker was not only meant to produce renewable energy, but do so in a way that would raise international public and media attention. A third teleology which was realised in the energy bunker was the creation of locally meaningful projects. The bunker project negotiated this third aim with the two afore mentioned by restoring the building and opening it for the public as an exhibition object.

Making sense of renewable energy production in the bunker as an 'exhibition object' has not been unchallenged. Other social worlds within the context of the project constantly contested this teleology. For Hamburg Energy and the municipal housing agency, renewable energy production should produce financial income or decrease tenements heating bills. Local residents have challenged the creation of innovative lighthouse projects. For them, urban development should improve those infrastructural institutions which are relevant in their everyday life.

Human actors, material artefacts and entities from the local environment are in diverse ways related to the IBA energy bunker project. The staff members of the IBA and of Hamburg Energie, politicians, and experts have contributed in different ways to the development and realization of the project. The context of the IBA influenced which actors were in what ways included in the project, while the integration of certain human actors shaped the project respectively. The integration of experts for example motivated the installation of a heat storage facility into the project. Local residents on the other hand do not appear as active participants in any energy production related activities. Instead they are relevant as consumers of the produced heat, as visitors of the bunker, and as beneficiaries of the IBA in general and the bunker project specifically. At the same time, local residents are essentially involved with all activities of the IBA, in that their engagement has prefigured its establishment.

The material features of the building, especially its mightiness and historical background were reasons why the project was envisioned to become a symbol of renewable energy production. The existence of the bunker and the space it offered were fundamental for the ability of the IBA to produce renewable energy in the district. The IBA provided a context within which the bunker was not only restored—making its interior available—but within which the created space inside the bunker became meaningful for the production of renewable energy. Likewise, the installation of a heat storage facility and of solar and photovoltaic panels can be attributed to the fact that for the IBA, the production of renewable energy is mainly made sense of as a way to attract international attention. Economic and technical considerations were subjected to this aim.

5 Conclusion

This thesis has analysed the contextuality of local renewable energy production. The underlying research interest was to study how renewable energy and the production of renewable energy are made sense of in different ways—not only within different projects, but also among different members of one project. Specifically, the analysis aimed to determine how energy production activities mutually are shaped by and shape the specific context in which they are enacted. Using the concept of social practices developed by Theodore Schatzki and Adele Clarke's situational analysis, this thesis' objective was to show how the organising elements of social practices—teleoaffectivities, general and practical understandings, and rules—are shaped by the context in which they are performed. The context in which these organising elements are made sense of consists of human actors, material artefacts, non-human actors, and things. Deriving from situational analyses, the study specifically focuses on how elements of social practices and entities of social orders are negotiated within certain situations. Focusing on negotiation implies an understanding of social practices and social orders as contested, ambiguous, heterogeneous, and unstable.

Analysing social practices of renewable energy production in the three different case studies CDT, KEBAP, and IBA has revealed a multitude of interests, ideas, knowledge, and norms with regard to how renewable energy is made sense of. In the CDT, energy production is a way to realise the community's interests. While this suggests that energy production is meant to increase the sustainability of the community, the production of renewable energy is also intended to provide an income stream for the CDT, which could be used to realise further community interests. The context within which renewable energy production is embedded in the CDT not only shapes the sense-making of energy production, but also has very practical outcomes. Thus, energy production in the CDT is based on volunteers, who plan, realise, and run the energy production as part of their private lives. This, on the one hand, limits the amount and level of knowledge, time and skills that can be raised for maintaining the system and, hence, influences the types of energy production technologies, which 'make sense' for and in the context of the CDT. Being realised by volunteers as part of their private lives, on the other hand, means that energy production in the CDT is always also a social activity. As such, it produces not only energy and a financial income, but also creates social contacts. Within the CDT, practical knowledge—for example about the interests of community members—is closely intertwined with theoretical knowledge about energy production technologies. Last but not least, energy production in the CDT is also always shaped by the local environment. As has been shown, (potential options for) energy production not only require suitable physical conditions, but have to negotiate different human ways of sense-making of diverse natural features. The high number of community members, which make sense of Comrie's local environment as 'unspoiled nature', is one crucial reason why the Trust restrains from the installation of wind turbines.

In KEBAP, the idea to produce renewable energy derived from the intention of the original core group to create a project that would challenge Vattenfall and Hamburg's government's energy policy. More specifically, the historical background of the project is a citizen initiative, which (successfully) aimed to prevent Vattenfall to install a new heat pipeline through Hamburg's inner city districts. The production of renewable energy in KEPAB had originally been a resistance strategy. This way of sense-making has significantly shaped the way energy production is meant to be realised in the project. The original idea of distributing heat via a micro-grid was being abandoned in favour of feeding heat into Vattenfall's long-distance grid. This option was chosen because it enabled the project to challenge Vattenfall—and especially its quasi-monopole of the long-distance grid—on practical and legal grounds. The possibility to challenge Vattenfall as a competitor on legal grounds was based on rules and conventions regulating the concept of 'free markets'. Making sense of energy production as a resistance strategy also influenced the perception of the bunker. A highly relevant feature of the bunker at that time was that it is located in a symbolically meaningful spot, close to the existing heat grid and more or less in the middle of Vattenfall's planned heat pipeline.

The IBA Hamburg had been set up as part of Hamburg's efforts to increase the city's attractiveness for an international audience. Deriving from this background, the key teleology of the IBA had been to create projects that would raise international public and media attention for the city. Because of its features, especially its historical background and its mightiness, the energy bunker was meant to become a lighthouse project of the IBA. After the publication of the fourth IPCC report in 2004, the team members of the IBA decided to make 'climate change and urban development' a key topic. Being realised within this topic, the energy bunker was meant to become a symbol for the production of renewable energies in the district. This intention has strongly shaped the project. For example, scientific knowledge and technical innovations have played a major role in designing and realising the energy bunker. The integration of a heat storage facility did not so much derive from technical requirements or economic considerations, but had been motivated by 'expert' advice. Likewise, the bunker as a highly symbolic and visible building was chosen as location for the production of energy, although this decision does not make sense from technological and economic points of view.

Energy production in the three cases is made sense of in very different ways. These ways of sense-making are not abstract ideas but significantly influence and shape the social practices of energy production. Sense-making of energy production has been shown to be closely related to the specific context of which it is an inherent part. Energy production practices, however, are not only shaped by the context within which they are enacted, but also influence the context respectively. In the CDT, energy production has, for example, severely influenced the practices enacted by and in the Trust. On the one hand, because of its inefficiency, the biomass boiler has increased instead of diminished the CDT's financial debts—hence inhibiting instead of subsidizing other activities in and of the Trust. On the other hand, the boiler generates heat, which enables the enactment of certain practices in the camp. Examples are a mushroom growing enterprise and a gym. These activities would not be possible without the existence of a heat production and distribution system.

Part of the motivation for the realization of the IBA energy bunker was the controversial situation caused by—on the one hand—the IBA's key issue of climate change and the derived strategies of carbon reductions through energy efficiency and energy production measures in the district and—on the other hand—an existing contract with Vattenfall because of which the district was meant to be provided with heat produced in a new coal power plant. The establishment of the IBA and its powerful position in the social arena of Wilhelmsburg's district politics altered the situation drastically. The IBA was able to push through its agenda and effect the cancellation of the contract. The subsequent production of heat in the bunker—and in other sites—has made redundant the necessity for an external heat provision. Heat production in the energy bunker hence can be understood as a negotiated order which, in particular because of its materialized character serves to momentarily stabilize the context.

For KEBAP it is not yet possible to establish how the planned energy production practices of the project might affect the context. Not only are the energy production plans of the project not yet realised, but the general situation of heat production and distribution in Hamburg's long distance heat grid is also extremely instable since the referendum forced Hamburg to re-municipalize the heat grid. The establishment of the project and its energy concept, however, already illustrate the dynamic interplay of social practices and context. The existence of the project and its obvious support through many of the district's residents enabled KEBAP to succeed against the private investor who aimed to purchase the bunker in order to transform it into apartments. The decision of the district's political parties to support KEBAP not only hindered this particular investor but also prevents other potential investors from bidding for the building, and hence created a negotiated order within and for the social arena which had evolved about the building. The case studies not only illustrate the dynamic interplay of social practices and the context within which these practices unfold but also exemplify that social practices of specific social worlds are both negotiated in and function as arguments in social arenas, i.e. the sites in which participants of different social worlds negotiate issues, perspectives, and problems. As such, arenas comprise of the '*interaction by social worlds around issues—where actions concerning those are being debated, fought out, negotiated, manipulated, and even coerced within and among the social worlds*' (Strauss 1993: 226). Actor groups do not enact social practices based on idiosyncratic interpretations of symbolic patterns of knowledge. Instead, as for example illustrated by the IBA, both general sense-making of Wilhelmsburg as a future low carbon district and decentralized heat production as a legitimate strategy are outcomes of negotiated orders which occur in social arenas.

Besides being describable with regard to certain core characteristics, the analysis of the three cases has emphasised the existence and relevance of heterogeneities and ambiguities within each project. Each project has to negotiate a range of different, sometimes conflicting teleologies, understandings, and knowledge. Most of these negotiations and conflicts are carried out in social arenas. Based on the theoretical framework, a number of social arenas was analytically differentiated in each project. Within the CDT, two different arenas have been identified as being of relevance for the production of renewable energy. Energy production is part of that social arena within which the Trust's sustainability agenda is being negotiated. As such, energy production has originally been made sense of as part of the CDT's carbon reduction strategies. After the Trust's sustainability agenda had come to be challenged by other social worlds in the CDT, the board and staff members decided to make its carbon reduction strategies henceforth intelligible as a way to save money. Another social arena within which energy production is being negotiated is created by the interactions of different social worlds with regard to Cultybraggan Camp and its usage. Within this social arena, the Heritage Group opposes the installation of energy production technologies in the camp, based on the group's interest to preserve the historical appearance of the camp. In KEBAP, different social worlds competed over the 'right' technology for the production of renewable energy. As has been shown, what was understood as the 'right' technology was closely related to the general aims and intentions of the respective group. Those people interested in a 'pure' local economy approach supported the installation of a woodchip gasifier. Their opponents argued for the importance of making KEBAP an actor in a future smart energy grid and, consequently, for the installation of a gas CHP plant. In the context of the IBA, the diverging interests of the social worlds of the IBA staff members, of Hamburg Energie, and the municipal social housing company caused severe conflicts, which at one point could only be solved by intervention of a high ranking political actor.

Social worlds are created through fluid, constantly ongoing negotiations within 'groups with certain commitments to certain activities, sharing resources of many kinds to achieve their goals, and building shared ideologies about how to go about their business' (Clarke 1991: 131). While conflicts and negotiations are mainly carried out among different social worlds in social arenas, complexity and ambiguity have been found to also be inherently part of every social world. Within the CDT, board members have been shown to have very different general understandings about renewable energy production. While most board and staff members do support sustainability, one board member does deny the legitimacy and relevance of the concept. For him, however, renewable energy production does make sense as a practice, which is more 'healthy' than conventional energy production. KEBAP exemplifies how the teleologies and understandings of social worlds might change over time. While the members of the project originally intended to challenge Vattenfall, this intention has been more or less abandoned in favour of a local economy approach. In both cases the specific consensus on energy production as a legitimate social practice within the project hence is not so much based on shared patterns of knowledge, but derives from the fact that renewable energy production was understood as legitimate practice with regard to different forms of sense-making.

Within an interview, a staff member of Hamburg Energie illustrates the existence of ambivalent ideas and understandings not only among different members of one social world, but within an individual person. Depending on the perspective assumed, the staff member was either in favour of the IBA's intentions of creating a lighthouse project, or discarded them as not compatible with Hamburg Energie's interests to create an economically viable energy project. The interview partner's contradictory positions showcase that sense-making of social practices is not stable, but instead might be characterized by highly ambivalent and potentially even irreconcilable ideas. The ambivalence shown to exist within social worlds and even individual persons draws attention to the fact that renewable energy production might be interpreted with regard to heterogeneous patterns of knowledge which are potentially negotiable (as sustainable or healthy) but which might also turn out to be incongruous (innovativeness versus economic). Different ways of sense-making exist at any time within social worlds. Furthermore, new ways of sense-making might enter into a social world at any point. Social practices of energy production hence appear not so much as stable or routinized social phenomena but as fragmented, contested and dynamic processes.

While this thesis has highlighted negotiation and instability, instances of stabilization can be found in all three cases. Stabilization occurs, first, in the form of institutionalization. Sustainability is institutionalized in the CDT's constitution, like local economy is in KEBAP's strategy paper or the creation of lighthouse projects in the memorandum for the IBA Hamburg. The existence of specific working groups or persons also institutionalise certain activities by making them permanent features of the respective projects, often ascribed with certain persons and responsibilities. Another type of stabilization—reification— is effected through the installation of technological equipment. The installation of solar panels, biomass boilers and other technologies materialises and thereby stabilises practices of energy production in the CDT and the IBA. However, when looking closer at these instances, their apparent stableness starts to dissolve. How something is interpreted or how activities are enacted by and in a specific working group might change over time. As illustrated by the CDT, sustainability was deleted from the Trust's vision after the strategy revision process. The IBA Hamburg, likewise, altered its social practices and ways of sense-making when it subsequently decided to integrate climate change into its agenda. Working groups change with regard to the people involved and the issues dealt with—for example in Comrie the renewable energy working group became the renewables energies working group when the members decided not only to deal with renewable energy production but associated issues like waste treatment and transport. After the energy cooperative had been founded, the energy working group in KEBAP was outsourced into the KEGA eG. Comrie also serves to illustrate the general instability of working groups in that recently (2017) the Renewables Working group has ceased to exist.

Likewise when looking at material artefacts—and especially sense-making of material artefacts, which has been my research interest—these were found to be subject to heterogeneous and changing ways of sense-making. The conflict about the 'right' energy production technology in KEBAP illustrates how material artefacts are related to heterogeneous ways of sense-making. The example of the IBA energy bunker highlights the dynamic character of these processes. The formerly derelict building only came to be made sense of as a potential lighthouse project for the production of renewable energies, when the members of the IBA put climate change onto their agenda. On a more general level, all three projects illustrate drastic changes of sense-making with regard to material artefacts. Each of the projects actively altered the sense-making of the sites within which their activities unfold from being (mostly/only) memorials of WWII to being sites of renewable energy production as well as other social and cultural activities. While instances of stabilisation hence might occur in each of the case studies, these instances become visible as outcome of continuously occurring negotiations within which processes of de- and re-stabilisation constantly ensure the existence of a 'dynamic stability'.

Contributing to the actual qualitative social science research debate, which has shifted its focus from methods of data gathering to methods of data analysis (Mey/Mruck 2014: 12), this thesis exemplifies the benefit of combining situational analysis—as a method of analysis—with practice theoretical approaches. Most scholars using practice theoretical approaches have so far only underlined the importance of using '*methodological techniques capable of observing what actually happens in the performance of practice such as ethnography, rather than relying solely on the results of either questionnaire surveys or interview*' (Hargreaves 2011: 84). Besides the nowadays conventional mentioning of coding procedures, only few practice theorists specifically explicate their methods of data analysis. This fallacy can in part be attributed to still existing differentiations between 'theoretical' and 'empirical' scientific work (Kalthoff 2008). In fact, whole books have been published, denouncing the chasm between qualitative empirics and theory (Kalthoff et al. 2008). Despite these efforts, scholars like Theodore Schatzki, Andreas Reckwitz, and Elizabeth Shove have developed and used highly sophisticated theoretical accounts and applied them in insightful ways, without theoretically elaborating upon their empirical methods. Addressing this weakness, this thesis aims to heighten awareness about and reflexivity with regard to the (qualitative) methods of data analysis in practice theory. Situational analysis developed by Adele Clarke provides an example of an explicit '*theory-method package*' (Clarke/Keller 2012: 37), which decidedly aims to tackle the chasm between qualitative theories and empirics.

The decision to employ a practice theoretical approach to analysing the case studies was not solely based on my research interest but also on methodological premises of qualitative social research. Qualitative research aims to reconstruct the constitution of social realities by reference to social actors' sense-making processes (Hitzler 2014: 64; Hollstein/Ullrich 2003: 35pp.). Thereby, qualitative social researchers are requested to take into account the contextuality of these sense-making processes (Hollstein/Ullrich 2003: 37). Sense-making is always embedded in certain contexts and can be understood only in relation to these contexts. Practice theoretical accounts are informed by cultural theory's conceptualisation of 'the social' as evolving from symbolic patterns of knowledge, which enable social actors to make sense of the world in a collectively shared manner (Reckwitz 2002: 245pp.).

Aiming to focus on the contextuality of social sense-making in social practices requires a method for data analysis, which makes the situation the '*key unit
of analysis*' (Clarke 2003: 559; 2012: 37). While Adele Clarke herself 'only'
makes use of situational analysis for her theoretical approach of social
worlds/arenas and negotiation of orders, her maps do not prescribe that something of which they visualize the included elements. Consequently, it has been
possible to use the method of situational analysis to analyse the contexts within
which social practices of energy and energy production unfold.

This last point also refers to a gap in the existing literature on community
energy. As has been criticised in the beginning, research on community energy
suffers from its homogenising and under-complex conceptualisation of community energy. On empirical grounds this thesis aimed to counter a tendency for
homogenisation which has been identified for much research within the field of
community energy studies. Even within qualitative research, a certain level of
homogenisation is a necessary prerequisite for scientific abstraction—especially
when researchers do not so much aim to study forms of subjective but of social
sense-making. It has been Adele Clarke's intention to enable scholars engage in
theorising, and hence abstractions, while focusing on complexity and ambiguity.
Making use of situational analysis consequently provides analytical tools which
enable insights into the fine differences among community energy projects. Instead of criticising scholars in the field of community energy studies for their
theoretical abstractions, I rather understand my thesis as adding some specifications to the existing debate.

One frequently instance of 'homogenisation' of community energy occurs
when scholars lump together these projects into categories like 'niche actors'
(Geels   2004,   2011,   Mautz/Rosenbaum   2012)   or   'challengers'
(Fligstein/McAdam 2011), which try to act at the margins of the energy system
(Markantoni 2016). For Scotland, Strachan and colleagues have found that despite recent policy changes, key political and economic features, such as market
support and planning arrangements, still favour large corporations (Strachan et
al. 2015; likewise Markantoni: 2016: 157; Bergmann et al. 2010). While new
instruments like FITs or the RHI have been created, these seem to '*trap community renewables in a dependence relationship with harder energy paths*' (Strachan et al., 2015: 106) and to maintain community energy projects within a niche
rather than to change the energy system in a sustainable way. According to literature, this position(ing) of local energy projects within the energy system affects their potential to mobilise resources (among others Markantoni 2016;
Walker 2008; Rogers et al. 2008; Hinshelwood 2001; Seyfang et al. 2013; Gubbins 2007). While I do agree with Markantoni and others about the subordination
of community energy, my case studies also indicate that both the reasons why

projects are situated at the margins as well as the ways how they perceive and respond to this position differ very much.

Confirming Markantoni (2016) and others' findings, the projects suffer from the structural conditions of the energy systems both in Germany and the UK. Quasi-monopoles of large energy companies over energy grids were obstacles for KEBAP and the IBA. In both cases, it would have been questionable, whether the projects could have been realised if they would have had to feed-in into a long-distance heat grid, operated by Vattenfall. Both projects, however, responded differently to this situation.

From its beginning, the members of KEBAP strived to challenge not only Vattenfall but the conventional energy system and Hamburg's energy policy. Especially in the beginning, most people who joined the project did so because they were motivated by the idea to challenge Vattenfall. While one motivation had been resistance against the Moorburg-Trasse, members also were keen to challenge the company's quasi-monopole over the heat grid. Setting up a project like KEBAP was not only a way to fight agains the monopole politically and symbolically, but also enabled the members to tackle this situation legally. Instead of being demotivated by Vattenfall's powerful position within the existing system, and the fact that the existing energy system has been shaped to accommodate the needs and requirements of the incumbent actors (Fligstein/McAdam 2011; Alle et al. 2017), this situation has been (and continues to be) a strong motivating factor for many participants of the project.

B3: 'And it is about power. This is just, well for KEBAP, what always motivates me. Because I believe that those power shifts as we have them in the energy sector—well in other sectors as well [...]—that has caused a concentration of power on a few, who form an ominous ['*unheilvoll*'] alliance with political actors. This is obvious in the energy sector' (KEBAP, Board Member 3: 57–62).

The members of KEBAP are aware of the efforts undertaken by conventional energy providers and political actors to keep decentralised renewable energy projects at the margins of the energy system. Instead of being put off by this situation, it is in fact the existence of perceived '*ominous*' alliances and hence the conditions of the energy system which motivated the participants to engage in this specific sector. KEBAP has been and continues to be motivated because of the structural conditions and vested power of and within the conventional energy system.

The staff members of the IBA, likewise, felt it to be necessary to develop alternatives to the existing energy system. Based on the insights of the fourth IPCC assessment report, the members of the IBA not only created an action plan to develop Wilhelmsburg into a carbon neutral district but also actively challenged an existing contract between the city of Hamburg and Vattenfall. By arguing that providing the district with power from a coal power plant would contradict the exemplary status of the IBA as an urban development process, showcasing the innovative potential of Hamburg, the IBA Hamburg deliberately made use of its position as a player who does not (normally) belong to the system. Making use of the high level of political support, deriving from its identity as a onetime and unique urban development process, the IBA did not tackle Vattenfall within the energy production system but repositioned the issue into the field (Fligstein/McAdam 2011) or regime (Geels 2004, 2011) of urban development. Hence, while the IBA Hamburg could be defined as a marginal actor within the energy system, this does not mean that the IBA has been marginalised. Instead, not having been a 'normal' actor of the energy system did bring certain advantages to the IBA.

In addition to this particular situation, the special character of the IBA generally enabled access to a range of additional resources. Besides political support and the creation of a unit in the ministry for urban development, specifically setup to coordinating communication and interaction with different public authorities for the IBA, the IBA also had access to funding which is not normally available to local renewable energy projects. While this situation enabled the realisation of the energy bunker, the particular context also affects the replicability of the energy bunker project.

PE: 'I agree with you; under the current framework conditions such projects are not
       replicable. Without massive financial assistance. But they show what is possible
       already now. What basically would be possible in Germany. In the end it's only
       about changing the legal framework conditions in a way that all the investment
       sources which the project got, are not necessary any more. So, to show it is not a
       technical problem, it is only a financial problem. And it will only be a financial
       problem as long as the framework conditions are as they are at the moment' (Po-
       litical Expert: 466–474).

The specific situation of the IBA with its core task to raise public and media attention for the innovative potential of the city means that the energy bunker could mobilize political, financial, and institutional (Schmid et al. 2016) resources not normally accessible for community energy projects. Additionally, because of the specific context within which it was realised, economic considerations were much less decisive for the project than is normally the case for energy production projects. Instead of generating financial income, the bunker realised the intention of creating a lighthouse project for renewable energy production in Hamburg. However, this situation also means that the energy bunker project is very likely to remain a lighthouse project—a situation which could be argued to contradict its lighthouse-character. Can a project that shines, but where nobody is able to follow the guide beam, be defined as a lighthouse project?! Hence, the energy bunker system has been realised outside the existing structures and vested power of and in the existing energy system.

Within the CDT, while many members are motivated by environmental objectives and would define themselves as political active, the CDT itself is not understood to be a political organisation. The CDT is not meant to be an energy project but the production of renewable energy is seen as a way to enable the sustainable development of the community. The members of the CDT, consequently, do not define themselves as actors in an existing or changing energy system, but much rather as actors in the developing system of community organisations. Instead of scrutinizing the energy system and their marginal position within it, the members of the CDT make use of newly created financial incentives in and for the renewable energy system in order to realise their community aspirations. Especially in the early years, the members did not perceive themselves as marginal(ized) actors in the energy system but much more as actors in a community development field, which have been able to gain from recent institutional and technical developments in the energy sector. By installing renewable energy production technologies, the CDT, however, has increasingly been drawn into the energy production system. Board and staff as well as the members of the renewable energy working group have been made aware of their marginal position and the corresponding lack of institutional and financial support mechnisms. Nevertheless, the members keep on practicing renewable energy production, and have been able to access sufficient volunteers and funding to increase their energy installations in a way suitable to the intentions of the project.[25] Energy production in the CDT is realised despite the structural conditions and vested power of and within the existing energy system. More to the point, because the CDT is focused on developing the community, its position in the energy system is not generally an issue of concern for and in the Trust.

While all three case studies can be defined as marginalised niche actors or challengers of the conventional energy system, the analysis of the case studies has shown that this situation might be perceived and responded upon very differently by the projects. Maybe more importantly, the case studies raise questions about whether the projects really correspond to the concept of niche actors or challengers of one particular field or system.

Within the analytical framework of the Multi-Level Perspective (Geels 2004, 2011; Foxon/Pearson 2011; Berkhout et al. 2003) or of Strategic Action Fields (Fligstein/McAdam 2011), IBA, KEBAP and the CDT would be conceptualized as niche or challenger projects within the existing energy system or

---

25 The CDT has been able to access funding in order to install the necessary piping to connect the biomass boiler to a number of huts.

field. The analysis of the case studies, however, illustrates not only the differences among the three projects but also has shown the diversity that can be found within the projects. In fact, each of the projects not only belongs to different niches (the CDT's and KEBAP's gardening activities make them part of a niche within the agricultural system, KEBAP and the IBA could be defined as participants in an urban living niche,…), but each project might also be part of different regimes or fields as well. When looking at the integration into the political system, the IBA Hamburg has been an incumbent (Fligstein/McAdam 2011) or regime actor. Because of high average levels of qualification and income among its members, it can be argued that the CDT is a player of a socio-economic regime or field. Likewise, participants in KEBAP have close relations to district regime actors like members of district party fractions or employees of the district authority. Having employed a practice theoretical account has helped to illuminate how in their daily activities, the project members entwine aspects of what in the MLP or SAF would be defined as different systems or fields.

Acknowledging the heterogeneity of community energy projects as niche actors within the energy system problematises the idea of 'the' energy transition. Questioning the idea of 'the' energy transition does not only mean to take into consideration the existence of different actors groups with varying motivations, interests, and resources (Alle et al. 2017; Schmid et al. 2016), or of different models of renewable energy production (Mautz/Rosenbaum 2012) but also the differences and ambiguities within just one actor group—community energy projects. Community energy groups, furthermore, not only differ with regard to what a certain project aims to achieve (Seyfang et al. 2013; Schmid et al. 2016; Strunz 2014; Fuchs/Hinderer 2014), or what motivates members to participate (Gormally et al. 2014; Bauwens 2016; Seyfang et al. 2013). Instead, it is argued here that understanding community energy projects as meshes of situated social practices draws attention to the multifold and dynamic interplay of intentions, emotions, knowledge, rules, human actors, material artefacts, and local environments. Community energy is thereby made visible as outcome of complex sociotechnical processes, within which no single element defines or shapes the reality of renewable energy production in isolation. Instead the position and meaning of each element is an outcome of its dynamic relations to other elements of the context. Understanding the complexity of energy production within one project means that one both has to analyse how the elements making up energy production within a certain case are related to one another in constantly shifting and changing ways. Developing such an understanding not only challenges simplistic notions of 'the' energy transition, by highlighting the diversity of energy transformations but also enables a dynamic perspective on community energy projects.

While this thesis has illustrated the situatedness of local renewable energy production, several aspects could not be covered. Differences between rural and urban contexts were not elaborated upon. Also, this thesis could not pay attention to the interplay of renewable energy production with other transformative activities. Due to its limited nature, a major shortcoming of this study is that it overlooks gender and other social categories. To further enhance knowledge about the specific characteristics of renewable energy production contexts, it would be interesting to study conventional energy production contexts and compare them to contexts of renewable energy production. Interesting issues that have not come up in the research include the meaning and relevance of local, regional, and transnational connections between transformative projects or actors within these relations. A highly relevant topic would be the analysis of situated renewable energy production practices in countries in the southern hemisphere. How is renewable energy production embedded into social contexts in localities that not only suffer from social-economic deprivation but are also highly vulnerable to the effects of climate change? This and further related questions need to be elaborated upon in future research.

Literature

AAE (2013): *Energiegenossenschaften. Bürger, Kommunen und lokale Wirtschaft in guter Gesellschaft.* Agentur für Alternative Energien e.V. Berlin. Online on: https://www.unendlich-viel-energie.de/media/file/34.AEE_DGRV_Energiegenossenschaften_2013_web.pdf, last checked: 26.08.2016.

Adams, Carol A.; Bell, S. (2015) Local energy generation projects: assessing equity and risks. In: *Local Environment* 20 (12): 1473-1488.

Adger, W. Neil (2006): Vulnerability. In: *Global Environmental Change* 16 (3): 268–281.

Alle, Katrin; Fettke, Ulrike; Fuchs, Gerhard; Hinderer, Nele (2017): Lokale Innovationsimpulse und die Transformation des deutschen Energiesystems. In: G. Fuchs (ed.): *Lokale Impulse für Energieinnovationen, Energie in Naturwissenschaft, Technik, Wirtschaft und Gesellschaft.* Wiesbaden: Springer: 1-25.

Allen, Joshua; Sheate, William R.; Diaz-Chavez, Rocio (2012): Community based renewable energy in the Lake District National Park – local drivers, enablers, barriers and solutions. In: *Local Environment* 17 (3): 261-280.

Avelino, Flor; Rotmans, Jan (2009): Power in transition: An interdisciplinary framework to study power in relation to structural change. In: *European Journal of Social Theory* 12 (4): 543–569.

Bauwens; Thomas (2016): Explaining the diversity of motivations behind community renewable energy. In: *Energy Policy* 93: 278-290.

Beck: Ulrich (2010): Climate for change, or how to create a green modernity? In: *Theory, Culture, and Society* 27 (2-3): 254-266.

Becker, Sören; Bues, Andrea; Naumann, Matthias (2016): Zur Analyse lokaler energiepolitischer Konflikte. Skizze eines Analysewerkzeugs. In: *Raumforschung und Raumordnung* 74 (1): 39–49.

Becker, Sören; Gailing, Ludger; Naumann, Matthias (2013): Die Akteure der neuen Energielandschaften – Das Beispiel Brandenburg. In: L. Gailing and M. Leibenath (eds.): *Neue Energielandschaften – Neue Perspektiven der Landschaftsforschung.* Wiesbaden: Springer: 19–31.

Becker, Sören; Gailing, Ludger; Naumann, Matthias (2012): Neue Akteurslandschaften der Energiewende. Aktuelle Entwicklungen in Brandenburg. In: *RaumPlanung* 162 (3): 42–46.

Berger, Peter L.; Luckmann, Thomas (1991 [1966]): *The social construction of reality. A treatise in the sociology of knowledge.* London: Penguin Books.

Bergman, Noam; Markusson, Nils; Connor, Peter; Middlemiss, Lucie; Ricci, Miriam (2010): *Bottom-up, social innovation for addressing climate change*. University of Oxford, School of Geography and the Environment. Oxford (Environmental Change Institute Working Paper).

Berkhout, Frans; Smith, Adrian; Stirling, Andy (2003): Socio-technological regimes and transition contexts. *SPRU Electronic Working Paper Series*. University of Sussex (SPRU – Science & Technology Policy Research, University of Sussex, 106).

Berlo, Kurt; Wagner, Oliver (2011): Zukunftsperspektiven kommunaler Energiewirtschaft. In: *RaumPlanung* 158/159: 236–242.

Bernard, H. Russell (2011): *Research methods in anthropology. Qualitative and quantitative approaches*. Lanham: AltaMira.

Berthon, Pierre; Ewing, Michael; Hah, Li Lian (2005): Captivating company: dimensions of attractiveness in employer branding. In: *International Journal of Advertising* 24 (2): 151-172.

Betsill, Michele M.; Bulkeley, Harriet (2006): Cities and the multilevel governance of global climate change. In: *Global Governance: A Review of Multilateralism and International Organisations* 12 (2): 141–159.

Bijker, Wiebe E.; Pinch, Trevor (1987): Preface to the anniversary edition. In: W. E. Bijker, T. P. Hughes and T: Pinch (eds.): *The Social construction of technological systems*. New directions in the sociology and history of technology. Cambridge: MIT Press: xi–xxxiv.

Blanchet, Thomas (2015): Struggle over energy transition in Berlin: How do grassroots initiatives affect local energy policy-making? In: *Energy Policy* 78: 246-254.

Bogusz, Tanja (2009): Erfahrung, Praxis, Erkenntnis. Wissenssoziologische Anschlüsse zwischen Pragmatismus und Praxistheorie – ein Essay. In: *Sociologia Internationalis* 47 (2): 197–228.

Bolton, Ronan P. G. (2011): *Socio-technical transitions and infrastructure networks. The cases of electricity and heat distribution in the UK*. PhD-Thesis. University of Leeds.

Bomberg, Elizabeth; McEwen, Nicola (2012): Mobilizing community energy. In: *Energy Policy* 51: 435–444.

Bongaerts, Gregor (2007): Soziale Praxis und Verhalten - Überlegungen zum Practice Turn in Social Theory. In: *Zeitschrift für Soziologie* 36 (4): 246–260.

Bontje, Marco; Musterd; Sako (2009): Creative industries, creative class, and competitiveness: Expert opinions critically appraised. In: *Geoforum* 40 (5): 843-852.

Boykoff, Maxwell T.; Goodman, Michael T.; Curtis, Ian (2010): Cultural politics of climate change: interactions in everyday spaces. In: M. T. Boykoff (ed.): *The politics of climate change. A survey*. London, New York: Routledge: 136–154.

Bracken, Louise J.; Bulkeley, Harriet A.; Maynard Claire M. (2014): Micro-hydro power in the UK: The role of communities in an emerging energy source. In: *Energy Policy* 68: 92-101.

Brand, Karl-Werner (2011): Umweltsoziologie und der praxistheoretische Zugang. In: M. Groß (ed.): *Handbuch Umweltsoziologie*. Wiesbaden: VS Verlag für Sozialwissenschaften: 173–198.

Breukers, Sylvia; Wolsink, Maarten (2007): Wind power implementation in changing institutional landscapes: An international comparison. In: *Energy Policy* 35 (5): 2737–2750.

Brickmann, Irene; Kropp, Cordula; Türk, Jana (2012): Aufbruch in den Alpen – Lokales Handeln für eine globale Transformation? In: G. Beck and C. Kropp (eds.): *Gesellschaft innovativ*. Wiesbaden: VS Verlag für Sozialwissenschaften: 65–83.

Bruckbauer, Daniel (ed.) (2003): *Sprung über die Elbe*. Internationale Entwurfswerkstatt = Leap across the Elbe. Hamburg.

Brulle, Robert J.; Dunlap, Riley E. (2015): Sociology and global climate change. In: (ibid.) (eds.): *Climate change and society. Sociological perspectives*. New York: Oxford University Press: 1–31.

Bryden, John; Geisler, Charles (2007): Community-based land reform: Lessons from Scotland. In: *Land Use Policy* 24 (1): 24–34.

Bucher, Rue; Strauss, Anselm (1961): Professions in process. In: *American Journal of Sociology* 66 (4): 325–334.

Burch, Sarah (2010): Transforming barriers into enablers of action on climate change: Insights from three municipal case studies in British Columbia, Canada. In: *Global Environmental Change* 20 (2): 287–297.

Buttel, Frederick H. (2010): Social institutions and environmental change. In: M. R. Redclift and G. Woodgate (eds.): *The international handbook of environmental sociology*. Cheltenham, Northampton: Edward Elgar Publishing: 33–47.

Callaghan, George; Williams, Derek (2014): Teddy bears and tigers: How renewable energy can revitalize local communities. In: *Local Economy – The Journal of the Local Economy Unit* 29 (6-7): 657-674.

Callander, Robin Fraser (1998): *How Scotland is owned*. Edinburgh: Canongate.

Callon, Michel; Law, John; Rip, Arie (eds.) (1986): *Mapping the dynamics of science and technology*. London: Palgrave Macmillan UK.

Cass, Noel; Walker, Gordon; Devine-Wright, Patrick (2010): Good neighbours, public relations and bribes: The politics and perceptions of community benefit provision in renewable energy development in the UK. In: *Journal of Environmental Policy & Planning* 12 (3): 255–275.

Charmaz, Kathy (2014): *Constructing grounded theory*. London, Thousand Oaks, New Delhi: Sage Publications.

Charmaz, Kathy (2015): Grounded Theory. In: J. A. Smith (ed.): *Qualitative psychology. A practical guide to research methods*. London, Thousand Oaks, New Delhi, Singapore: Sage Publications: 53-84.

Chenevix-Trench, Hamish; Philip, Lorna J. (2001): Community and conservation land ownership in highland Scotland: A common focus in a changing context. In: *Scottish Geographical Journal* 117 (2): 139–156.

Clark, Duncan; Chadwick, Malachi (2011): *The rough guide to community energy*. London: Rough Guides Ltd.

Clarke, Adele E. (1991): Social Worlds/Arenas Theory as Organisational Theory. In: A. L. Strauss and D. R. Maines (eds*.): Social organisation and social process. Essays in honor of Anselm Strauss*. New York: A. de Gruyter: 119–158.

Clarke, Adele E. (2005): *Situational Analysis: Grounded theory after the postmodern turn*. Thousand Oaks/London/New Delhi: Sage.

Clarke, Adele E. (2003): Situational Analyses: Grounded theory mapping after the postmodern turn. In: *Symbolic Interaction* 26 (4): 553–576.

Clarke, Adele E.; Friese, Carrie (2007): Grounded theorizing using situational analysis. In: A. Bryant (eds.): *The SAGE handbook of grounded theory*. Los Angeles: Sage Publications: 363–397.

Clarke, Adele E.; Keller, Reiner (2012): *Situationsanalyse. Grounded theory nach dem Postmodern Turn*. Wiesbaden: Springer.

Condon, Patrick M.; Cavens, Duncan; Miller, Nicole (2009): *Urban planning tools for climate change mitigation*. Cambridge: Lincoln Institute of Land Policy (Policy focus report/Lincoln Institute of Land Policy).

Cubasch, Ulrich; Wuebbles, Donald; Chen, Daliang; Facchini, Maria C.; Frame, Davind; Mohawald, Natalie; Winther, Jan-Gunnar (2013): Introduction. In: Intergovernmental Panel on Climate Change (eds.): *Climate Change 2013: The physical science basis. Contribution of Working Group I to the Fifth Assessment Report of the Intergovernmental Panel on Climate Change*. Cambridge, New York: Cambridge University Press: 119–158.

Desfor, Gene (2011): *Transforming urban waterfronts. Fixity and flow*. New York: Routledge.

Devine-Wright, Patrick; Fleming, Paul D.; Chadwick, Henry (2001): Role of social capital in advancing regional sustainable development. In: *Impact Assessment and Project Appraisal* 19 (2): 161–167.

Devine-Wright, Patrick (2005): Beyond NIMBYism: towards an integrated framework for understanding public perceptions of wind energy. In: *Wind Energy* 8 (2): 125–139.

Devine-Wright, Patrick (2007): *Reconsidering public attitudes and public acceptance of renewable energy technologies: a critical review*. Working Paper 1.4. School of Environment and Development. University of Manchester, Online on: http://www.sed.manchester.ac.uk/research/beyond_nimbyism/, last checked: 25.08.2016.

DeWald, Kathleen M.; DeWald, Billie R.; Wayland, Coral (1998): Participant observation. In: R. Bernard (ed.): *Handbook of methods in anthropology.* Walnut Creek: AltaMira: 259–299.

Dey, Ian (1993): *Qualitative Data Analysis: A User-Friendly Guide for Social Scientists.* London, New York: Routledge.

DGRV (2016): *Energiegenossenschaften. Ergebnisse der DGRV-Jahresumfrage (zum 31.12.2015).* Deutscher Genossenschafts- und Raiffeisenverband e.V. Berlin. Online on: http://www.genossenschaften.de/sites/default/files/Auswertung%20Jahresumfrage_0.pdf, last checked: 25.08.2016.

Diaz-Bone, Rainer (2012): Review Essay: Situationsanalyse – Strauss meets Foucault? In: *Forum Qualitative Sozialforschung* 14 (1): n.p.

Diesendorf, Mark (2011): Redesigning Energy Systems. In: J. S. Dryzek, R. B. Norgaard and D. Schlosberg (eds.): *Oxford handbook of climate change and society.* Oxford, New York: Oxford University Press: 561–578.

Dooley, Larry M. (2002): Case study research and theory building. In: *Advances in Developing Human Resources* 4 (3): 335–354.

Dryzek, John S.; Norgaard, Richard B.; Schlosberg, David (2011): Climate change and society: approaches and responses. In: J. S. Dryzek, R. B. Norgaard and D. Schlosberg (eds.): *Oxford handbook of climate change and society.* Oxford, New York: Oxford University Press: 3–17.

Dunlap, Riley E. (2010): Social institutions and environmental change. In: M. R. Redclift and G. Woodgate (eds.): *The international handbook of environmental sociology.* Cheltenham, Northampton: Edward Elgar Publishing: 33–47.

Dunlap, Riley, E.; Catton, William R. (1979): Environmental Sociology. In: *Annual Review of Sociology* 5: 243-273.

Dunlap, Riley E.; McCright, Aaron M. (2015): Challenging climate change: the denial countermovement. In: R. E. Dunlap and R. J. Brulle (eds.): *Climate change and society. Sociological perspectives.* New York: Oxford University Press: 300–332.

Eagle, Robin; Aled, Jones; Greig, Alison (2017): Localism and the environment: A critical review of UK Government localism strategy 2010-2015. In: *Local Economy* 32 (1): 55-72.

Ehrhardt-Martinez, Karen; Schor, Juliet B.; Arahamse, Wokje; Alkon, Alison H.; Axsen, Jonn; Brown, Keith; Shwom, Dale S.; Wilhite, Harold (2015): Consumption and climate change. In: R. E. Dunlap and R. J. Brulle (eds.): *Climate change and society. Sociological perspectives.* New York: Oxford University Press: 93–126.

Ek, Kristina (2005): Public and private attitudes towards "green" electricity: the case of Swedish wind power. In: *Energy Policy* 33 (13): 1677–1689.

Elsen, Susanne (2014): Genossenschaften als transformative Kräfte auf dem Weg in die Postwachstumsgesellschaft. In: C. Schröder and H. Walk (eds.): *Genossenschaften und Klimaschutz. Akteure für zukunftsfähige, solidarische Städte.* Wiesbaden: Springer: 31–47.

Ellsworth-Krebs, Katherine; Reid, Louise (2016): Conceptualising energy prosumption: Exploring energy production, consumption and microgeneration in Scotland, UK. In: *Environment and Planning A* 48 (10): 1988-2005.

Engels, Anita (2010): CO2-Märkte als Beitrag zu einer carbon-constrained business world. In: H.-G. Soeffner (ed.): *Unsichere Zeiten. Herausforderungen gesellschaftlicher Transformationen.* Wiesbaden: Springer: 411-423.

Engels, Anita (2015): Stand und Entwicklung einer nachhaltigen und interdisziplinären Transformationsforschung im internationalen Vergleich. „Transformationsforschung: Ist der Weg das Ziel?" Tagung des Clusters Transformationsforschung der Heinrich-Böll Stiftung. Heinrich Böll Stiftung. Heinrich Böll Stiftung. Berlin, 15.11.2015.

Engels, Anita; Knoll, Lisa; Huth, Martin (2008): Preparing for the 'real' market: national patterns of institutional learning and company behaviour in the European Emissions Trading Scheme (EU ETS). In: *European Environment* 18 (5): 276–297.

Feldman, Martha S.; Orlikowski, Wanda J. (2011): Theorizing practice and practicing theory. In: *Organisation Science* 22 (5): 1240–1253.

Flick, Uwe (2009): *An introduction to qualitative research.* London: Sage Publications.

Flick, Uwe; Kardoff, Ernst von; Steinke, Ines (2004): What is qualitative research? An introduction to the field. In: (ibid.) (eds.): *A companion to qualitative research.* London: Sage Publications: 3–12.

Fligstein, Neil; McAdam, Doug (2011): Towards a general theory of strategic action fields. In: *Sociological Theory* 29 (1): 1-26.

Florida, Richard (2003): Cities and the creative class. In: *City & Community* 2 (1): 3–19.

Fluehr-Lobban, Carolyn (2014): Ethics. In: R. Bernard (ed.): *Handbook of methods in cultural anthropology.* London: Rowman & Littlefield: 173–201.

Flyvbjerg, Bent (2006): Five misunderstandings about case-study research. In: *Qualitative Inquiry* 12 (2): 219-245.

Forman, Alister (2017): Energy justice at the end of the wire: Enacting community energy and equity in Wales. In: *Energy Policy* 107:(649-657).

Foxon, Timothy J.; Pearson, Peter (2008): Overcoming barriers to innovation and diffusion of cleaner technologies: some features of a sustainable innovation policy regime. In: *Journal of Cleaner Production* 16 (1): 148–161.

Foxon, Timothy J.; Pearson, Peter (2011): Transition pathways for a UK low carbon electricity system. Exploring roles of actors, governance and branching points. Working Paper for 2nd International Conference on Sustainability Transitions: "Diversity, Plurality and Change: Breaking New Grounds in Sustainability Transitions Research", University of Lund.

Frater, Harald; Podbregar, Nadja; Lohmann, Dieter (2008): Neuer Streit um Atomenergie. In: (ibid.) (eds.): *Wissen Hoch 12.* Berlin, Heidelberg: Springer: 218–227.

Fröhlich, Jannes (2011): Instrumente der regionalen Raumordnung und Raumentwicklung zur Anpassung an den Klimawandel. Hamburg: HafenCity Universität, Stadtplanung und Regionalentwicklung (Neopolis working papers, 10).

Frondel, Manuel (2012): Der Rebound-Effekt von Energieeffizienz-Verbesserungen. In: *Energiewirtschaftliche Tagesfragen* 62 (8): 12–17.

Fuchs, Gerhard; Hinderer, Nele (2014): Sustainable electricity transitions in Germany in a spatial context: between localism and centralism. In: *Urban, Planning and Transport Research* 2 (1): 354–368.

Fuller, Sara; Bulkeley, Harriet A. (2013): Energy justice and the low-carbon transition: assessing low-carbon community programmes in the UK. In: K. Bickerstaff, G. Walker, H. Bulkeley (eds.) *Energy Justice in a Changing Climate: Social Equity and Low-Carbon Energy*. London: Zed Books: 61-78.

Gailing, Ludger; Röhring, Andreas (2015): Was ist dezentral an der Energiewende? Infrastrukturen erneuerbarer Energien als Herausforderungen und Chancen für ländliche Räume. In: *Raumforschung Raumordnung* 73 (1): 31–43.

Geels, Frank W. (2004): From sectoral systems of innovation to socio-technical systems. In: *Research Policy* 33 (6-7): 897–920.

Geels, Frank W. (2010): Ontologies, socio-technical transitions (to sustainability), and the multi-level perspective. In: *Research Policy* 39 (4): 495–510.

Geels, Frank W. (2011): The multi-level perspective on sustainability transitions: Responses to seven criticisms. In: *Environmental Innovation and Societal Transitions* 1 (1): 24–40.

Geels, Frank W.; Schot, Johan (2007): Typology of sociotechnical transition pathways. In: *Research Policy* 36 (3): 399–417.

Giddens, Anthony (1984): *The constitution of society. Introduction of the theory of structuration*. Berkeley: University of California Press.

Glaser, Barney G.; Strauss, Anselm L. (1973 [1967]): *The discovery of grounded theory. Strategies for qualitative research*. Chicago: Aldine Publications.

Goldman, Michael; Schurman, Rachel A. (2010): Closing the ‚great divide‘: New social theory on society and nature. In: *Annual Review of Sociology* 26: 563-584.

Goldthau, Andreas (2014): Rethinking the governance of energy infrastructure: Scale, decentralization and polycentrism. In: *Energy Research & Social Science* 1: 134–140.

Gormally, Alexandra M.; Pooley, Colin G.; Whyatt, Duncan.; Timmis, Roger J. (2014) "They made gunpowder … yes down by the river there, that's your energy source": attitudes towards community renewable energy in Cumbria. In: *Local Environment* 19 (8): 915-932.

Gov.UK: Renewable Heat Incentive. Online on: https://www.gov.uk/government/policies/increasing-the-use-of-low-carbon-technologies/supporting-pages/renewable-heat-incentive-rhi, last checked:15.05.2014.

Gram-Hanssen, Kirsten (2010): Residential heat comfort practices: understanding users. In: *Building Research & Information* 38 (2): 175–186.

Grin, John; Rotmans, Jan; Schot, Johan (2010a): Conclusion: How to understand transitions? How to influence them? Synthesis and lessons for further research. In: J. Grin, J. Schot and J. Rotmans (eds.): *Transitions to sustainable development.* New York, Abingdon: Routledge: 320–337.

Grin, John; Schot, Johan; Rotmans, Jan (eds.) (2010b): *Transitions to sustainable development.* New York, Abingdon: Routledge.

Gronow, Antti (2012): From Habits to Social Structures. Pragmatism and Contemporary Social Theory. Frankfurt/M: Peter Lang.

Guba, Egon G.; Lincoln, Yvonna S. (1994): Competing paradigms in qualitative research. In: N. K. Denzin and Y. S. Lincoln (eds.): *Handbook of qualitative research.* Thousand Oaks: Sage Publications: 105–117.

Gubbins, Nicholas (2007): Community Energy in Practice. In: *Local Economy* 22 (1): 80–84.

Haggett, Claire; Creamer, Emily; Harnmeijer, Jelte; Parsons, Matthew; Bomberg, Elizabeth (2013): Community energy in Scotland. The social factors for success. University of Edinburgh. Online on: http://www.climatexchange.org.uk/files/4413/8315/2952/CXC_Report_-_Success_Factors_for_Community_Energy.pdf, last checked: 26.08.2016.

Haraway, Donna (1988): Situated Knowledges: The Science Question in feminism and the privilege of partial perspective. In: *Feminist Studies* 14 (3): 575–599.

Hargreaves, Tom (2011): Practice-ing behaviour change: Applying social practice theory to pro-environmental behaviour change. In: *Journal of Consumer Culture* 11 (1): 79–99.

Hargreaves, Tom; Haxeltine, Alex; Longhurst, Noel; Seyfang, Gill (2011): Sustainability transitions from the bottom-up. Civil society, the multi-level perspective and practice theory. Norwich: CSEGRE (CSERGE working paper, 2011-01).

Hargreaves, Tom; Hielscher, Sabine; Seyfang, Gill; Smith, Adrian (2013): Grassroots innovations in community energy: The role of intermediaries in niche development. In: *Global Environmental Change* 23 (5): 868–880.

Harlan, Sharon L.; Pellow, David N.; Roberts, J. Timmons; Bell, Sharon E.; Holt, William G.; Nagel, Joanne (2015): Climate Justice and Inequality. In: R. E. Dunlap and R. J. Brulle (eds.): *Climate change and society. Sociological perspectives.* New York: Oxford University Press: 127–163.

Harnmeijer, Anna; Harnmeijer, Jelte; McEwen, Nicola; Bhopal, Vijay (2012): A Report on Community Renewable Energy in Scotland. Scene Connect Report. Sustainable Community Network; Wageningen University; University of Edinburgh. Online on: http://static1.squarespace.com/static/536b92d8e4b0750dff7e241c/t/53f2251ce4b04928a223d78f/1408378140506/SCENE_Connect_Report_Scotland.pdf, last checked: 26.08.2016.

Hawkey, David; Webb, Janette; Winskel, Mark (2013): Organisation and governance of urban energy systems: district heating and cooling in the UK. In: *Journal of Cleaner Production* 50: 22–31.

Heins, Bernd; Alscher, Stefan (2013): Change Agents. Pioniere des Wandels als Akteure für Klimaschutz und Energiewende. In: P. Schweizer-Ries, J. Hildebrand and I. Rau (eds.): U*mweltpsychologische Untersuchung der Akzeptanz von Maßnahmen zur Netzintegration Erneuerbarer Energien in der Region Wahle – Mecklar (Niedersachsen und Hessen)*. Abschlussbericht. Saarbrücken: 119–134.

Heiskanen, Eva; Johnson, Mikael; Robinson, Simon; Vadovics, Edina; Saastamoinen, Mika (2010): Low-carbon communities as a context for individual behavioural change. In*: Energy Policy* 38 (12): 7586–7595.

Hielscher, Sabine; Seyfang, Gill; Smith, Adrian (2013): Grassroots innovation for sustainable energy. Exploring niche-development processes among community-energy initiatives. In: M. J. Cohen, H. Szejnwald Brown and P. J. Vergraft (eds.): *Innovations in Sustainable Consumption. New Economics, Socio-technical Transitions and Social Practices*. Cheltenham, Massachusetts: Edward Elgar Publishing: 133–157.

Hillebrandt, Frank (2014): *Soziologische Praxistheorien. Eine Einführung.* Wiesbaden: Springer.

Hinshelwood, Emily (2001): Power to the People: community-led wind energy - obstacles and opportunities in a South Wales Valley. In: *Community Development Journal* 36 (2): 96–110.

Hitzler, Ronald (2014): Wohin des Wegs? In: G. Mey and K. Mruck (eds.): *Qualitative Forschung.* Wiesbaden: Springer: 55–72.

Hoffman, Matthew (2013): Why community ownership? Understanding land reform in Scotland. In: *Land Use Policy* 31: 289–297.

Hollstein, Bettina; Ullrich, Carsten G. (2003): Einheit trotz Vielfalt? Zum konstitutiven Kern qualitativer Forschung. In: *Soziologie* 32 (4): 29–43.

Holton, Judith (2010): The coding process and its challenges. In: *The Grounded Theory Review* 9 (1): 21–40.

Huener, Uli; Bez, Michael (2015): Erneuerbare Energien als Grundlage für Prosumer-Modelle. In: C. Herbes and C. Friege (eds.): *Marketing Erneuerbarer Energien.* Wiesbaden: Springer: 335–358.

Kalthoff, Herbert (2008): Einleitung: Zur Dialektik von qualitativer Forschung und soziologischer Theoriebildung. In: H. Kalthoff, S, Hirschauer and G, Lindemann (eds.): *Theoretische Empirie.* Frankfurt/Main: Suhrkamp: 8-33.

Herbert Kalthoff, Stefan Hirschauer and Gesa Lindemann (eds.) (2008): *Theoretische Empirie.* Frankfurt/Main: Suhrkamp

Kardorff, Ernst von (1995): Qualitative Sozialforschung. Versuch einer Standortbestimmung. In: U. Flick, E. von Kardorff, H. Keupp, L. von Rosenstiel and S. Wolff (eds.): *Handbuch Qualitative Sozialforschung. Grundlagen, Konzepte, Methoden und Anwendungen.* Weinheim: Beltz: 3–8.

Kern, Kristine; Bulkeley, Harriet (2009): Cities, Europeanization and multi-level governance: governing climate change through transnational municipal networks. In: *Journal of Common Market Studies* 47 (2): 309–332.

Kleining, Gerhard (1995): Methodologie und Geschichte qualitativer Sozialforschung. In: U. Flick, E. von Kardorff, H. Keupp, L.von Rosenstiel and S. Wolff (eds.): *Handbuch Qualitative Sozialforschung. Grundlagen, Konzepte, Methoden und Anwendungen.* Weinheim: Beltz: 11–22.

Klemisch, Herbert (2014): Energiegenossenschaften als regionale Antwort auf den Klimawandel. In: C. Schröder and H. Walk (eds.): *Genossenschaften und Klimaschutz.* Wiesbaden: Springer: 149–166.

Knieling, Jörg; Fröhlich, Jannes; Grieving, Steffan; Kannen, Andreas; Morgenstern, Nelly; Moss, Timothy; Ratter, Beate; Wickel, Martin (2011): Planerisch-organisatorische Anpassungspotentiale an den Klimawandel. In: H. von Storch and M. Claussen (eds.): *Klimabericht für die Metropolregion Hamburg.* Berlin, Heidelberg: Springer: 231–270.

Knoblauch, Hubert (2005): Focused ethnography. In: *Forum Qualitative Sozialforschung* 6 (3). Online on: http://nbn-resolving.de/urn:nbn:de:0114-fqs0503440, last checked: 27.08.2016.

Knoblauch, Hubert (2014): Qualitative Methoden am Scheideweg. In: G. Mey and K. Mruck (eds.): *Qualitative Forschung.* Wiesbaden: Springer: 73–85.

Knorr-Cetina, Karin (1999): Epistemic cultures. How the sciences make knowledge. Cambridge, Massachusetts: Harvard University Press.

Kocka, Jürgen (1990): Neue Energien im 19. Jahrhundert: zur Sozialgeschichte der Elektrizitätswirtschaft. In: E. Gröbl-Steinbach (ed.): *Licht und Schatten. Dimensionen von Technik, Energie und Politik.* Wien: Böhlau: 17–31.

Kousky, Carolyn; Schneider, Stephen H. (2003): Global climate policy: will cities lead the way? In: *Climate Policy* 3 (4): 359–372.

Kress, Michael; Rubik, Frieder; Müller, Ria (2014): Bürger als Träger der Energiewende. In: *Ökologisches Wirtschaften* 29 (1): 14.

Kunze, Conrad (2011): *Soziographie ländlicher Energieprojekte. Eine vergleichende explorative Untersuchung über ländliche partizipative Initiativen zur Entwicklung regionaler Energie-Infrastrukturen mittels regenerativer Energien am Beispiel von sieben Kommunen in einem neuen Bundesland.* Dissertation. Universität Cottbus.

Lamnek, Siegfried (1995): *Qualitative Sozialforschung. Methoden und Techniken.* Weinheim: Beltz.

Landwirtschaftskammer Hamburg (2009): *Studie zum Biomassepotential in der Freien und Hansestadt Hamburg.* Hamburg. Online on: http://www.hamburg.de/contentblob/3978800/data/biomassestudie.pdf, last checked: 29.10.2015.

Leaney, Victoria; Jenkins, Dilwyn; Rowland, Andy; Gwilliam, R.; Smith, D. (2001): Local and community ownership of renewable energy power production: Examples of wind turbine projects. In: *Wind Engineering* 24 (4): 215-226.

Legard, Robin; Keegan, Jill; Ward, Kit (2003): In-depth interview. In: J. Ritchie and J. Lewis (eds.): *Qualitative research practice. A guide for social science students and researchers.* London, Thousand Oaks, Sage Publications: 138-169b.

Leggewie, Claus; Welzer, Harald (2010): Another "Great Transformation"? Social and cultural consequences of climate change. In: *Journal of Renewable Sustainable Energy* 2 (3): 31009-1 – 31009-13.

Leibenath, Markus (2013): Landschaften unter Strom. In: L. Gailing and M. Leibenath (eds.): *Neue Energielandschaften – Neue Perspektiven der Landschaftsforschung.* Wiesbaden: Springer: 7–15.

Li, Wen; Birmele, Janine; Schaich, Harald; Konold, Werner (2013): Transitioning to community-owned renewable energy: Lessons from Germany. In: *Procedia Environmental Sciences* 17: 719–728.

Lorenzoni, Irene; Nicholson-Cole, Sophie; Whitmarsh, Lorraine (2007): Barriers perceived to engaging with climate change among the UK public and their policy implications. In: *Global Environmental Change* 17 (3-4): 445–459.

Lüders, Christian (2013): Beobachten im Feld und Ethnographie. In: U. Flick, E. von Kardoff and I. Steinke (eds.): *Qualitative Forschung. Ein Handbuch.* Reinbek bei Hamburg: Rowohlt: 384–401.

Luterbacher, Urs; Sprinz, Detlef F. (2001): *International relations and global climate change.* Cambridge, Mass: MIT Press.

Lutzenhiser, Loren; Shove, Elizabeth (1999): Contracting knowledge: the organisational limits to interdisciplinary energy efficiency research and development in the US and the UK. In: *Energy Policy* 27 (4): 217–227.

Mackay, H.; Gillespie, G. (1992): Extending the social shaping of technology approach: Ideology and appropriation. In: *Social Studies of Science* 22 (4): 685–716.

Markantoni; Marianna (2016): Low carbon governance: Mobilizing community energy through top-down support? In: *Environmental Policy and Governance* 26: 155-169.

Markard, Jochen; Truffer, Bernhard (2008): Technological innovation systems and the multi-level perspective: Towards an integrated framework. In: *Research Policy* 37 (4): 596–615.

Markert, Margret (2008): Der Sprung über die Elbe - Wilhelmsburgs Weg in die Mitte der Stadt. In: Geschichtswerkstatt Wilhelmsburg Honigfabrik e.V. und Museum Elbinsel Wilhelmsburg e.V. (eds.): *Wilhelmsburg: Hamburgs große Elbinsel.* Hamburg: Medien-Verlag Schubert: 191–205.

Martin-Brelot, Helene; Grossetti, Michel; Eckert, Denis; Gritsai, Olga; Kovács, Zoltán (2010): The spatial mobility of the ‚creative class‘: A European perspective. In: *International Journal of Urban and Regional Research* 34 (4): 854-870.

Mautz, Rüdiger; Rosenbaum, Wolf (2012): Der deutsche Stromsektor im Spannungsfeld energiewirtschaftlicher Umbaumodelle. In: *WSI-Mitteilungen* 12: 85-93.

Mautz, Rüdiger; Byzio, Andreas; Rosenbaum, Wolf (2008): *Auf dem Weg zur Energie-wende; die Entwicklung der Stromproduktion aus erneuerbaren Energien in Deutschland.* Göttingen: Universitätsverlag Göttingen.

McCright, Aaron M.; Dunlap, Riley E. (2003): Defeating Kyoto: The Conservative Movement's Impact on U.S. Climate Change Policy. In: *Social Problems* 50 (3): 348–373.

McCrone, David (2017): *The new sociology of Scotland.* London/Thousand Oaks/New Delhi/Singapore: Sage.

McCrone, David (2005): Devolving Scotland. In: A. Balfour, D. McCrone and G. Reid (eds.): *Creating a Scottish Parliament.* Edinburgh: Finlay Brown: 5–33.

Mey, Günter; Mruck, Katja (2014): Qualitative Forschung: Analysen und Diskussionen. In: G. Mey and K.Mruck (eds.): *Qualitative Forschung.* Wiesbaden: Springer: 9–31.

Meyer, Christian; Meier zu Verl, Christian (2014): Ergebnispräsentation in der qualitati-ven Forschung. In: N. Baur and J. Blasius (eds.): *Handbuch Methoden der empiri-schen Sozialforschung.* Wiesbaden: Springer: 245–257.

Middlemiss, Lucie (2008): Influencing individual sustainability: a review of the evidence on the role of community-based organisations. In: *International Journal of Envi-ronmental Science and Development* 7 (1): 78.

Middlemiss, Lucie; Parrish, Bradley D. (2010): Building capacity for low-carbon com-munities: The role of grassroots initiatives. In: *Energy Policy* 38 (12): 7559–7566.

Miller, Jody; Glassner, Barry (2010): The "inside" and the "outside": finding realities in interviews. In: D. Silverman (ed.): *Qualitative research.* London, Thousand Oaks, New Delhi: Sage Publications: 131–148.

Moloney, Susie; Horne, Ralph E.; Fien, John (2010): Transitioning to low carbon com-munities—from behaviour change to systemic change: Lessons from Australia. In: Energy Policy 38: 7614-7623.

Moore, Michele-Lee; Tjornbo, Ola; Enfors, Elin; Knapp, Corrie; Hodbod, Jennifer; Bag-gio, Jacopo A. (2014): Studying the complexity of change: toward an analytical framework for understanding deliberate social-ecological transformations. In: *Ecology and Society* 19 (4): 54.

Moss, Timothy; Becker, Sören; Naumann, Matthias (2014): Whose energy transition is it, anyway? Organisation and ownership of the Energiewende in villages, cities and regions. In: *Local Environment* 20 (12): 1–17.

Munday, Max; Bristow, Gill; Cowell, Richard (2011): Wind farms in rural areas: How far do community benefits from wind farms represent a local economic develop-ment opportunity? In: *Journal of Rural Studies* 27 (1): 1–12.

Murphy, Joseph (2010): At the Edge: Community ownership, climate change and energy in Scotland. In: Joseph Rowntree Foundation briefing paper: community assets.

Naus, Joeri; Spaargaren, Gert; van Vliet, Bas J.M.; van der Horst, Hilje M. (2014): Smart grids, information flows and emerging domestic energy practices. In: *Energy Policy* 68: 436–446.

Neukirch, Mario (2014): Konflikte um den Ausbau der Stromnetze: Status und Entwicklung heterogener Protestkonstellationen. SOI Discussion Paper 2014/1, Stuttgarter Beiträge zur Organisations- und Innovationsforschung. Stuttgart.

Nolden, Colin (2013): Governing community energy—Feed-in tariffs and the development of community wind energy schemes in the United Kingdom and Germany. In: *Energy Policy* 63: 543–552.

Norberg, Jon; Cumming, Graeme S. (2008): *Complexity theory for a sustainable future.* New York: Columbia University Press.

Norgaard, Kari M. (2006): "We Don't Really Want to Know": Environmental Justice and Socially Organised Denial of Global Warming in Norway. In: *Organisation & Environment* 19 (3): 347–370.

Nyborg, Sophie; Røpke, Inge (2013): Constructing users in the smart grid—insights from the Danish eFlex project. In: *Energy Efficiency* 6 (4): 655–670.

Olausson, Ulrika (2009): Global warming—global responsibility? Media frames of collective action and scientific certainty. In: *Public Understanding of Science* 18 (4): 421–436.

Ott, Eckhard; Wieg, Andreas (2014): Please, in My Backyard – die Bedeutung von Energiegenossenschaften für die Energiewende. In: C. Aichele and O. D. Doleski (eds.): *Smart Market.* Wiesbaden: Springer: 829–841.

Otto, Antje; Leibenath, Markus (2013): Windenergielandschaften als Konfliktfeld. Landschaftskonzepte, Argumentationsmuster und Diskurskoalitionen. In: L. Gailing and M. Leibenath (eds.): *Neue Energielandschaften – Neue Perspektiven der Landschaftsforschung.* Wiesbaden: Springer: 65–75.

Peck, Jamie (2005): Struggling with the creative class. In: *International Journal of Urban and Regional Research* 29 (4): 740-770.

Perrow, Charles; Pulver, Simone (2015): Organisations and markets. In: R. E. Dunlap and R. J. Brulle (eds.): *Climate change and society. Sociological perspectives.* New York: Oxford University Press: 61–92.

Peters, Michael; Fudge, Shane; Sinclair, Philip (2010): Mobilising community action towards a low-carbon future: Opportunities and challenges for local government in the UK. In: *Energy Policy* 38 (12): 7596–7603.

Pinch, Trevor J.; Bijker, Wiebe E. (1984): The social construction of facts and artefacts: Or how the sociology of science and the sociology of technology might benefit each other. In: *Social Studies of Science* 14 (3): 399–441.

Prus, Robert C. (1996): *Symbolic interaction and ethnographic research. Intersubjectivity and the study of human lived experience.* Albany: State University of New York Press.

Przyborski, Aglaja; Wohlrab-Sahr, Monika (2014): *Qualitative Sozialforschung. Ein Arbeitsbuch.* Oldenbourg: De Gruyter.

Pulla, Venkrat (2014): Grounded theory approach in social research. In: *Space and Culture* 2 (3): 15–23.

Reckwitz, Andreas (2002): Toward a theory of social practices: A development in culturalist theorizing. In: *European Journal of Social Theory* 5 (2): 243–263.

Reckwitz, Andreas (2003): Grundelemente einer Theorie sozialer Praktiken. Eine sozialtheoretische Perspektive. In: *Zeitschrift für Soziologie* 32 (4): 282–301.

Reckwitz, Andreas (2004): Die Entwicklung des Vokabulars der Handlungstheorien. Von den zweck- und normorientierten Modellen zu den Kultur- und Praxistheorien. In: M. Gabriel (ed.): *Paradigmen der akteurszentrierten Soziologie.* Wiesbaden: VS Verlag für Sozialwissenschaften: 303–328.

Reusswig, Fritz (2010): Klimawandel und Gesellschaft. Vom Katastrophen- zum Gestaltungsdiskurs im Horizont der postkarbonen Gesellschaft. In: M. Voss (ed.): *Der Klimawandel.* Wiesbaden: VS Verlag für Sozialwissenschaften: 75–97.

Reuveny, Rafael (2007): Climate change-induced migration and violent conflict. In: *Political Geography* 26 (6): 656–673.

Richardson, Rudy; Kramer, Eric H. (2006): Abduction as the type of inference that characterizes the development of a grounded theory. In: *Qualitative Research* 6 (4): 497–513.

Rogers, Jenny C.; Simmons, Eunice A.; Convery, Ian; Weatherall, Andrew (2012): Social impacts of community renewable energy projects: findings from a woodfuel case study. In: *Energy Policy* 42 (2): 239-247.

Rogers, Jenny C.; Simmons, Eunice A.; Convery, Ian; Weatherall, Andrew (2008): Public perceptions of opportunities for community-based renewable energy projects. In: *Energy Policy* 36 (11): 4217–4226.

Røpke, Inge (2009): Theories of practice - New inspiration for ecological economic studies on consumption. In: *Ecological Economics* 68 (10): 2490–2497.

Rosa, Eugene A.; Rudel, Thomas K.; York, Richard; Jorgenson, Andrew K.; Dietz, Thomas (2015): The human (anthropogenic) driving forces of global climate change. In: R. E. Dunlap and R. J. Brulle (eds.): *Climate change and society. Sociological perspectives.* New York: Oxford University Press: 32–60.

Rost, Dietmar (2015): *Konflikte auf dem Weg zu einer nachhaltigen Energieversorgung - Perspektiven und Erkenntnisse aus dem Streit um die Carbon Capture and Storage-Technology (CCS).* Kulturwissenschaftliches Institut Essen (KWI). Essen.

Rotmans, Jan; Loorbach, Derk (2009): Complexity and transition management. In: *Journal of Industrial Ecology* 13 (2): 184–196.

Rouse, Joseph (2007): Practice theory. In: S. Turner and M. Risjord (eds.): *Philosophy of anthropology and sociology,* Amsterdam: Elsevier: 499–540.

Rutherford, Jonathan; Coutard, Olivier (2014): Urban energy transitions: places, processes and politics of socio-technical change. In: *Urban Studies* 51 (7): 1353–1377.

Rutland, Ted; Aylett, Alex (2008): The work of policy: Actor networks, governmentality, and local action on climate change in Portland, Oregon. In: *Environment and Planning D* 26 (4): 627–646.

Schaefer Caniglia, Beth; Brulle, Robert J.; Szasz, Andrew (2015): Civil society, social movements, and climate change. In: R. E. Dunlap and R. J. Brulle (eds.): *Climate change and society. Sociological perspectives.* New York: Oxford University Press: 235–268.

Schäfer, Hilmar (2010): Bourdieu gegen den Strich lesen - eine poststrukturalistische Perspektive. In: H. Schäfer, D. Suber and S. Prinz (eds.): *Pierre Bourdieu und die Kulturwissenschaften. Zur Aktualität eines undisziplinierten Denkens.* Konstanz: UVK: 63–88.

Schäfer, Hilmar (2012): Kreativität und Gewohnheit. Ein Vergleich zwischen Praxistheorie und Pragmatismus. In: U. Göttlich and R. Kurt (eds.): *Kreativität und Improvisation. Soziologische Positionen.* Wiesbaden: Springer: 17–43.

Schatzki, Theodore R. (2003): A New societist social ontology. In: *Philosophy of the Social Sciences* 33 (2): 174–202.

Schatzki, Theodore R. (2010): Materiality and social life. In: *Nature and Culture* 5 (2): 123–149.

Schatzki, Theodore R. (2011): *Where the Action Is* (On Large Social Phenomena Such as Sociotechnical Systems) (Sustainable Practices Research Group Working Paper Series, 1).

Schatzki, Theodore R. (1996): *Social practices. A Wittgensteinian approach to human activity and the social.* New York: Cambridge University Press.

Schatzki, Theodore R. (1997): Practices and actions - A Wittgensteinian critique of Bourdieu and Giddens. In: *Philosophy of the Social Sciences* 27 (3): 283–308.

Schatzki, Theodore R. (2002): *The site of the social. A philosophical account of the constitution of social life and change.* University Park: Pennsylvania State University Press.

Schatzki, Theodore R. (2005a): Introduction - Practice Theory. In: T. R. Schatzki, K. Knorr-Cetina and E. von Savigny (eds.): *The practice turn in contemporary theory.* London, New York: Routledge: 10–23.

Schatzki, Theodore R. (2005b): Practice mind-ed orders. In: T. R. Schatzki, K. Knorr-Cetina and E. von Savigny (eds.): *The practice turn in contemporary theory.* London, New York: Routledge: 50–63.

Schatzki, Theodore R. (2010): Materiality and social life. In: *Nature and Culture* 5 (2): 123-149.

Schatzki, Theodore R. (2012): Primer on practices. In: J. Higgs, R. Barnett, S. Billett, M. Hutchings and F. Trede (eds.): *Practice-based education: perspectives and strategies.* Rotterdam, Boston, Taipei: Sense Publications: 13–26.

Schatzki, Theodore R. (2013): Timespace and the organisation of social life. In: E. Shove, F. Trentmann and R. Wilk (eds.): *Time, consumption and everyday life. Practice, materiality and culture.* London: Bloomsbury: 35–48.

Schleicher-Tappeser, Ruggero (2012): How renewables will change electricity markets in the next five years. In: *Energy Policy* 48: 64–75.

Schmid, Eva; Knopf, Brigitte; Pechan, Anna (2016): Putting an energy system transformation into practice: The case of the German Energiewende. In: *Energy Research & Social Science* 11: 263–275.

Schneidewind, Uwe; Singer-Brodowski, Mandy (2014): *Transformative Wissenschaft. Klimawandel im deutschen Wissenschafts- und Hochschulsystem.* Weimar: Metropolis.

Schreurs, Miranda A. (2008): From the Bottom Up: Local and Subnational Climate Change Politics. In: *The Journal of Environment & Development* 17 (4): 343–355.

Schumann, Diana; Pietzner, Katja; Esken, Andrea (2010): Umwelt, Energiequellen und CCS. regionale Einstellungsunterschiede und -veränderungen in Deutschland. In: *Energiewirtschaftliche Tagesfragen* 60 (5): 52–56.

Schweizer-Ries, Petra; Hildebrand, Jan; Rau, Irina (eds.) (2013): *Umweltpsychologische Untersuchung der Akzeptanz von Maßnahmen zur Netzintegration Erneuerbarer Energien in der Region Wahle – Mecklar.* Abschlussbericht. Forschungsgruppe Umweltpsychologie. Saarbrücken.

Seyfang, Gill; Hielscher, Sabine; Hargreaves, Tom; Martiskainen, Mari; Smith, Adrian (2014): A grassroots sustainable energy niche? Reflections on community energy in the UK. In: *Environmental Innovation and Societal Transitions* 13: 21-44.

Seyfang, Gill; Park, Jung Jin; Smith, Adrian (2013): A thousand flowers blooming? An examination of community energy in the UK. In: *Energy Policy* 61: 977–989.

Seyfang, Gill; Haxeltine, Alex (2012): Growing grassroots innovations: exploring the role of community-based initiatives in governing sustainable energy transitions. In: *Environment and Planning C* 30 (3): 381–400.

Seyfang, Gill; Smith, Adrian (2007): Grassroots innovations for sustainable development: Towards a new research and policy agenda. In: *Environmental Politics* 16 (4): 584–603.

Shove, Elizabeth (2003): Users, technologies, and expectations of comfort, cleanliness, and convenience. In: *Innovation: the European Journal of Social Science Research* 16 (2): 193-206.

Shove, Elizabeth (2010): Beyond the abc: Climate change policy and theories of social change. In: *Environment and Planning A* 42 (6): 1273-1285.

Shove, Elizabeth; Pantzar, Mika (2005): Consumers, Producers and practices: Understanding the invention and reinvention of Nordic walking. In: *Journal of Consumer Culture* 5 (1): 43–64.

Shove, Elizabeth; Trentmann, Frank; Wilk, Richard (2013): Introduction. In: (ibid.) (eds.): *Time, consumption and everyday life. Practice, materiality and culture.* London: Bloomsbury: 1–13.

Shove, Elizabeth; Walker, Gordon (2014): What is energy for? Social practice and energy demand. In: *Theory, Culture & Society* 31 (5): 41–58.

Shove, Elizabeth; Walker, Gordon (2007): CAUTION! Transitions ahead: politics, practice, and transition management. In: *Environment and Planning* 39 (4): 763–770.

Shove, Elizabeth; Watson, Mathew; Hand, Martin; Ingram, Jack (2007): *The design of everyday life*. Oxford, New York: Berg.

Silverman, David (1994*): Interpreting qualitative data. Methods for analysing talk, text, and interaction*. London, Thousand Oaks: Sage Publications.

Silverman, David (2013): *Doing qualitative research: A practical handbook*. London, Thousand Oaks, New Delhi: Sage Publications.

Slocum, Rachel (2004): Polar bears and energy-efficient lightbulbs: Strategies to bring climate change home. In: *Environment and Planning D* 22 (3): 413–438.

Smith, Adrian; Hargreaves, Tom; Hielscher, Sabine; Martiskainen, Mari; Seyfang, Gill (2016): Making the most of community energy: Three perspectives on grassroots innovation. In: *Environment and Planning A* 48 (2): 407-432.

Smith, Paul J. (2007): Climate change, mass migration and the military response. In: *Orbis* 51 (4): 617–633.

Soeffner, Hans-Georg (2006): Wissenssoziologie und sozialwissenschaftliche Hermeneutik sozialer Sinnwelten. In: D. Tänzler (ed.): *Neue Perspektiven der Wissenssoziologie*. Konstanz: UVK: 51–78.

Soeffner, Hans-Georg (2014): Interpretative Sozialwissenschaft. In: G. Mey and K. Mruck (eds.): *Qualitative Forschung*. Wiesbaden: Springer: 35–53.

Sommer, Bernd; Schad, Miriam (2014): Change Agents für den städtischen Klimaschutz. Empirische Befunde und praxistheoretische Einsichten. In: *GAIA - Ecological Perspectives for Science and Society* 23 (1): 48–54.

Sovacool, Benjamin (2014): What are we doing here? Analyzing fifteen years of energy scholarship and proposing a social science research agenda. In: *Energy Research & Social Science* 1: 1–29.

Sovacool, Benjamin; Ryan, Sarah; Stern, Paul C.; Janda, Katy; Rochlin, Gene; Spreng, Daniel; Pasqualetti, Marting; Wilhite, Harold; Lutzenhiser, Loren (2015): Integrating social science in energy research. In: *Energy Research & Social Science* 6: 95–99.

Spittler, Gerd (2001): Teilnehmende Beobachtung als Dichte Teilnahme. In: *Zeitschrift für Ethnologie* 126 (1): 1–25.

Steffen, Will (2011): A truly complex and diabolical policy problem. In: J. S. Dryzek, R. B. Norgaard and D. Schlosberg (eds.): *Oxford handbook of climate change and society*. Oxford; New York: Oxford University Press: 21–37.

Stephan, Benjamin; Rothe, Delf; Methmann, Chris (2015): The third side of the coin: Hegemony and governmentality in global climate politics. In: H. Bulkeley and P. Newell (eds.): *Governing climate change*. London: Routledge: 59–76.

Stern, Nicholas (2006*): The economics of climate change: The Stern review*. Cambridge: Cambridge University Press.

Strachan, Peter A.; Cowell, Richard; Ellis, Geraint; Sherry-Brennan, Fionnguala; Toke, David (2015): Promoting community renewable energy in a corporate energy world. In: *Sustainable Development* 23: 96-109.

Strauss, Anselm (1978): A social world perspective. In: *Studies in Symbolic Interaction* 1: 119–128.

Strauss, Anselm (1993): *Continual permutations of action*. New York: Aldine de Gruyter.

Strauss, Anselm; Corbin, Juliet (1996): *Grounded Theory: Grundlagen Qualitativer Sozialforschung*. Weinheim: Beltz.

Strengers, Yolande (2012): Peak electricity demand and social practice theories: Reframing the role of change agents in the energy sector. In: *Energy Policy* 44: 226–234.

Strengers, Yolande; Maller, Cecily (2011): Integrating health, housing and energy policies: social practices of cooling. In: *Building Research & Information* 39 (2): 154–168.

Strübing, Jörg (2005): *Pragmatistische Wissenschafts- und Technikforschung. Theorie und Methode*. Frankfurt/M, New York: Campus.

Strübing, Jörg (2008): Pragmatismus als epistemische Praxis. Der Beitrag der Grounded Theory zur Empirie-Theorie Frage. In: H. Kalthoff (ed.): *Theoretische Empirie. Zur Relevanz qualitativer Forschung*. Frankfurt/M: Suhrkamp: 279–310.

Strübing, Jörg (2013): *Qualitative Sozialforschung. Eine komprimierte Einführung für Studierende*. München: Oldenbourg.

Strübing, Jörg (ed.) (2014): *Grounded theory*. Wiesbaden: VS Verlag für Sozialwissenschaften.

Strunz, Sebastian (2014): The German energy transition as a regime shift. In: *Ecological Economics* 100: 150–158.

The Scottish Government (2009): *Climate change delivery plan. Meeting Scotland's statutory climate change targets*. Edinburgh.

Toffler, Alvin (1980): *The third wave*. New York: William Morrow.

Toke, David (2007): Wind power, governance and networks. In: Joseph Murphy (ed.): Governing technology for sustainability. London, Sterling: *Earthscan*: 168–181.

trend:reserach; Leuphana Universität Lüneburg (2013): *Definition und Marktanalyse von Bürgernenergie in Deutschland*. Studie im Auftrag der Initiative "Die Wende - Energie in Bürgerhand" und der Agentur für Erneuerbare Energien. Leuphana Universität Lüneburg, Bremen/Lüneburg. Online on: http://www.buendnis-buergerenergie.de/fileadmin/user_upload/downloads/Studien/Studie_Definition_und_Marktanalyse_von_Buergerenergie_in_Deutschland_BBEn.pdf,   last checked: 09.12.2015.

Unger, Jochem; Hurtado, Antonio (2013): *Energie, Ökologie und Unvernunft*. Wiesbaden: Springer.

Unruh, Gregory C. (2000): Understanding carbon lock-in. In: *Energy Policy* 28 (12): 817–830.

Urry, John (2010): Consuming the planet to excess. In: *Theory, Culture & Society* 27 (2-3): 191–212.

Urry, John (2011): *Climate change & society*. Cambridge, Malden: Polity Press.

Urry, John (2014): The problem of energy. In: *Theory, Culture, and Society* 31 (5): 3-20.

van der Horst, Dan (2007): NIMBY or not? Exploring the relevance of location and the politics of voiced opinions in renewable energy siting controversies. In: *Energy Policy* 35 (5): 2705–2714.

Volbers, Jörg (2015): Theorie und Praxis im Pragmatismus und in der Praxistheorie. In: T. Alkemeyer, V. Schürmann and J. Volbers (eds.): *Praxis denken*. Wiesbaden: Springer: 193–214.

Voss, Martin; Schildhauer, Sascha (2016): Zivilgesellschaftliche Partizipation im Klimawandel. In: C. Besio and G.Romano (eds.): *Zum gesellschaftlichen Umgang mit dem Klimawandel*. Baden-Baden: Nomos: 117–148.

Walker, Gordon (2008): What are the barriers and incentives for community-owned means of energy production and use? In: *Energy Policy* 36 (12): 4401–4405.

Walker, Gordon; Cass, Noel (2007): Carbon reduction, 'the public' and renewable energy: engaging with socio-technical configurations. In: *Area* 39 (4): 458–469.

Walker, Gordon; Devine-Wright, Patrick (2008): Community renewable energy: What should it mean? In: *Energy Policy* 36 (2): 497–500.

Walker, Gordon; Devine-Wright, Patrick; Hunter, Sue; High, Helen; Evans, Bob (2010): Trust and community: Exploring the meanings, contexts and dynamics of community renewable energy. In: *Energy Policy* 38 (6): 2655–2663.

Walker, Gordon P.; Hunter, Sue; Devine-Wright, Patrick; Evans, Bob; Fay, Helen (2007): Harnessing community energies. Explaining and evaluating community-based localism in renewable energy policy in the UK. In: *Global Environmental Politics* 7 (2): 64–82.

Warren, Charles R.; McFadyen, Malcolm (2010): Does community ownership affect public attitudes to wind energy? A case study from south-west Scotland. In: *Land Use Policy* 27 (2): 204-213.

Watson, Mike (2001): *Year zero. An inside view of the Scottish Parliament*. Edinburgh: Polygon at Edinburgh.

WBGU (2011): *Welt im Wandel: Gesellschaftsvertrag für eine große Transformation.* Zusammenfassung für Entscheidungsträger. Berlin: Wissenschaftlicher Beirat der Bundesregierung Globale Umweltveränderungen.

Webb, Janette (2015): Improvising innovation in UK urban district heating: The convergence of social and environmental agendas in Aberdeen. In: *Energy Policy* 78: 265-272.

Webb, Janette (2012): Climate Change and Society: The chimera of behaviour change technologies. In: *Sociology* 46 (1): 109–125.

Weingart, Peter; Engels, Anita; Pansegrau, Petra (2000): Risks of communication: discourses on climate change in science, politics, and the mass media. In: *Public Understanding of Science* 9 (3): 261–283.

Welzer, Harald; Soeffner, Hans-Georg; Giesecke, Dana (2010): KlimaKulturen. In: (ibid.) (eds.): *KlimaKulturen. Soziale Wirklichkeiten im Klimawandel.* Frankfurt/M, New York: Campus: 7–19.

Westley, Frances; Olsson, Per; Folke, Carl; Homer-Dixon, Thomas; Vredenburg, Harrie; Loorbach, Derk (2011): Tipping toward sustainability: Emerging pathways of transformation. In: *AMBIO* 40 (7): 762–780.

Weyer, Jürgen (2010): Einleitung zum Plenum: Klimawandel und nachhaltige Energieversorgung. Transformationen und sozialer Wandel. In: H.-G. Soeffner (ed.): *Unsichere Zeiten. Herausforderungen gesellschaftlicher Transformationen.* Wiesbaden: Springer: 385-386.

Wightman, Andy (1999): *Scotland. Land and power: the agenda for land reform.* Edinburgh: Luath Press.

Williams, Robin; Edge, David (1996): The social shaping of technology. In: *Research Policy* 2: 865-899.

Wilson, Charlie; Hargreaves, Tom; Hauxwell-Baldwin, Richard (2015): Smart homes and their users: a systematic analysis and key challenges. In: *Personal and Ubiquitous Computing* 19 (2): 463–476.

Winskel, Mark (2007): Multi-level governance and energy policy. Renewable energy in Scotland. In: J. Murphy (ed.): *Governing technology for sustainability.* London, Sterling, VA: Earthscan: 182–219.

Wise, R.M.; Fazey, I.; Stafford Smith, M.; Park:E.; Eakin, H.C.; van Archer Garderen, E.R.M.; Campbell, B. (2014): Reconceptualising adaptation to climate change as part of pathways of change and response. In: *Global Environmental Change* 28: 325–336.

Wüstenhagen, Rolf; Wolsink, Maarten; Bürer, Mary Jean (2007): Social acceptance of renewable energy innovation: An introduction to the concept. In: *Energy Policy* 35 (5): 2683–2691.

www.altona.info.de (5.3.2011): *Hamburger Urteil zur Fernwärmetrasse Moorburg wegweisend.* Online on: http://www.altona.info/2010/03/05/hamburger-urteil- zurfernwarmetrasse-moorburg-wegweisend/, last checked: 16.8.2017.

www.bbk.bund.de: *Baulicher Bevölkerungsschutz*. Online on: http://www.bbk.bund.de/DE/AufgabenundAusstattung/BaulicherBevoelkerungsschutz/baulicherbevoelkerungsschutz_node.html, last checked: 29.10.2015.

www.bbk.bund.de: *Schutzbauten - Entwicklung bis 2007*. Online on: http://www.bbk.bund.de/DE/AufgabenundAusstattung/BaulicherBevoelkerungsschutz/Schutzbauten-Entwicklungbis2007/schutzbauten-entwicklung-bis2007_node.html, last checked: 29.10.2015.

www.bmub-bund.de: *Die Deutsche Klimaschutzpolitik. Bundesministerium für Umwelt, Naturschutz, Bau und Reaktorsicherheit*. Online on: http://www.bmub.bund.de/themen/klima-energie/klimaschutz/nationale-klimapolitik/klimapolitik-der-bundesregierung/, last update 09.04.2014, last checked: 27.08.2016.

www.dtascot.org.uk: *What is a Development Trust?* Online on: http://www.dtascot.org.uk/content/what-is-a-development-trust, last checked: 28.08.2016.

www.erneuerbare-energien.de: *Entwicklung der erneuerbaren Energien in Deutschland im Jahr 2016*. Bundesministerium für Wirtschaft und Energie. Online on: http://www.erneuerbare-energien.de/EE/Navigation/DE/Service/Erneuerbare_Energien_in_Zahlen/Entwicklung_der_erneuerbaren_Energien_in_Deutschland/entwicklung_der_erneuerbaren_energien_in_deutschland_im_jahr_2015.html, last update February 2016, last checked: 27.08.2016.

www.genossenschaften.de: *Was ist eine Genossenschaft?* Online on: https://www.genossenschaften.de/was-ist-eine-genossenschaft, last checked: 16.8.2017.

www.gov.uk: 2010 to 2015 *government policy: greenhouse gas emissions*. Committee on Climate Change. Online on: https://www.gov.uk/government/publications/-2010-to-2015-government-policy-greenhouse-gas-emissions/2010-to-2015-government-policy-greenhouse-gas-emissions, last checked: 27.08.2016.

www.iba-hamburg.de: *The IBA Consulting Committee on Climate and Energy*. Online on: http://www.iba-hamburg.de/en/story/actors/iba-consulting-committee-on-climate-and-energy.html,last checked 29.8.2017.

www.nucleus.iaea.org/Pages/inres.aspx: *international nuclear and radiological event scale*. Online on: https://nucleus.iaea.org/Pages/inres.aspx, last checked: 28.10.2015.

www.moorburgtrasse-stoppen.de (21.2.2014): *Moorburgtrasse ist gestoppt! Endlich endgültig!* Online on: http://moorburgtrasse-stoppen.blogspot.de/2014/02/moorburgtrasse-ist-gestoppt-endlich.html, last checked: 16.8.2017.

www.moorburgtrasse-stopppen.de (30.11.2011): *Moorburgtrasse ist zum 2. Mal gestoppt, jedoch (noch) nicht verhindert*. Online on http://moorburgtrasse-stoppen.blogspot.de2011/11/moorburgtrasse-ist-zum-2-mal-gestoppt.html, last checked: 16.8.2017.

www.moorburgtrasse-stoppen.de (29.6.2011): *4250 Einwendungen sind ein deutliches politisches Signal*. Online on http://moorburgtrasse-stoppen.blogspot.de/2011/07/4250-einwendungen-sind-ein-deutliches.html, last checked: 16.8.2017.

www.unser-hamburg-unser-netz.de (28.20.2015). Warum nicht 25,1%? Online on: www.unser-netz-hamburg.de/fragenantworten/25prozent/, last checked 16.8.2017.

Yildiz, Özgür (2013): Energiegenossenschaften in Deutschland. Bestandsentwicklung und institutionenökonomische Analyse. In: *Zeitschrift für das gesamte Genossenschaftswesen* 63 (3): 173–186.

Yin, Robert (1981a): The case study as a serious research strategy. In: *Science Communication* 3 (1): 97–114.

Yin, Robert (1981b): The case study crisis: some answers. In: *Administrative science quarterly* 26 (1): 58–65.

Zahle, Julie (2012): Practical knowledge and participant observation. In: *Inquiry* 55 (1): 50–65.

Zimmer, René; Kloke, Sarah; Gaedtke, Max (2012): *Der Streit um die Uckermarkleitung – Eine Diskursanalyse*. Unabhängiges Institut für Umweltfragen e. V. Berlin (UfU-Paper, 3/12).